GLENDALE PUBLIC LIBRARY
GLENDALE, CALIFORNIA

REFERENCE

Not to be taken from the Library

The Tincal Trail
A History of Borax

The Tincal Trail
A History of Borax

N. J. Travis
and
E. J. Cocks

Harrap   London

2-96- EQ. 7.00

First published in Great Britain 1984
by HARRAP LIMITED
19–23 Ludgate Hill, London EC4M 7PD

© *R.T.Z. Borax Ltd* 1984

All rights reserved. No part of this
publication may be reproduced in any
form or by any means without the prior
permission of Harrap Limited

ISBN 0 245-53798-8

Designed by Robert Wheeler
Printed and bound in Great Britain
by the Pitman Press, Bath

They crossed a snowy pass in cold moonlight, when the lama, mildly chaffing Kim, went through up to his knees, like a Bactrian camel – the snow-bred, shag-haired sort that come into the Kashmir Serai. They dipped across beds of light snow and snow-powdered shale, where they took refuge from a gale in a camp of Tibetans hurrying down tiny sheep, each laden with a bag of borax.

Rudyard Kipling
'Kim', Chapter XIII

Contents

Preface	xi
Introduction	xv
Part I The development of an industry	1
1 Early history	3
2 How the borax trade developed	14
3 Where North American borax began	38
4 Colemanite and consolidation in the USA	60
5 Redwood and Sons, Unlimited	69
6 Smith goes to London	74
7 The effects of the Borax Consolidated merger	85
8 Borax in Britain after the merger	100
9 China – new adventures in old haunts	110
10 Colemanite rivalries in the USA	114
11 The Tonopah and Tidewater Railroad saga	125
12 The fall of F. M. Smith	139
13 The First World War	146
14 The challenge from Lake Borax	154
15 Steve Mather goes to Washington	163
16 1914–26: The end of an era in Death Valley and a new discovery	168
17 The aftermath of the discovery of sodium borate at Boron	185
18 The start of potash-mining in America, and Borax before the Second World War	197
19 The Second World War	208
20 The post-war world	222
21 The end of underground mining at Boron and the beginning of U.S. Borax	228
Part II South America	235
Part III The pursuit of borax in Turkey	255
Glossary	291
Index	294

Illustrations

Borax-refining in the sixteenth century	*32*
Early alternative alchemical symbols for borax	*32/33*
Borax minerals: tincal, colemanite, ulexite, kernite	*32/33*
Clear Lake, Cal.: first borax discovery in N. America	*32/33*
Death Valley, Cal.	*32/33*
1876: an imaginative borax advertisement	*33*
Mule Team transport, about 1888	*48*
Smith's letter-heading as an independent agent of 'The San Francisco Board'	*48*
Professor Theophilus Redwood	*48/49*
Redwood's opinion on 'Preservitas'	*48/49*
'Redwoods Unlimited'	*48/49*
Early trademarks	*48/49*
The Founding Fathers of the future Borax Consolidated	*48/49*
A selection of early 'Borax' letterheads	*49*
The Mule Team leaving Borate, *c.* 1895	*96*
'Old Dinah', an 1894 traction engine	*96*
'Francis', a tank engine on the Borate and Daggett Railroad	*96/97*
The Railroad, approaching the mine at Borate	*96/97*
John Ryan, about 1890	*96/97*
On the move again in Death Valley	*96/97*
Daggett plant, *c.* 1906	*97*
The Memorandum and Articles of Association of Borax Consolidated, 1899	*97*
Lila C mine, Death Valley	*112*
Lang mine, Ventura County	*112*
One of the earliest advertisements for Smith Bros. Borax	*112/113*
An early (*c.* 1890) advertising theme of The Borax Company, Paris	*112/113*
From a brochure of 1896 entitled '200 Borax Best Recipes'	*112/113*
Early 20 Mule Team Borax advertisements	*112/113*
Pacific Coast Borax refinery at Bayonne, N.J., about 1900	*113*
Bilingual share certificates	*113*
Borax refineries at Belvedere and Barcelona	*144*
The Tonopah and Tidewater Railroad timetable	*144/145*
No. 1 locomotive on the Tonopah and Tidewater	*144/145*
The Bullfrog Goldfield Railroad	*144/145*
A Bullfrog Goldfield Railroad share certificate	*144/145*
The dramatic news of Smith's financial collapse	*145*
Smith starts all over again....	*145*
Smith towards the end of his career	*160*
Smith's cabin in Nevada	*160*
Smith's mansion in San Francisco	*160*
View of the mining camp at Ryan	*160/161*
Death Valley Railroad	*160/161*

A poker game in the Borate company store	*160/161*
The Death Valley Brass Band	*160/161*
Horace Albright and Steve Mather with 'National Park Service No. 1'	*161*
Tribute to Steve Mather	*161*
Clarence Rasor	*224*
John Suckow	*224*
Boron, Cal.: borax-mining before and after 1957	*224/225*
Furnace Creek Inn, Death Valley	*224/225*
A group at the opening of the Pacific Coast Borax site at Wilmington in 1923	*224/225*
Lord Clitheroe, chairman of Borax (Holdings) Ltd, speaking in Death Valley in 1960	*225*
The boron high-energy fuels of the 1950s	*240*
Ronald Reagan in the television series 'Death Valley Days'	*240*
The mine at Tincalayu, Argentina	*240/241*
The borate of lime deposit at Salar de Ascotan, Chile	*240/241*
Pandermite ore, Sultan Tchair, Turkey	*240/241*
Across the fields of Anatolia to the mine	*240/241*
Borates at work in the modern world	*241*

Maps

I	Tibet, India and adjoining states showing the tincal trade	*20/21*
II	'Soffioni' of the Maremma, Tuscany	*25*
III	The borate deposits of the Western United States	*39*
IV	The 'Marsh' borate deposits of Nevada	*43*
V	The 20-Mule Team routes in the Death Valley area	*55*
VI	Borate deposits, mines and railroad systems in the Calico Mountains	*61*
VII	Railways in the area of Death Valley and Searles Lake	*129*
VIII	The Kramer borate district, California	*176/177*
IX	The borate areas of South America	*239*
X	Turkey: The borate areas of North-West Anatolia	*259*

Preface

The story of man's struggle to ease and enhance his life by overcoming the harsh impact of his environment, and by using this success to brighten his outlook, can be related to many different branches of art and science. The history and development of agriculture, architecture, medicine, engineering, transport, industry, painting, sculpture, and music is in each case a long process of the accumulation by man of the knowledge, methods and skills which together compose what we know as civilization.

This book is derived from a curiosity to see what emerges from the choice of a narrow subject – namely, the story of a somewhat obscure mineral called tincal and its geological allies, from which the principal chemical derivative is borax. The search for these minerals took men to places which were remote and sometimes politically volatile – Tibet, the Maremma district of Tuscany, the Andes of South America, the Anatolian plain of Turkey and the far west of the United States – and from which were supplied the centres of civilization and industry where borax was used. For centuries borax was an exotic and expensive commodity greatly valued by those who worked with the precious metals gold and silver, and for a long time its source was a well-kept secret. Later, those who made glass, pottery and enamels found that borax enhanced the quality and versatility of their products, and production spread throughout industrial Europe. Eventually, in the latter part of the nineteenth century, when the United States added greatly to the source of supply and to the world industrial scene, borax became much cheaper, and has found new and expanding roles in the technology of the last hundred years.

The preservation of company archives in London and California, containing much primary source material and covering the last hundred years and more, has provided a fertile area for research into the history of the borax industry. It has also provided its fair share of colourful personalities, and makes some contribution to the contemporary history of the places where these events took place.

Following a brief introduction, the book is divided into three parts. Part I covers the history and development of borax chronologically from the earliest times to approximately 1960. Parts II and III, which relate respectively to South America and Turkey, make use of the extensive material available to describe contemporary life and times in these areas, giving a somewhat broader picture than matters relating to borax alone.

What has happened since 1960 will perhaps be part of some future history of the borax industry; but since that time there have been changes in ownership and name of three of the principal companies involved in this history, and the following explanation will help the reader to identify past with present.

In 1956 Borax Consolidated, Ltd, the London-based public company founded in 1899, and whose history forms a substantial part of this book, became Borax (Holdings) Ltd.

In 1968 Borax (Holdings) Ltd agreed to a merger with The Rio Tinto-Zinc Corporation Ltd, and following this the company Borax (Holdings) Ltd became RTZ Borax Limited.

Following the formation of Borax (Holdings) Ltd in 1956, the interests of Borax Consolidated, Ltd in borax and potash in North America were brought together in a new operating company called United States Borax and Chemical Corporation. Thus the Pacific Coast Borax Co. and the United States Potash Co., which had previously developed as separate companies, now became part of a single corporation. Today the United States Borax and Chemical Corporation, whose headquarters are in Los Angeles, is a wholly owned member of the RTZ Borax Group.

The American Potash and Chemical Corporation and its predecessor the American Trona Corporation, which pioneered the technology that enabled the brine in Searles Lake, California, to be used successfully for the production of chemicals – including potash, borax and soda ash – became part of the Kerr McGee Corporation of Oklahoma in 1967. These operations continue today as part of the Kerr McGee Chemicals Corporation.

West End Chemical Company, formed in 1920, and also established on Searles Lake as producers of borax and soda ash, was merged into the Stauffer Chemical Company in 1956. These operations are today part of the Kerr McGee Chemicals Corporation, which acquired them from Stauffer in 1974.

The authors wish to thank all those who have helped in the production of this book. There are the many members of the staff (too numerous to mention individually) of RTZ Borax Ltd and its predecessor companies in England and those of the United States Borax and Chemical Corporation in California, who have helped to make available letters, documents and photographs and to answer specific questions. Particular thanks go to:

Dr Robert P. Multauf of the Smithsonian Institution, Washington, D.C. and former Director of its Museum of History and Technology, has given helpful advice and generously supplied a number of important references concerning the early history of borax.

Mr James Gerstley, former President of United States Borax and Chemical Corporation, and Mr Norman Pearson, former Director of Borax (Holdings) Ltd, have commented most helpfully on certain sections.

The archives at Borax House, London, and at US Borax's office in Los Angeles have been the principal source of photographs, but others have been supplied by the Royal Botanic Gardens, Kew; the Pharmaceutical Society, London; Oakland City Museum, California; the Department of the Interior (National Park Service), Washington, D.C.; and Dr George Hildebrand, Arizona.

Mr J. M. Hooper of the Stauffer Chemical Company kindly supplied an unpublished account (dated 1956) of the company's work in the borax industry by Albert J. Steiss and Vincent H. O'Donnell.

We are grateful for the library services supplied in London by the British Library, the Royal Society, the Royal Geographical Society and India Office; and also the Bancroft Library in California.

The maps are the work of Mrs Tricia Weber, a member of the staff of the U.S. Borax Exploration Department, and for these we are extremely grateful. We also thank Mrs Vivienne Waller for undertaking the extensive secretarial and typing work.

Abbreviations

The following are abbreviations in the references and relate to documents in the archives of Borax House, Carlisle Place, London S.W.1, used as primary sources:

M.D. — These copies of letters written by the Managing Directors of Borax Consolidated, Ltd. from London between 1902 and 1960 are bound in 58 volumes numbered serially.

R.C.B. — These are letters in files containing the incoming letters to R. C. Baker (the Managing Director of Borax Consolidated, Ltd) from 1899 to 1929, originating from centres of borax activity, particularly North and South America, Europe and Turkey.

F.M.S. — A collection of letters written by F. M. Smith, mostly in manuscript and between the years 1898 and 1904. Other F. M. Smith letters used are in the archives of U.S. Borax in Los Angeles.

F.M.S. Ann. Rep. — These are Annual Reports written by F. M. Smith as Joint Managing Director of Borax Consolidated, Ltd covering the activities of the Company in the United States from 1900 to 1912.

P.C.B. — Pacific Coast Borax – a collection of files containing letters exchanged between the management in the United States and that of Borax Consolidated in London between 1911 and 1960.

Minute Book — Refers to Minutes of Board Meetings of Borax Consolidated, Ltd except when stated otherwise.

Introduction

Borax – what is it, and what is it for?
This is always the question implied by any relatively unfamiliar substance. For instance, playgoers today no doubt find the following conversation in Noël Coward's *Private Lives* (1930) somewhat obscure:

(The scene is Amanda's flat in Paris)
ELYOT (*as they dance*): Quite a good floor, isn't it?
AMANDA: Yes, I think it needs a little borax.
ELYOT: I love borax.

This passage, written fifty years ago, refers to a bygone age when boracic crystals (called 'spangles') spread on the dance-floor helped to make you glide.

To most of us in Europe, borax is little more than a memory of boracic lint or ointment or a white powder bought in the chemist's shop. In the United States it is a different story. For nearly a hundred years generations of mothers have had borax on the shopping list and in the kitchen cupboard, using it for a multitude of household tasks from washing woollens and babies' diapers to killing cockroaches.

Borax is a white crystalline substance, known chemically as sodium tetraborate decahydrate ($Na_2B_4O_7.10H_2O$); it occurs in nature as a mineral called tincal. Sometimes the word borax has been borrowed for other minerals and salts containing the element boron, but today boron minerals and compounds are more often known generally as borates. Other borate minerals which occur naturally, and which are of commercial interest, are colemanite (calcium borate) and ulexite (sodium calcium borate), and each form part of the mystery and history of borax and of the story of the 'Tincal Trail'.

For nearly a thousand years prior to the nineteenth century, borax was used almost entirely in metalwork as a flux – that is, a substance which assists melting and fusion. It was expensive, and this confined it largely to the precious-metal trade. Gold- and silver-smiths and jewellers used it as a soldering agent and in the refining of metals and assaying of ores. In these early times the quantities used were small; it was traded as an exotic commodity in the same category as spices, and it was associated with particular trade routes, cities and seaports. Secrecy surrounded its method of production, and its source remained a mystery until well into the second half of the eighteenth century.

Today world production of borax minerals has risen from about ten thousand tons a year a century ago to over three million tons today. To find out where all this goes we need to look beyond the home and to the industries in which borax, boric acid and other boron compounds are vital components, although invisible as such in the end-products which contain them.

Borax products are bulk-shipped in 30,000-ton ocean liners, destined for markets throughout the industrial world. Glass, glass-fibre, enamel and ceramics production are parts of a family of related technologies where borax plays a special part in creating end-products with desired characteristics. For example, it is borax in heat-resistant glass that prevents cracking. The additive in detergents which makes laundry 'whiter than white' is the chemical sodium perborate ($NaBO_3.4H_2O$), which is made from borax and hydrogen peroxide. Borax and boron products feature in a multitude of industrial processes and products – fire retardants, wire-drawing, electronic capacitors, nylon-manufacture, and shields for nuclear reactors.

In the field of aerospace and electronics also, boron plays its part. Insulating tiles were fitted to the underside of the Space Shuttle *Columbia* to resist the intense heat generated from re-entry to the earth's atmosphere. These were prepared with a borosilicate (i.e., boron-containing silica) glaze especially designed for the purpose, which succeeded in its vital role. High-purity boron tribromide is used in making semiconductors where boron plays an essential role in enabling the familiar silicon chip to work its modern magic in computers and pocket calculators. While concentrated borax is toxic to vegetation, and is used to kill unwanted plants, its absence in a soil can cause vulnerability to disease and low yield in a crop. In very small amounts boron is an essential 'trace element' for most plants, and to remedy its deficiency in soil, borates are mixed into fertilizers to promote healthy plant life. Traces of borates (mostly as tourmaline and datolite) are widely dispersed in rocks and soil in nature, and boron forms about 0·001 per cent of the earth's crust. However, concentrated mineral deposits from which borax can be obtained occur in few places on earth, and are mainly of volcanic origin.

PART ONE
The development of an industry

1

Early history

The distant past

Just how long borax has been used by man is a question unlikely to be resolved; the same problems arise as in the attempts to trace the history of salt, soda, saltpetre, potash, nitre and similar materials. In all these cases it is often difficult to know whether a name which we attribute to a substance today of known chemical identity is the same substance as one mentioned in an early book or manuscript; and borax is no exception to the confusion surrounding the early history of such substances.

According to legend, the Babylonians brought borax from the Far East more than four thousand years ago to be used by the goldsmiths,[1] and works of reference have frequently cited the ancient Egyptians as users of borax in metallurgy, medicine and mummification, but none of this can be substantiated. There were several types of embalming in ancient Egypt (which depended on how much one wished to spend), but the essence of all of them was the gutting of the corpse and its pickling for at least thirty days in natron, which was the name given to the impure soda (sodium carbonate and bicarbonate) found in the dry lake deposits in Egypt. It was used, together with common salt and gypsum, in cleansing and food-preservation as well as in embalming, and its use can be traced to the Tasian period of Egyptian civilization in the fifth millennium B.C.[2] However, no one has suggested or detected the presence of borax in these Egyptian lake deposits, and perhaps the reason for this confusion arises from the later Arabic classification of minerals (see page 5) which included natron as a 'borax'.

The Coptic Papyrus, dated about the tenth century A.D., which was found at Meshaikh in 1892 and which is now at Cairo, records many medical prescriptions. Much of it relates to diseases of the eye, and Armenian borax (*paurak armenei*) is among the materials mentioned.[3] Although Coptic medicine relied mostly on the practice of these ancient Egyptian Christians, the papyrus also contains information from Greek and Arabic sources, so that this cannot be taken as evidence of the use of borax in ancient Egypt.

The nineteenth-century novelist Bulwer-Lytton, in describing the gladiatorial arena in *The Last Days of Pompeii*, makes one of his characters say,[4] 'Panso regrets nothing more than that he is not rich enough to strew the arena with borax and cinnabar, as Nero used.' This reference is said to have been taken from Suetonius' 'Life of Caligula', but neither in this nor in any other of Suetonius' *Lives of the Caesars* can any mention of borax be traced, and Lytton must have found this somewhere else.

Those who have written about the derivation of the word borax from other languages have sometimes helped to perpetuate beliefs about the civilizations which may have used it. For example, Chaptal, the eminent French chemist, writing in 1804, says:[5] 'The nitron baurak of the Greeks, the borith of the Hebrews, the baurack of the Arabians, the boreck of the Persians, the burack of the Turks, the borax of the Latins, all appear to express one and the same substance, the borate of soda.' But this is not so. The word borith appears in early translations of the Bible, along with the substance nether, and these words have undergone many changes in successive translations. In the New English Version (1970), Jeremiah 2 xxii reads: 'Though you wash yourself with soda and do not stint the soap. . . .' Soda and soap have been substituted for borith and nether. Borith and cinders of borith (mentioned in Malachi 3 ii) almost certainly referred to the vegetable alkali obtained in the form of potash (potassium carbonate) from the ashes of burnt wood or other inland vegetation or in the form of soda ash (sodium carbonate) by burning certain plants that grow by the sea in Sinai and other parts of the Holy Land. Nether (also called natron) was natural soda imported from Egypt. These substances were used for cleansing purposes, and it seems that neither soap nor borax were known in ancient Palestine.

The frequently quoted and specific references to the use of borax in ancient Babylonian and Egyptian and more recent Greek and Roman times need examination against the broader background of early chemical and technical knowledge. In attempting to distinguish fact from fiction, it is useful to start with the knowledge and the records left by Islamic civilization and the writers in Arabic between the eighth and tenth centuries A.D. These writings (which were later translated into Latin and relayed to Europe by various routes) were the foundations of alchemy, and the prelude to the science of chemistry. The appearance of Arabic words which translate to borax commences with a collection of works known as the Jabir Corpus, which contain those of the best-known Islamic chemist, Jabir ibn Hayyan (c. A.D. 722–817) and his contemporaries.[6] Boraxes are mentioned as a class of substances separate from salts, alums, vitriols and others, but little can be learnt about sources or users. Burraq, bora, burak, baurach are words which began to appear at that time, and which later were all given the single Latin form 'borax', and there is slender evidence upon which to decide when or whether these names described the substance we know as borax today. These words from which borax is derived are varying transliterations of the Arabic word meaning to glitter or shine, and probably indicate the white and crystalline properties of such mineral salts.

The Islamic chemists who followed Jabir (Latin form, Geber) make it clear that boraces (boraxes) at that time referred to a family of mineral substances, and some really definitive writing is first found in the works of Al Razi (known as Rhazes in the Latin West). Razi (c. A.D. 865–925) was a Persian who lived in Bagdad; he practised as a physician, and wrote widely on alchemy and other scientific subjects. He was the first to classify substances into the now well-known animal, vegetable and mineral categories, and he subdivided minerals into six different types. These were: Spirits (which sublime on heating) – mercury, sulphur and sal-ammoniac; Bodies – the metals gold, silver, copper, tin, iron and lead; Stones – malachite, turquoise, pyrites, glass, mica, gypsum, etc.; Salts – rock-salt, lime, potash and kali (calcined wood ashes); Vitriols – green, yellow, white (alum); Boraces (boraxes) – tincar, goldsmith's borax, bread borax, natron, borax of Zarawand and *buraq al garb*.[7]

Natron was the impure natural soda found in dry lakes in Egypt. It is crude sodium sesquicarbonate (a mixture of carbonate and bicarbonate), in recent times given the name trona, and was probably the 'bread borax' used in baking, which gave a sought-after shiny appearance to the loaf. Razi refers to the use of borax for softening metals, which indicates that goldsmith's borax could be the borax of today, but he describes it as white and similar to the efflorescence found on walls – which is indicative of saltpetre (also used as a flux by goldsmiths). The borax of Zarawand was similar, but reddish in colour. Tincar was described by Razi as the 'artificial borax', made by boiling 'buraq' with quali salt (probably soda) and buffalo milk, and indicates a refined product. Buraq al garb has been identified as the gum of the willow or acacia tree, and seems out of place as a borax.

Tincar and tincal are the Western words derived from the Persian word *tinkar*.[8] Later tincal became the name given to the natural mineral product from which borax was refined, and of Razi's boraxes, tincar is the one perhaps most likely to be a refined sodium borate. The alchemy of Islam was mostly derived from Greek sources, through translators in Egypt, Syria and Persia, and it might be expected that the earlier history of borax would be explained in the scientific writings of Greek or earlier civilizations. However, the writers of Greece and Rome in the last centuries B.C., such as Aristotle, Theophrastus, Hippocrates, Dioscorides, Pliny, Galen, Zosimos and others who summarized and systemized the discoveries of previous civilizations and added their own scientific observations, have nothing to say which indicates they used or even knew the substance borax. Theophrastus (372–287 B.C.) refers to the 'common bastard emeralds' found in copper-mines of Cyprus and used for soldering gold which were called 'chrysocolla' (the Greek word for gold-glue).[9] Later the elder Pliny (A.D. 23–79) describes the soldering material chrysocolla, which was obtained from verdigris or from copper mines.[10] According to Chaptal (1790),[11] a similar substance was prepared by the goldsmiths themselves – 'It does not appear that Borax was known to the Ancients. The chrysocolla of which Dioscorides speaks was nothing but an artificial solder composed by the goldsmiths themselves with the urine of children and rust of copper, which were beaten together in a mortar of the same metal.'

Thus chrysocolla appeared in various forms, crystals of copper carbonate ('bastard emeralds'), malachite, verdigris, copper scale, all of which were easily reduced by a flame to metallic copper and which formed a low melting alloy with the gold surfaces and joined them together. This technology, which was used in Greek times and probably earlier, was later superseded by the use of borax. However, for over two hundred years from the sixteenth century onward, writers often used chrysocolla and borax synonymously, which has added to the confusion about the origins of borax, and makes it difficult to decide when it was first used in the goldsmiths' workshops.

From Islam to Europe

Medieval Europe had its own skilled craftsmen in such fields as dyeing, glass-making and metal-working, but chemistry and the associated technology were imports from the civilization of Islam. Before the twelfth century the only contact between the Arab and the Christian civilizations had been the Crusades, and this was not conducive to the transmission of learning. However, soon after A.D. 1100 European scholars began to discover that Islam contained learning and wisdom unknown to Europe. The first point of such contact was in Sicily, but it was in Spain where the exchange was greatest, and the Muslim colleges and libraries in many Spanish cities were opened to Christian scholars and students, who diligently translated the Arabic into Latin. The science of chemistry was thus brought to Europe, and Arab writings were found to contain not only the science of Islam, but also a record of what had come to them from Greek and other civilizations. Robert of Chester, Gerard of Cremona and Adelard of Bath were all translators who enabled chemistry to emerge in Europe, but most of their work contained nothing new, nor was any critical faculty applied. As the Latin language had no adequate technical vocabulary, Arabic and Hebrew words were adapted, and borax emerged with many others. These Latin translators stimulated other new works covering wide areas of scientific and technical knowledge, and a whole range of Latin works attributed to the Islamic chemists Jabir, Al-Razi, Avicenna and others appeared; many of these were the works of Moorish alchemists and originated later in Western Islam and Spain. For example, *De Aluminibus et Salibus* ('On alums and salts'), believed to be a translation by Gerard of Cremona (A.D. 1114–87), covers similar matters to Al Razi, but with certain differences; boraxes are no longer a class of minerals, and borax is called a salt; sal-ammoniac is no longer a spirit, and is also now a salt.[12]

These and later works spread the knowledge of borax and its use as a flux for soldering, and some time before 1500 the Venetians had managed to establish a monopoly in the borax trade in Europe, which continued for over two hundred years.

Agricola, first in his *De Natura Fossilium* (1546) and later in his *De Re Metallica* (1556), refers always to 'chrysocolla which the Moors called Borax';[13] however, his descriptions of its use in the assaying and refining of gold and silver leave no doubt that he was referring to the substance we know as borax. *De Re Metallica* is a magnificently illustrated book which remained the most authoritative source of information on mining and metallurgy for two centuries, but on the matter of borax it helped to perpetuate some confusion. Agricola visited the mines of central Europe as a doctor; but he was also a scholar, and developed a great interest in the arts and technology of mining, metallurgy and minerals, and reported what he observed and what he was told. He probably visited Venice, and the woodcut illustration depicts borax-refining with realistic accuracy, but his description of the process, which says borax was made from nitrum and urine, shows that he was completely misled by those who informed him. ('Nitrum' as used by Agricola meant soda.)

Biringuccio, master-craftsman in metal-work from Siena, in his work on metallurgy *Pyrotechnica* (1540), states[14] that borax could also be made from sal-ammoniac (ammonium chloride) and rock alum, while the versatile Andreas Libavius (*c.* 1540–1616), doctor of medicine, professor of poetry and history at Jena, and teacher of chemistry, who published in 1597 the *Alchemia* (generally recognized as the first systematic textbook, and a leading source of information on contemporary chemistry), described what he calls Venetian borax as made from milk, honey, crocus (yellow oxide of iron), saltpetre and alkaline lye, boiled and evaporated.[15] All of which makes no sense, and must certainly have suited the Venetians, who did not disclose their trade secrets to scholars or to itinerant authors, in order to have them disseminated by the new invention of the printing press. Thus their source of material and their process remained a well-kept secret.

In 1640 Alonzo Barba, a Spanish priest in the mining area of Peru, wrote *El Arte de los Metales*, mainly about gold and silver. The work was important and kept secret in Spain, but the British Ambassador, the Earl of Sandwich, managed to get a copy and translated it in 1669, wherein it is said:[16] 'Borax (which is called by the Spaniards Chrysolica and Anticar) is an artificial sort of nitre, made of urines stirred together in the heat of the sun, in a copper pan, with a ladle of the same until it thicken and coagulate, although others make it of sal-ammoniac and alum'. In 1671 Webster's *History of Metals* refers to 'Chrysocolla or Native Borax' which he says[17] 'is digged up in Hungaria, Bohemia at Goldberg in Silesia and occurs with copper'. Both are clearly describing the chrysocolla of Greek and Roman times.

It is easy to visualize how this confusion between chrysocolla, borax and other metal fluxes such as saltpetre and sal-ammoniac continued for so long. Chrysocolla meant gold-glue, and European craftsmen who bought these substances thought of them as 'chrysocollas', some like borax being better and more expensive than others. Appearance, price and performance (but not chemical composition) were the criteria by which these fluxes were identified in the bazaars and markets, and adulteration of borax with salt or alum was not uncommon.

It has been suggested that 'during the thirteenth century Marco Polo introduced borax into Europe from the Far East'.[18] English translations of his travels make no reference to borax, and indeed if he had brought to Europe everything attributed to him he could have been heavily overburdened. The Venetians obtained their supplies of tincal via Alexandria and the Levant, and it seems likely that it came by the long-established Arab trade routes by sea along the coast of India and the Horn of Africa. Marco Polo may well have traded in tincal, but the notion that he introduced it to Europe as a new product is in all probability a fable. Scholars are increasingly dubious about the authenticity of Marco Polo's travels as an account of personal experience, and it has been suggested he may not have reached China at all.[19]

Throughout history practically every substance known to man has been tried in medical recipes; staunch beliefs resulted, and the written word passed down often went unquestioned. One of the earliest references to borax in English literature is to be found at the end of the fourteenth century in the works of Chaucer, who took a keen interest in the art of alchemy. In the Prologue of the *Canterbury Tales* he says of the Somptner:

> Of his visage children were aferd.
> There nas quyk-silver, litarge, ne brymstoon,
> Boras, ceruce, ne oille of tartre noon;
> Ne oynement that wolde clense and byte,
> That him myghte helpen of his whelkes white,
> Ne of the knobbes sittynge on his chekes.*

The Somptner or Summoner was a much-feared official, whose job it was to summon those required before the ecclesiastical courts, where not only clerics but

---

*In modern English: 'His face scared the children. There was no quicksilver, litharge, brimstone, borax, white lead, oil of tartar, or ointment that would clean and purge, that might help free him from his white blotches or the knobs on his cheeks.'

also lay persons could be dragged for all manner of offences, such as heresy. Although the more usual view is that his complaint was due to overeating, one editor of Chaucer asserts that his position gave him an unfair advantage over the girls, and his face suffered from a complaint that even borax could not cure. Borax is also mentioned in a short epitome of the art of alchemy in Chaucer's Canon's Yeoman's Tale (lines 233–8).

To summarize, the use of borax by the goldsmiths of Europe followed soon after the translations of Arab chemistry and technology became available in the twelfth and thirteenth centuries. Therefore it seems probable that one of the boraxes referred to by Razi and others was sodium borate, and that its use first became known in the civilization that began around Mecca and Medina, which was governed from Bagdad from A.D. 762 by the Abbassid Caliphs, the most famous of whom was Harun-ar-Rashid, under whom Islamic civilization reached its zenith.

Investigations of the origins of the borax trade must also include India and China, where agreement about the names of substances and their identity present even more difficulties than have been described already. Alchemy and associated technology were mostly carried to India from the Arabic world; however, passages of the Hindu manuscript *Susrata Samhurta* – which is an extensive account of medical matters – were translated from Sanskrit into Arabic in the eighth century A.D. for the benefit of the Abbassid Caliphs. *Susrata* mentions various alkaline substances, one of which is *tankana*,[20] the equivalent of the Persian word *tinkar*. This it recommends for use internally for subduing phlegm and for increasing the appetite, which is an unlikely role for the borax of today. The date of the manuscript is uncertain – perhaps between the first and sixth century A.D. – and the same difficulty of identifying *tankana* with sodium borate arises as in the case of the Islamic boraxes, especially in the absence of other early Indian references.

Early information about borax in China is scarce. Lorca (1916) asserts that 'borax was first found in Asia more than 2000 years B.C.: the Chinese used it to solder metals and in the preparation of glazes'.[21] There is no evidence to support this early date, or its use in the great glazed ceramic ages of China, which started in about 500 B.C. Cibot, a French missionary in Peking, together with others carried out an extensive historical investigation of scientific matters appearing in Chinese literature and their application to the arts, and this was published in 1786. In an article on borax[22] he concludes that it was unknown in ancient China, a view now generally accepted. Cibot's comments underline some of the difficulties of interpretation – for example, although by the eighteenth century *pong-cha* was well established as the Chinese name for borax, in the eleventh century *pong* was also the name of a plant used in medical remedies. A pharmacopoeia of about A.D. 970 mentions *pong-cha* (today usually written *pheng-sha*), and the more general references to its use as a flux in metalwork from the tenth century A.D. onward clearly refer to borax of today.[23] The first Arab ships dropped anchor in Canton in A.D. 714 and thereafter trade and cultural communication between China and the Islamic world developed rapidly. A mosque was erected in the cosmopolitan city of Sian in the eighth century, and in the Sung period (A.D. 960–1279) the Arabs in Canton were said to exceed a hundred thousand. Borax was probably first brought to China by Arab traders during the latter part of the T'ang dynasty (A.D. 618–907).

Borax and the scientists

In the period which followed the Renaissance the progress of science in Europe was spectacular. The work of Galileo, Kepler and Newton gave a new vision and a satisfying explanation of the physical world. Alchemy was superseded by the experimental science of chemistry, culminating in the last quarter of the eighteenth century with what has been described as the 'chemical revolution', heralded by such names as Lavoisier and Berthollet in France and Priestley and Cavendish in Britain. A new, convincing explanation of fire and combustion followed the discovery of the gas oxygen, air was identified for the first time as a mixture of gases, and hydrogen and oxygen were shown to be the basic ingredients of water. However, in spite of all these scientific advances the composition of borax remained unresolved throughout the eighteenth century.

In 1702 Wilhelm Homberg observed that when borax was heated with green vitriol (ferrous sulphate) a new substance was obtained which he named sedative salt (later to be known as boric acid).[24] Homberg was chemistry instructor and later physician to the Duke of Orleans, who had a splendidly equipped laboratory. His discovery of sedative salt (which he thought contained sulphur) started many attempts to resolve the chemical composition of borax. The results were frustrating to those involved, and gave rise to many false conclusions.

Of the known substances in the eighteenth century, borax was one of the most resistant to classification and examination with the methods available at the time. Dr Samuel Johnson, who was quite a dabbler in chemistry, published his famous *Dictionary* in 1756, and the entry for borax reads thus: 'Borax (borax, low latin) an artificial salt prepared from sal-ammoniac, nitre, calcined tartar, sea-salt and alum, dissolved in wine. It is principally used to solder metals, and sometimes as an uterine ingredient in medicine.' This included even more substances unrelated to borax than was usual in describing its manufacture! Poor Dr Johnson, he probably found them all in references by apparently reliable authors, and included some wine for good measure. If challenged, we may be sure his comment would have been equal to the occasion.

Many distinguished eighteenth-century chemists blunted their wits and experimental skills on the composition of borax. A few of those who recorded their observations (some at great length) are N. Lemery (Paris),[25] C. J. Geoffrey[26] (Paris), Théodore Baron[27] (Paris), J. H. Pott[28] (Berlin), and J. G. Model[29] (St Petersburg). Lemery (1715) described the preparation of sedative salt by treating borax with sulphuric acid. Geoffrey (1732) showed that borax has the same basis as Glauber's salt and soda (a similarity explained later by the fact they are all sodium salts) and he also described the resulting green flame when a solution of sedative salt in alcohol was ignited, thus establishing what has ever since been a field test for borax and borate minerals. The mineral sample is treated with sulphuric acid and alcohol; if a borate is present boric acid (formerly known as sedative salt) is formed in situ, and the alcohol burns with a green flame.

Baron (1747) showed that borax was formed by the combination of marine alkali (soda) with sedative salt. In the same period Pott and Model wrote extensively but added nothing to the unsolved question as to what was the sedative salt part of borax, and they concluded it was a 'peculiar substance'. The green-flame test for borax caused some further confusion. Chemists associated the green flame with copper, and thus the old confusion between borax and the copper compound chrysocolla (see p.5) would not go away. L. C. Cadet, a contemporary of Lavoisier in Paris, who published a paper on borax in 1766,[30] felt certain that because of the green flame copper must be one of the constituents, and that the acid content of borax was *l'acide marin* (hydrochloric acid). Pierre Macquer, Professor of Chemistry at the Jardin du Roi in Paris, summed up the situation very well in his *Chemical Dictionary*, which was the first ever reference book of this kind when published in 1766. Referring to borax, he said:[31]

> We are far from knowing as much concerning borax as is desirable. We are even ignorant of its origin.... It is brought from the East Indies in a state which only requires slight purification... but it is not yet known whether this matter be a natural or an artificial substance, nor whence, nor how it is obtained. Our ignorance concerning borax is certainly owing to the interest which they who make a lucrative commerce of it have to keep everything concerning its origin secret.

After referring to Baron's discovery that borax consisted of alkali and sedative salt, Macquer concluded his review with irony and the simple truth:

> On the nature of borax, nothing more remains to have all the knowledge we can desire, than to discover what this sedative salt is.

About this time there were numerous claims that borax had been made by new methods, and in 1768 the Society of Arts in London offered a prize for a 'genuine borax' made from British materials or a satisfactory substitute. No prize was awarded, but many concoctions were suggested.

In 1777 Francesco Hoefer, chemist at the court of the Duke of Tuscany, reported the presence of sedative salt in natural waters from Monterotundo, in the Maremma area of Tuscany, which he detected by the green-flame test.[32] The attention of chemists was then focused on the thought that this was the 'natural' product from which borax was derived and borax was perhaps after all an 'artificial' product made from it by some secret process in the Far East or one still being guarded by the Dutch refiners, who now controlled the European trade.

Chemists now gave increasing attention to the nature of sedative salt, which was shown to be an acid from which a whole series of neutral salts in addition to borax could be obtained with bases such as lime, magnesia, potash, ammonia and the metal oxides of zinc, lead, cobalt, etc. Lavoisier had not included it in his own researches, but he dealt with borax with clarifying simplicity when he published his famous *Traité de Chimie* in 1789:

> Borax is a neutral salt with excess of base consisting of soda, partly saturated with a peculiar acid, long called Homberg's sedative salt, now the boracic acid.... The boracic radical is hitherto unknown, no experiments having, as yet, been able to decompose the acid.[33]

This put the boracic part of borax where it belonged – i.e. among the unknown – but by implication it suggested the presence of a chemical element not yet isolated and a prize still to be won.

It thus turned out that Homberg's sedative salt was neither a salt nor a sedative of any value medically. Chemists had referred to the inadequacy of the name for some time, and when Guyton de Morveau, Lavoisier and their collaborators published new chemical nomenclature (*Méthode de Nomenclature Chimique*) in 1787 'sel sedatif' was renamed 'acide boracique' (boracic acid).[34]

The beginnng of the nineteenth century saw a remarkable new tool being applied to chemical research, namely the electric current, which was generated from the Voltaic pile (a battery). Humphry Davy was one of the first to recognize its possibilities and won a prize awarded by Napoleon for 'the best work on the galvanic fluid'. England and France were engaged in war, and there was some criticism of Davy, who travelled to France to receive his medal. However, Davy expressed his own view, saying: 'If the two countries or governments are at war, the men of science are not. That would indeed be civil war of the worst description. We should rather through the instrumentality of men of science, soften the asperities of national hostility.' What would he have thought of the march of progress into the twentieth century?

Working at the Royal Institution, Davy discovered several new elements, potassium and sodium, followed by barium and strontium in quick succession. In his Bakerian lecture on 19 November 1807 he reported[35] that he had obtained 'a dark coloured combustible matter by the action of electricity, on moist boracic acid but the researches upon the alkalies have prevented me pursuing this fact, which seems to indicate decomposition.' Some months later he produced the same substance by heating potassium and boracic acid in a tube, and found that his new product burned brightly in oxygen and returned to boracic acid. In a later lecture on 15 December 1808 he concluded: 'There is strong reason to consider the boracic basis as metallic in nature, and I venture to propose for it the name "boracium".' About the same time Gay-Lussac (professor of chemistry at the Ecole Polytechnique) and his assistant, Thenard, independently obtained the same substance, also using potassium and boracic acid.[36] This new chemical element was the clue to the elusive question of the composition of the substance. Priority was claimed on both sides of the Channel, each with their dates and evidence.

Davy was a man in a hurry, a remarkable inventor with a strong desire to be first. However, just at this time he was ill for some months, and it is not surprising that, with the multiplicity of work and duties he had on hand, the records of his experiments on this matter are not as well dated and precise as might be expected and his reporting was delayed. Thus the award of priority for the discovery of boron has never been resolved to the satisfaction of all concerned. Boron was certainly discovered independently in London and Paris in a climate of intense scientific and national rivalry, 1808 being the year that Wellington landed in Portugal and the Peninsular War began, and the commencement of Napoleon's decline. However, Gay-Lussac was a man of broad and generous outlook, and on the death of Sir Joseph Banks in 1820, when a contest for the Presidency of the Royal Society seemed likely, he strongly supported the nomination of Davy, who was elected.

The French gave the new element the name *bore* and changed the name of *acid boracique* to *acid borique*. Four years later Davy said that boracium was more analagous to carbon than to the metals, and the name in Britain was changed from boracium to boron. 'Bore' was considered appropriate to the French language, but not suitable for English, but about the name of the acid the two sides of the Channel agreed to differ, and boracic acid continued into the twentieth century, gradually giving way in the English-speaking world to boric acid.

The discovery of the new element boron unravelled the mystery of the composition of borax, and it also opened the door for science to pursue the problems set by boron's refusal to fit tidily into the theoretical picture awaiting the arrival of the new element with the atomic number five. The story of the continuing challenge of boron chemistry can be found elsewhere, and since the Second World War the Nobel Prize for chemistry has been awarded twice for advances made in this area of research, to William N. Lipscomb (Harvard University, 1976) and to Herbert C. Brown (Purdue University, 1979).

REFERENCES

1. *Encyclopaedia Britannica*: 15th Edn., Macropaedia, 3.44c.
2. Lucas, Alfred: *Ancient Egyptian Materials and Industries*, 3rd Edn. (Arnold, London, 1948), p.303. Also R. P. Multauf: *Origins of Chemistry* (Oldbourne, London, 1966), p.17.
3. The Coptic Papyrus refers to *paurak armenei* – lines 24, 26, p.89. See also Partington: *Origins and Developments of Applied Chemistry* (Longmans, London, 1935), Vol. 1. pp.193–4.
4. Lytton, Bulwer: *The Last Days of Pompeii* (Nelson, 1902), p.467.
5. Chaptal, J. A.: *Chimie Appliquée aux Arts* (Paris, 1807). Vol. IV, p.246.
6. Kraus, Paul: *Jabir-ibn-Hayyan*, Institut Français d'archéologie orientale (Cairo, 1942–3).
7. Stapelton, H. R., Azo, R. P. and Hidagat, Hussin M.: *Memoirs of the Asiatic Society of Bengal*, 'Chemistry in Iraq and Persia in the Tenth Century A.D.'. Vol. 8 (1927), pp.321–2, also p.247.
8. Laufer, B.: Sino-Iranian Nat. History Series 51, No. 3 (Chicago, 1919), p.503.
9. Theophrastus (c. 372–287 B.C.) *History of Stones*. Translated from the Greek by John Hill (London, 1764), pp.67–71.
10. Pliny: *Historia Naturalis* (Venice, 1469), Ch. xxv, 28; Ch. xxxiii, 26–9.
11. Chaptal, J. A.: *Elemens de Chimie* (Montpelier, 1790), Vol. I, p.241. (A translation.)
12. Multauf, R. P.: *The Origins of Chemistry* (Oldbourne, London, 1966), pp.160–1.
13. Agricola, G.:
    (i) *De Natura Fossilium* (Basel, 1546), pp. 206–8.
    (ii) *De Re Metallica* (Basel, 1556).
    English translation by H. C. and L. H. Hoover (London, 1912): Borax in gold-smelting, pp.444, 457, 464; assaying, pp.245–6 and manufacture, p.560.
14. Biringuccio: *Pyrotechnica*, 2nd Edition (Venice, 1550), Ch. XI, p.38.
15. Libavius, A.: D.O.M.A. *Alchemia* (1st edition 1597), 2nd edition Frankfurt (1606), p.178.
16. Barba, A.: *Arte de los Metales* (1st edition 1640 (Spanish): English translation, Earl of Sandwich (London, 1669), Ch. viii, p.30.
17. Webster, J.: *Metallographia, or A History of Metals* (London, 1671), Ch. xxviii, p.344.
18. *Encyclopaedia Britannica*: 15th Edn., Macropaedia, 3.44c.
19. Craig, C. (Far East Department, Victoria and Albert Museum): supplement to *The Times* 14 April 1982.
20. 'Susrata Samhurta', in the *Subasthana*, Ch. 46, pp.354, 359. See also Bose, Sen, Subb, *A Concise History of Science in India* (New Delhi, 1971).
21. Lorca: *Indistria del Borax* (Chile, 1916).
22. Cibot, P. M.: 'Note sur le borax', in *Mémoires concernant l'histoire des sciences . . . des Chinois, par les missionaires de Peking*, Vol. XI (1786), pp.343–6.
23. Needham, J.: *Science and Civilisation in China* (Cambridge, 1959), Vol. 3, p.662 quotes the Jih Hua Pen Tshao Pharmacopoeia (c. A.D. 970) as the earliest reference to borax in China.
24. Homberg, W.: *Mémoires de l'Académie des Sciences* (Paris, 1702), pp.50–2.
25. Lemery, N.: *Cours de Chimie* (Paris, 1715), 11th Edition – see 4th English Edition (London, 1720), pp.302–5. This was the last edition of this work published in Lemery's lifetime, and contains a revised section on borax.
26. Geoffrey, C. J.: 'Nouvelles expériences sur le borax', in *Mémoires de l'Académie des Sciences* (1732), p.398.
27. Baron. T.: 'Expériences pour servir à l'analyse du borax', in *Mémoires présentés par divers savants de l'Académie des Sciences de l'Institut de France* (1750) Vol. 1, pp.295–8 (read 1747) and 447–77 (read 1748).
28. Pott, J. H. L.: (1) *Observationum et Animadversionum Chymicarum* (Berlin, 1471) Vol. II. (2) *Dissertations Chimiques*, translated by Demachy (Paris, 1759), Vol. ii, p.319.
29. Model, J. G.: 'Von den Bestandtheilen des Boraxes' (*Hamburzischen Magasin*, 1747), Vol. 14, pp.473–521.
30. Cadet, L.C.: 'Expériences sur le borax', in *Mémories de l'Académie des Sciences* (1766), p.365.
31. Macquer, P. J.: *Dictionnaire de Chimie* (Paris, 1766). See under Borax.
32. Hoefer, F.: *Memoria sopra il sale secativo naturale della Toscana e del borace che con quello si compone* (Florence, 1778).
33. Lavoisier, A. L.: *Traité de Chimie* (Paris, 1789), Vol. I, p.267; also Robert Kerr's English translation *Elements of Chemistry* (Edinburgh, 1799), p.316.
34. Guyton de Morveau, Lavoisier et al.: *Méthode de Nomenclature Chimique* (Paris, 1787) – table opposite p.100.
35. Davy, H.: *Phil. Trans.* (1) (1808) Vol. 98, p.43: Bakerian Lecture (read 19 Nov. 1807). (2) (1808) Vol. 98, p.343: paper entitled 'Electrochemical Researches' (read 30 June 1908). An added

footnote describes result of heating boracic acid with potassium in a gold tube – 'borate of potash was formed and a black substance'. (3) (1809) Vol. 99, p.75: Bakerian Lecture (read 15 Dec. 1808) describes experiments made in spring and summer – Davy proposes name 'boracium' for the new substance.

36. J. L. Gay-Lussac and L. J. Thenard announced the discovery of boron in the *Moniteur* of 15 and 16 November 1808, and said their experiments were made on the 21st of June 1808 and first announced in July.

2

How the borax trade developed

Refining borax

To return to matters of trade and commerce, by the end of the seventeenth century the Dutch had acquired both the secret of the source of tincal and a knowledge of the refining process, so that in a relatively short time Amsterdam had displaced Venice as the centre of borax-production. In 1807 Chaptal, the French chemist, wrote:[1] 'The process of refining borax was long confined to the Venetians, but the commerce with the Levant having been interrupted, in consequence of the war carried on for such a length of time between the Turks and Persians, the trade fell into the hands of the Dutch who continued to monopolise it.' The Dutch maintained monopoly and secrecy for about a hundred years, although by the middle of the eighteenth century the East Indies and Persia were being vaguely specified as the sources of tincal.

France was the most advanced country in industrial chemistry in the eighteenth century, and was incensed by the Dutch borax monopoly. By the 1770s the French had developed a source of tincal in India, and although their product was at first inferior to the Dutch they soon produced competitively. The first borax refinery in Britain which can be identified dates from 1798, when Luke Howard started production, based on his own experiments, at the chemical works of Allen and Howard at Plaistow in Essex.[2] Peter Shaw – well known for his lectures on chemistry to the medical profession, and who later became physician to both George II and George III – wrote in 1741 that borax-refining has recently been started in England,[3] but this probably refers to some activity that did not survive or flourish.

Luke Howard, FRS (1772–1864), was a man of many interests. Besides manufacturing a variety of chemicals required by pharmacy and the arts, he became famous as the 'father of meteorology', and his work was acknowledged by his election to the Royal Society.[4] Goethe wrote a poem to 'The Honourable Memory of Howard' in which he acclaimed the 'cloud scientist', who was first to 'give precision to the imprecise, confine it, name it tellingly'. Later Goethe got in touch with Howard and asked that he might be told more about him. The modest Quaker thought the letter was a hoax, but was persuaded to respond in an autobiographical letter written in 1822. In accounting for himself, his beliefs and actions he refers to his failure as a scientist to publish anything on the subject of chemistry. My response, he said, is 'short and decided. *C'est notre métier*, we have to live by the practice of chemistry as an art, and not by exhibiting it as a science.'

Howard believed firmly in competition as a catalyst to progress in the manufacturing arts, and went on to explain that he had chosen meteorology as a science suitable for publication. He spent a considerable time in central Europe after the Napoleonic wars, and pioneered the concept of relief work to be done by actually going to the areas of distress, and thus started the association of the Quaker movement with this practice.

Allen and Howard became Luke Howard and Sons in 1807, and later the well-known chemical company Howards of Ilford (which today is part of Laporte Industries Ltd), and the refining of borax was a main part of their business for a hundred and forty years.

Today it may seem surprising that the process of converting tincal to borax – which as Macquer remarked requires only slight purification – could be kept a secret and have caused so many problems to those who attempted it. However, as later discussion of the source and production of tincal will show, this commodity was far from consistent, containing all kinds of impurities which varied from shipment to shipment as well as a coating of animal fats which was used to stop the crystalline tincal getting wet on the marathon journey from the East.

An early detailed account of refining borax was written by the French chemist M. Demachy (1773),[5] which he obtained from someone who had observed the practice of the Dutch refiners, and he concludes by saying: 'The Dutch withheld a secret from our observer which they say is essential to the purification', but he does not speculate on what this might be. The problems of good refining seem to have been related to size and colour of the borax crystals, and the Dutch produced large crystals which were the hallmark of genuine borax of high purity. Less effective refining gave small crystals with a yellow colour imparted by the grease with which the tincal was covered. Demachy says that clarification of the hot solution of borax is performed by 'the aid of the white of an egg or its equivalent of slaked lime and slate'. Slow cooling in copper pots, he says, produces large crystals with the bluish colour 'which, in the borax business, is always good to find'.

Those aware of the problems of introducing a new form of product to a traditional market will find interest in the following comment by Professor Sheridan Muspratt of Liverpool, relating to the efforts of a newly established French refinery to compete with Dutch borax. Although the French were producing large crystals of fine appearance, Muspratt writes:[6]

> It happened that commerce, accustomed to finding in Dutch borax . . . crystals with the edges broken by long transport, did not approve the new product. . . . It was necessary to imitate the Dutch article, and for that purpose to round the edges of the crystals by placing them in casks, which were made to turn on their axes.

The problems of operation of chemical processes today based on raw materials of consistent quality, monitored by chemical analysis, bear little resemblance to those which confronted the borax industry in earlier times. They had to deal with raw materials of unknown chemical composition, which had to be bought in advance at enormous commercial risk, with the likely possibility of adulteration and treachery. Appearance and performance were often the only methods of assessment, and it must have been quite hard for a new producer to remain profitable long enough to avoid bankruptcy through raw-material difficulties.

Even in the middle of the nineteenth century trading in tincal was still a hazardous business. An account taken from *Travels in Search of New Trade Products* by Arthur Robottom[7] describes the gamble which trading in tincal as late as around 1850 could present:

> ... Tincal has always been a favourite article of commerce with me, and it was one of the first things I did business with in Calcutta. I have followed its ups and downs, gains and losses, through all my commercial career. I became connected with some native merchants, who shipped me some dreadful rubbish. They sent a very large quantity to my consignment by the vessel *David Begg*, and another large lot by the *Lord Clive*. When the overland samples came to hand, I found the tincal to be composed of about half sand, mixed with ghee oil and other impurities, and I anticipated a terrible loss on these shipments. There were about 80 tons in one of the vessels, and [I] drew for it at about £160 per ton. The value in the London market for such rubbish was only about £20 per ton. On looking in *The Times* I found that both the vessels were lost off the Cape of Good Hope and, being insured for over the invoice cost, I came out with a gain instead of a loss.

Chemical analysis of some tincal samples shipped from India in about 1900 shows a relatively high grade of product,[8] which had had a substantial amount of impurities removed before shipment. Upgrading in India probably took many different forms and became more a feature of the trade from about 1840 onward, when the new competitive material from Italy made it no longer possible to sell the crude tincal that had been shipped in former times. However, Robottom's description above indicates that upgrading and quality control had not become much of a feature of the tincal trade even in the middle of the nineteenth century.

Tincal – the secret of its source

Although it had been observed for centuries that tincal came from the East, the places and methods by which it was produced remained a carefully guarded secret by those involved in producing borax.

By the seventeenth century its origin had begun to attract the attention of European travellers. These early reports associated its source with various trade routes, commercial trading centres and seaports. Jean B. Tavernier, who made several journeys to Asia and India, found borax being sold in 1631 in the port of Surat in Western India.[9] He said it came from near-by Ahmedabad, and suggested it was brought there from 'Thibet or Kathiawar'. However, the latter produced salt but never borax.

A little later (Sir) John Mandelso, Ambassador of the 'Duc de Holstein en Muscovie et Perse' made an extensive voyage starting from Ispahan in Persia in January 1638 and ending in Holland in 1640.[10] He visited India, China, Japan and the Dutch East Indies, and returned home via the Cape of Good Hope. He was a keen observer of matters relating to local manufacture and trade, and while in Northern India he learnt that borax was found in 'the Province of Purbet under the rule of Raja Biberon towards Great Tartary', and that silver, musk and copper also came from the same area. The borax, he said, grew like coral in the river Jankenekar, and it was traded in the bazaars of Guzarate (now Gujarat). Mandelso also reported on many other substances; he found saltpetre being made in India, China and Madagascar, but he heard nothing of the production of borax within

China or Persia. These early indications were noted by the scientific world, but no positive steps to explore the source of borax were taken until over a century later.

John Hill in his translation (1746) of *The History of Stones* by the Greek writer Theophrastus refers to borax in a footnote[11] as 'salt made by the evaporation of an ill-tasting and foul water, of which there are springs in Persia, Muscovy and Tartary'; however, J. G. Model writing in 1762 in St Petersburg refers specifically to tincal from Tibet.[12] Demachy, the French chemist, reports in 1773 a conversation with M. Durabec,[13] a merchant in Tranquebar (which is a port south of Madras), who said that somewhere in Tibet there was a lake called Necbal, from the bottom of which borax was dredged. At the same time he also noted that a French naval officer, the Marquis of Beauvau, who had resided for some time in Tranquebar, said that borax was extracted from a lake forty miles away. Saltpetre – which was a vital component of explosives, and always in inadequate supply within the Great Powers of Europe – was produced near Tranquebar, and the Marquis probably confused this with borax, although Tranquebar had been a centre for Arab traders dealing with the Coromandel Coast, on their way to and from China, since the time borax was first of interest to the Arab world, and it is possible that some sorting and upgrading of the crude tincal took place there.

A few years later Professor Kirwan in Dublin wrote,[14] 'It is now beyond doubt that it [borax] is a natural production since Mr Grill Abrahanson sent some to Sweden in 1772 in crystalline forms as dug out of the earth in Thebet' and that the chemist von Engerstrom had identified it. However, the first attempt to seek definite information about the location of tincal and the method of obtaining it was initiated by the Royal Society in London, and as a result two papers were presented to the Society on 17 May 1787, and later published in the *Philosophical Transactions*.[15] The first was a report to Dr Gilbert Blane, FRS, from his brother William, who had travelled some two hundred miles north-east of Lucknow to Betowle, where he met 'some of the wild and unsettled mountaineers from the Borax country'. Blane was told that thirty days north of Betowle (now Butwal in Nepal), in the Kingdom of Jumlate in Tibet, there was a lake six miles in diameter, so hot that the hand could not be held in it, and that during the snowy season borax was crystallized from the liquors near the edge of the lake, which were fed into reservoirs six inches deep.

The other response to the Royal Society came from the Father Prefect of the Italian Capuchin missions in Tibet, Joseph de Rovato, who was based at Patna in India. He obtained an interview through the brother of the King of Nepal with a native from the borax country, who told him that borax came from the Province of Marme, 28 days north of Nepal and 28 days west of Lhasa, where there is a valley containing two villages, Scierugh and Kangle, whose inhabitants lived solely by the sale of borax. The report indicated that the borax was collected from the bottom of a series of lakes, which men entered and then felt and broke the borax with their feet and collected it with shovels. This paper was presented to the Royal Society by its President, Sir Joseph Banks, on the same day as that from Mr Blane.

It is not surprising that information from Tibet about tincal is scarce. In spite of Marco Polo's account and the fame of Prester John, there is no authentic record that Tibet had been visited by any European before the seventeenth century. In 1774 Warren Hastings organized an expedition to the Tissoolumbo monastery near Shigatze, and before this European contact had been confined to the Jesuit and Capuchin missionaries. There is no account of the 1774 expedition, but in 1783 Hastings organized a second mission to the Court of Teshoo Lama (Tashi Lama) at Tissoolumbo and Lt Turner and Mr Saunders, a surgeon, reported extensively on their visit. At that time Tibet was ruled by the Teshoo Lama as Regent, as the Dalai Lama at Lhasa was still a minor; however, neither expedition was permitted to visit Lhasa. Saunders described what he heard about tincal (*tshale* in Tibetan) as follows:[16]

> The lake from whence tincal and rock salt are collected is about fifteen days journey from Tissoolumbo, and to the northward of it. The tincal is deposited, or formed in the bed of the lake. Although tincal has been collected from this lake for a great length of time, the quantity is not perceptibly diminished, and as the cavities made by digging it soon wear out or fill up, it is an opinion with the people that the formation of fresh tincal is going on. They have never yet met it on dry land or high situations, but it is found in the shallowest depths and the borders of the lake. From the deepest parts of the same lake they bring rock-salt, which is not found in the shallows. The waters of the lake rise and fall very little: it is at least twenty miles in circumference, and is frozen over for a great part of the year.

Geographically this points to the area of Lake Tengri-Nur. The full report of this mission was not published until 1800 by the then Captain Turner, who had carefully recorded what he had seen and heard regarding trade in Tibet and its neighbouring countries.[17] His report probably indicates the pattern of trade as it had been for several hundred years, and the products which Tibet imported from and exported to its three neighbouring territories, Nepal, Bengal and China, were noted. Exports were listed thus: to Nepal, rock salt, tincal and gold dust; to Bengal, musk, tincal and gold dust; to China, gold dust, diamonds, pearls, coral, musk, woollen clothes and lamb-skins. The trade with China was by barter carried on in the garrison town of Silling (Sining) on Tibet's eastern frontier. Tincal was not included in the China trade, indicating that it found its way to the outside world through India.

Cibot, the French misisonary in Peking, who wrote about the same time,[18] says Tibet was known to be the source of borax, and that it also came from Hainan, the offshore island south of China; he could find no reference in Chinese scientific literature to the production within the provinces of China itself. The mistaken idea of borax in Hainan no doubt arose from the presence of Hainan on the trade route from the west to the southern Chinese port of Canton, which was a centre of the borax trade.

The borax on sale in Peking was in a purified form: there is no reference to refining in China, and this again implies that the crude tincal from Tibet did not come across the Chinese frontier. Cibot refers to the merchants of Yunnan and Szechwan who offered a borax from Tibet, but he was uncertain how it came to them and thought it might be counterfeit, as their price was so cheap that it could not be reconciled with the cost of transporting it long distances. It seems that the small quantities of borax used in China came almost entirely by sea.

Since 1720 Tibet had been a dependency of China, but the civil and religious government was left to the Tibetans, and with few exceptions their frontiers were closed to all but Chinese officials. Those who crossed into Tibet did so at their peril, and many lost their lives. Lhasa the 'Forbidden City' became a goal for individuals motivated by the explorer's urge to be first, or by religious fervour to carry Christianity there or by the wish of foreign Powers, particularly Britain and Russia, to gather military and political intelligence.

The maps of the world showed Tibet as a blank white space, and to remedy this a British army officer, Captain T. G. Montgomerie, who was attached to the Survey of India at Dehra Dun, hit upon the idea of training hand-picked men to enter Tibet.[19] Nian Singh, who came from Milam – close to the frontier with Tibet and in the shadow of Nanda Devi (25,645 ft), the tenth highest mountain in the world – was perhaps the most famous of these spies. He crossed the frontier in 1865, disguised as a Buddhist pilgrim and carrying a compass in his prayer-wheel, a sextant in a secret compartment in his baggage, a thermometer in his staff and a supply of mercury in a cowrie shell which he poured into his pilgrim's bowl to provide the artificial horizon for his observations. He covered 1,200 miles on foot, counting each of his calibrated steps on the beads of his Buddhist rosary, and returned to India one and a half years later. He not only entered the 'Forbidden City', he also took tea with the Dalai Lama and was shown round the Royal Palace. He witnessed the public execution of a Chinese who had entered the city without permission, a fate which would have certainly befallen him if he had been detected.

On a subsequent visit Nian Singh made his way to the Tibetan gold-field Thok-Jahlung, which he estimated to be at 16,330 ft, and gave a fascinating description of the mining community living in yak-hair tents some eight feet *below ground*. Nian Singh reported he had seen from a mountain-top the 'borax lake', Bul-Tso, near Lake Tengri-Nur. However, 'bul' is the Tibetan for soda, and there is nothing to indicate specifically that this was a main borax-producing area. Nian Singh also noted in his diary that there was a large market at Babuk, north of Katmandu, on the frontier between Nepal and Tibet, where salt, wool, felt and borax were brought from Tibet prior to being carried to Nepal and adjacent territories.

The main purpose of these British spies (called Pundits) was to obtain topographical and political intelligence, and their lives were in constant danger. It is therefore not surprising that their reports did little to sort out details concerning materials like salt, soda and borax – gold was another matter.

The area of Tibet in which borax has been reported embraces a strip stretching east–west for about a thousand miles through the centre and north of the Himalayas, starting at Lake Tengri-Nur (Namar-Tso) about a hundred miles north of Lhasa and ending beyond its frontier in the Province of Ladakh (now in Kashmir), and mostly at an altitude of 13,000–15,000 ft.[20] There is little to indicate specifically where most of the tincal was obtained within this huge area, and those receiving it in India had probably no idea where it had originated. Lake Yamdok Cho (sometimes referred to as Lake Palte) in S.E. Tibet, south of Lhasa, seems to be a place from which borax was obtained since time immemorial, and north of Lhasa the large lake Tengri-Nur is also mentioned. Farther west, reported sources are the lakes Purang-Chaki and Rudok, the latter producing a high-quality borax known as Chu Tsale (water borax).

Eastern Tibet was probably the main source of tincal for a long time, and Arthur Robottom, the commodity trader referred to earlier, wrote an interesting account of the Tibetan scene in the period 1840–50.[21] He specifically refers to the Lake Yamdok Cho, south-west of Lhasa, and describes the passage to India as follows:

> Tincal is brought into the town in Tibet and sold at the bazaars there. Sheep owners buy it from the small dealers, and put it into their saddlebags, placing as much as from 30 to

Map I

40 lb in each bag, or about equal in weight to the animal that is carrying it. Some few goats are also employed in this trade.

When they have received their loads they are started on their wearisome journey, many of the flocks carrying nothing but tincal. Each sheep-driver carries a distaff and bobbins, and as they travel along every bit of wool that falls from the sheep, or that stick to the thorny bushes with which the sheep may come in contact, is carefully collected. The wool thus gathered is spun into yarn or strong thread and then woven into cloth; which in its turn is made into bags. These are covered outside with sheepskins to prevent the tincal from getting wet, and also to protect the woollen bags from getting torn by the thorny bushes. Numbers of the sheep and goats die on the road and their flesh is always eaten by the drivers. From 800 to 1,000 sheep constitute a drove.

The animals are driven from seven to nine miles a day and it takes from six to eight weeks for the journey from the starting point to Moradabad. They travel through a pass of the Himalayan Mountains, about 100 miles north of Almira. In one part of the pass there is no grass for a distance of 25 miles. The young underwood has to be cut down for the leaves to feed the sheep. Some points of the pass are 15,000 ft above the sea level.

The tincal is sold in the bazaars at Moradabad by the bootus* to native dealers and it is sent from thence to Calcutta.

The western approaches to Tibet are sealed off by the formidable Karakoram and Ladakh mountain ranges, and it is in Ladakh on the Kashmir frontier with Tibet that the most western of the Himalayan borax deposits occur. While Tibet was closed to them in the nineteenth century, there were a number of intrepid Europeans who explored right up to the frontiers, and the borax deposit in Ladakh finds a place in several accounts. Major Cunningham (Bengal Engineers) and Dr Thomson (botanist) visited Ladakh in 1845, and Cunningham described the area thus:[22]

> The hot springs of Puga, I have myself examined, occur in a rivulet called Rulang-Chu [Radlang river] for a length of about two miles. The springs vary in strength from gentle to bubbling to strong ebullition and the temperature varies from 80° to 145°.

This river, he says, joins the Indus on its left bank ten miles from Puga (literally, a hole), which he placed at 33°12′N and 78°16′E, at a height of 15,264 ft (about seventy miles south-east of Leh).

The air in this volcanic area bears the sulphurous smell of hydrogen sulphide, and the hot springs throw up boric acid. A mixture of borate of soda and other salts is deposited along both banks of the river, and also sulphur as pure transparent crystals. Cunningham saw no human activity when he was there in September, but he heard that the shepherds were still coming in the summer months and collecting relatively small quantities of borax.

Dr H. von Schlagentweit from the Munich Academy visited Ladakh, Nepal and Bhutan to investigate the borax deposits in 1857, but like others he could not get to Tibet. He describes the hot springs at Puga[23] and the narrow gorges where borax occurs in the river at an average thickness of about 3 ft, presenting a strange and striking land formation with a wavy surface with large isolated borax accumulations in the shape of ninepins. At the time of his visit the production of borax from this area had almost ceased and the summer residences built for the visiting caravans of those who came to collect the borax were now in ruin; the buildings had moats and walls but no roofs.

---

*Also referred to as Bhots, a race which predominates on the Indian side of the Himalayas in Nepal, Sikkim and Bhutan. (See note 23.)

Production at Puga, it seems, was never large, and von Schlagentweit met caravan leaders who had earlier brought tincal from Eastern Tibet through Nepal, Bhutan and Assam and who told him this had been the main source, which was later to be confirmed by Burrard and Hayden, who visited Ladakh and Tibet in 1907.[24]

By the middle of the nineteenth century Britain was the largest borax-consuming country in the world, and a new source of production in Italy – a development to be described later – had captured the British market. The loss of Indian trade to Italy was not to the liking of the British Government, and the effectiveness with which they policed the British Empire in order to promote trade is shown by the suggestion of the India Office in 1855 to Lord Dalhousie, Viceroy of India, that the feasibility of restoring the tincal trade be given attention.[25] This resulted in some renewed activity in the Ladakh area on the Kashmir–Tibet frontier, but it is not surprising (taking into account the altitude of 15,000 ft, and the general inaccessibility of the area) that not much occurred.

However, the first half of the 1880s saw a revival in the tincal trade, due to government effort in the North-West Provinces to improve the quality of the product, and perhaps due also to shortages in the European market. At this time Jagadhi, a small place 37 miles south-east of Ambala, had become a centre for the tincal trade, where the mineral was sorted and upgraded before being taken by bullock cart to Furuckabad or Mirzapur, and thence to Calcutta by the river Ganges.

Later a small refinery was established at Jagadhi which supplied crystal borax to India and made some shipments to China. But this seems to have been a short-lived activity, as by 1890 borax from Europe of superior quality was beginning to be landed and sold in India at prices which gave the Indian refined borax – with its transport problems, and irregularity of supply and quality – little chance of survival in the twentieth century.

It seems that the tincal trade through western Tibet into north-west India outlasted the more direct supply route from east Tibet through Sikkim and Bhutan to Calcutta, which had ceased during the second half of the nineteenth century. At this time the Tibetan frontiers with Sikkim, Bhutan and Assam were certainly a hostile scene and they saw little in the way of trade in the nineteenth century; however, a passage by Chaptal, written about the tincal trails at the end of the eighteenth century, may explain the reason. He says two kinds of crude borax came from the East to Europe:[26]

> The one is brought by sea from Gomnon and Bengal, the other by land from Bender to Bassy, to Ispahan, and even Gihlan, whence it is sent by the Caspian Sea to Astracan, and thence by land to Petersburg, and from this last place to different parts of Europe. The first is in a very impure state, and held in little estimation; but the second, which is brought by the caravans of the East, is composed of hard and greenish crystals.

No nineteenth-century report about borax in Tibet adds much to earlier ones, and by the time Tibet's frontiers were opened following the controversial Younghusband expedition – which fought its way to Lhasa in 1904 – the tincal trade had virtually ceased. The later explorers and visitors who might have taken some interest were mostly unaware of tincal or its history. Just how this material situated at 15,000 ft in the icy wastes of Tibet first found its way to the Islamic civilization and the Mediterranean area remains a mystery, and there is no plausible explanation of how it could have happened.

There are many references in the literature of minerals and travel to other sources of borax; Egypt, India, Ceylon and Persia are all mentioned. In Egypt borax was traded throught the port of Alexandria, but there was no production there. Ports in India and Ceylon also played a part in the Tibetan tincal trade, and were wrongly thought to be where the substance was found.

Persia has often been mentioned as a source of tincal. The word is itself Persian, and it was Islamic chemists, mostly Persians, who made the virtues of borax known. However, there is little to support the idea that tincal was found there. John Mandelso (referred to earlier) and Jean Chardin (1674),[27] who both paid particular attention to minerals in Persia, saw nothing of borax there. The chemists Bergman and Chaptal repeat earlier references to borax being found in caves in Persia,[28] which sounds like confusion with the chrysocolla which was found in copper mines. However, after it was established in the 1780s that tincal came from Tibet, Persia soon ceased to be mentioned in the literature. There have been no reports of sodium borate being found in any lake deposits there. In 1871 borax mineral, mainly ulexite (see p.27), was discovered in West Kerman, in south central Persia, and some small shipments were made to Europe prior to the First World War, but little activity has ever taken place, and Persia (now Iran) has no commercial source of borates.[29] Earlier Persia stretched far to the east, and the belief by the western world that this was the land from which tincal came is not difficult to understand.

Italy – a new European borax source

After Tibet, the next phase of world borax production took place in Italy. The steam vents of Tuscany issuing from the ground, known locally as *soffioni*,* had been observed for centuries on the slopes of the mountains which separate the valleys of the rivers Cecina and Cornia, south of Volterra, and north of Massa Maritima. This Tuscan area (known as the High Maremma) is an awesome place, and commonly thought to be that described by Dante at the end of the 7th Canto of the *Inferno*, as the place of punishment of those whose crime had been the indulgence of angry passions. In their natural state the soffioni have the appearance of white and shifting clouds, issuing violently and noisily from crevices in the ground, impregnating the air with the smell of sulphur.

As mentioned earlier, in 1777 Hoefer, a chemist from Cologne in the service of the Duke of Tuscany, detected boric acid in the muddy blue water of the lagoons in this area,[30] although it was not until about 1820 that a combination of commercial drive and technical skill began a business which was soon to bring the Tibetan tincal trade in Europe almost to an end. The key figure in the Italian scene was a young French émigré from Leghorn, Francesco Lardarel, who always

---

*The Tuscan soffioni have characteristics in common with the geysers of Iceland, North and South America, New Zealand and the Azores, but they differ in the fact that the latter shoot out alternating jets of steam and boiling water, while the soffioni eject almost exclusively and continuously super-heated steam mixed with small quantities of gases. The average composition of the steam and gases mixture is about as follows: steam 95·5 per cent and carbon dioxide 4·3 per cent, while the remaining 0·2 per cent is a mixture of hydrogen sulphide, nitrogen, ammonia, combustible gases (methane and hydrogen), rare gases (helium, argon, neon) and boric acid (present to the extent of about 0·03 per cent).

seems to have been one step ahead of his rivals. Some reports[31] indicate he dealt ruthlessly with those who were his initial collaborators in order to establish his ownership of the company which was to become 'Lardarello'. On the other hand, the energy, tenacity and confidence shown by Lardarel in the early years was undoubtedly responsible for his ultimate success. With due secrecy he had acquired land in the area, long before the viability of the process was worked out.

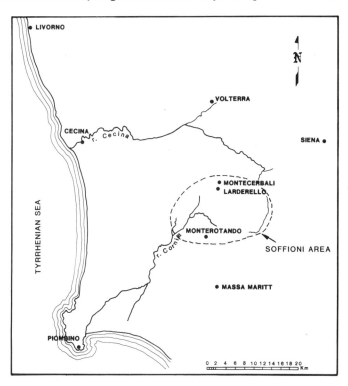

Map II

Lardarel was under thirty when he took up the challenge of converting an idea which had been discussed for forty years into a productive industry in a hostile and barren area. It was a question of how to utilize a natural phenomenon consisting of steam, boiling water and toxic gas, in an area where roads and transport systems, drinking water, indigenous labour and living quarters were all totally absent. Vagabonds and robbers watching the routes were an additional hazard, and caravans of mules and later wagon trains had to be protected with guns at the ready.

The initial method of concentration of the boric acid solution using wood or coal as fuel soon created problems of supply and costs, and work could not be sustained, as most of the timber in the area had been consumed. Harnessing steam from the soffioni as a source of energy was an essential step in the continuing production of boric acid and borax at a cost which could compete.[32] Boring into a soffione was a difficult operation, and explosions and risks to human life delayed progress. Professor Gasseri and an engineer, Manteri – neither of whom were associated with Lardarel – developed the first successful drilling methods, and their results were published in 1840. However, it was Lardarel who first made commercial use of the technology.

Since that time the technique of drilling boreholes at increasing depths into the soffioni has continued to progress. By 1905 the first power station was started at Lardarello, and throughout the twentieth century this area has continued to develop as a source of power-generation, employing much of the drilling technology developed in the oil industry. Later the production of boric acid was dwarfed in relative importance by power-generation, but the former industry continued for well over a hundred years until it was phased out in the late 1950s, and borax and boric acid production within Lardarello was replaced by a traditional process based on imported ore.

During the ten years 1818 to 1828 production averaged only about 75 tons annually, but thereafter it soon reached 1,000 tons, and began to make a serious impact on the Tibetan tincal trade. By 1860 annual production had reached 2,000 tons, and by the end of the century it was about 3,000 tons.

The crude boric acid was shipped in 10-cwt casks for refining or conversion to borax at Leghorn; it was also exported to new European borax-refiners who based their business on this competitive Italian source of supply. Lardarello established a refinery in Marseilles, and in 1829 Messrs Gerber and Thilo were using Tuscan boric acid in Hamburg, and seem to have been the first German borax-refiners.

In 1836 Edward Ward started refining Tuscan boric acid at Old Swan in Liverpool. The plant was sold to Howards in 1862, and was then acquired in turn by Coghill & Sons in 1874, refiners who at the end of the century were among the largest in England. In 1876 William Arbuthnot also commenced refining Tuscan boric acid at Stroud in Gloucestershire; later this became the Boron Products Company Ltd.

Lardarel died in Florence in 1858 at the age of sixty-eight and the industry which was largely established by his efforts and initiative was of tremendous benefit to the Grand Duchy of Tuscany at a time when the economy was largely agricultural and commercial. In 1846 the Grand Duke Leopold II named the place at the centre of the activity, and close to Montecerboli, Lardarello. This spot had developed during Lardarel's life into a model industrial centre, with its church, schools, a music centre and housing for employees. The company provided a free health insurance scheme and pensions for employees and their widows – an unusual situation, particularly in the mid-nineteenth century – and a few years after Lardarel's death the new borate industry of Tuscany proudly entered the young kingdom of Italy. By the end of the century boric acid was being produced in about ten different places in the Maremma district of Tuscany, seven of which belonged to Lardarello.

The age of expansion – new tincal trails in the Old World and the New

In the middle of the nineteenth century the world-wide increase in demand for cheaper sources of borax coincided with the age when steam had revolutionized communication and transport on land and sea. Within a short space of time mining of borax was being established commercially in new and distant places and the third phase of its history had begun. The Andes of South America, the Anatolian Plain of Turkey and the Californian and Nevada deserts of the western United States were the new areas of activity, and by the end of the century these, with Italy, were the principal producing areas of the world.

South America

The earliest report of borates in South America came in 1787 from a physician,

Dr Carrere, employed in the mining area of Potosí, Bolivia.[33] Nevertheless, it was not until 1852 that South American borates were first mined commercially, in Chile, and production there did not develop on a large scale until after 1880. By the end of the century mining was taking place in both Chile and Peru. Production in Argentina came later and was insignificant in the nineteenth century, while in Bolivia borates were not mined until the twentieth century, and then only for a short time (1901–5). The borate ore from South America was referred to as borate of lime. It is a sodium calcium borate, which was named ulexite in 1849 after examination by the German chemist Georg Ulex (1811–83).

The borate of lime districts of South America are like those in Tibet in that they are of volcanic origin at high altitudes of about 15,000 ft, and they exist only between the latitudes of 15° and 18° South. The Andes mountains on the Pacific side show three distinct levels of plateaux – the lower running parallel to the ocean between 2,500 and 3,000 ft above the sea, while above this is another level plain at 7,000 to 8,000 ft. A third, at 13,000 to 15,000 ft, is an immense level plain – almost a barren waste – which is totally lacking in vegetation. Above these plateaux there are many volcanic peaks, 3,000 to 5,000 ft above the high, level plain in which all the well-developed deposits of borate of lime are found. Borate of lime deposits are found at the lower levels, but seem to have been derived geologically from those formed at the highest level.

On the eastern side of the Andes, in Argentina, the borate of lime deposits occur at about 12,000 to 14,000 ft, and are seldom in the stratified form of the west. Here the ulexite mineral is in lumps about the size of a potato and had to be dug up and separated from the sand, so that it is more difficult and costly to produce.

The borates from the Andes were frequently carried long distances by llamas, burros or mules to narrow-gauge railways, and on to such ports as Antofagasta and Iquique in Chile and Mollendo in Peru. These mines provided an opportunity to compete with Italian borax, and this produced numerous new refiners in Europe based on South American borates.

Turkey

The mining of borates started in Turkey in 1865, where the ore deposit was situated about forty miles south of the port of Panderma on the Sea of Marmara, at Sultan Tchair, near Susurluk. The mineral was known locally as pandermite, which is a form of calcium borate also called priceite. This Asia Minor mineral, containing about 46·6 per cent boric oxide (B_2O_3), was consistently richer than any other boron-bearing ore that has been mined commercially; and this deposit was mined continuously until supplies were exhausted in 1955.

There has been speculation about the use of pandermite as a source of borax in Roman times. Early civilizations of the Eastern Mediterranean certainly depended heavily on Anatolia for minerals such as marble and gypsum, and these were extracted from this area. However, there is no evidence that the Romans obtained borax from any quarter, and if they had mined pandermite and developed a process for its conversion to borax (an unlikely event), this would have been a known and adequate source for the Arab civilization and the Western nations which followed it, and the marathon haul of tincal from Tibet would have been unnecessary. Outcroppings of pandermite occurred at the surface; it is a beautiful white stone more pure in colour than marble, but not so close in grain or so susceptible to high polish. It is easily carved, and was used in place of alabaster or marble for making images and statues. It was probably used locally for this purpose in Roman, Greek and earlier times, but not for borax.

A delightful story of how the French first came to mine pandermite turns on its proximity to marble in this area.[34] Henri Groppler, a Polish refugee, was carrying on a failing marble business at Bebek on this coast, and during a visit to Paris to try to drum up financial support he called on Desmazures, a French engineer with whom he had once been associated in contructing lighthouses on the Marmara coast. He gave him as a souvenir some small rough statuettes which had been carved by his workmen from certain unknown material which from time to time was picked out of the gypsum that was used for polishing the marble. The unknown substance somewhat resembled lumps of hard chalk. The statuettes stood on Desmazures's desk for a long time, but since, like Groppler, he was a dabbler in chemistry, the day came when he wondered about the composition of the unknown substance. An analysis showed that it contained a high percentage of borax.

Desmazures at once cabled Groppler to ask for fuller particulars about the area in which the figurines had been produced, and suggested that if possible Groppler should buy the ground concerned at once. The latter had by this time given up his unprofitable marble business, and was unable to give any exact information as to the provenance of the gypsum and the strange substance from which the statuettes had been made. Desmazures nevertheless left Paris for Turkey without delay, and he and Groppler, through the help of a local inhabitant, traced the source of the gypsum, and also the stone of the statuettes.

They obtained a *permis* from the Sultan's Government in 1865 and started mining the pandermite ore, which was then dispatched to the borax-refinery, which Desmazures had built at Maisons Lafitte, near Paris. The 'permis' was exchanged for a regular 'firman' of concession granted by the Imperial Ottoman Government in 1887 for fifty years, and in the same year the Desmazures and Groppler enterprise was acquired by a new company called The Borax Company, registered in London. In 1889 the Société Lyonnaise de Borate de Chaux, with refineries in Lyons and Vienna, acquired a concession in Turkey on a site adjacent to The Borax Company. The further history of these two companies will be recounted later.

The United States

At almost the same time that these events took place in Turkey borates were discovered for the first time in the USA in 1856 in mineral springs in Tehama County in California, and were first produced commercially at Clear Lake, Lake County (about ninety miles north of San Francisco) in 1864.[35] More important discoveries were made in Nevada and in California in the Death Valley and Searles Lake areas in the 1870s, and the story of the development of these discoveries is the subject of the next chapter.

Borax in use – before the nineteenth century

Precious metals

As mentioned earlier, before the eighteenth century borax was mainly used as a flux in metalwork; jewellers working in gold and silver found it the best soldering agent, and were among the few able to afford it. In the sixteenth century no field of applied science was more advanced or more precise in its quantitative methods than the assaying of precious metals. The methods described by Biringuccio (1540), Agricola (1556) and Ercker (1574) were substantially those which lasted until the present century,[36] and the use of borax to remove the final traces of impurities from the molten metal has remained a general practice in assaying and refining ever since.

Johann Cramer, who taught assaying in Leyden and London before returning to Germany, described in a Latin work dated 1739 the knowledge of the art and its related chemistry.[37] He discusses borax along with nitre, common salt, sal-ammoniac (ammonium chloride) and sandiver (salts from the scum produced in glass-manufacture) as a class of substances for removing impurities and aiding fusion. About borax he says:

> It is therefore very detrimental when these viler metals [i.e., copper, iron, tin and lead] are mixed in ever so small a quantity with gold and silver. For in the melting there comes upon the surfaces light scoriae, in which gold and silver is retained as in spunges. . . . To remove these inconveniences, borax is added as it helps the melting of metals . . . bringing the whole mass to a quick fusion, causes the metals to sink to the bottom without loss and vitrifies the lightest scoriae, throwing them to the surface. . . . Borax, by covering the surface of metal tortured in the fire, as it were, by a very thick glass, defends it against the combined force of fire and air.
>
> It is likewise expedient to rub with borax crucibles in which precious metals are to be melted. . . . It is a chief point never to neglect this particular, when any little mass of gold or silver is to be melted for a second time. Observe, however, that if you melt gold with borax, you must add to it a little nitre or sal ammoniac, but not both together or they would make a detonation. For borax alone makes gold pale, but it recovers its colour again by means of nitre or sal ammoniac.

By 1800 a range of fluxes which contained varying amounts of borax were in use in assaying metal ores (which also included base metals such as copper, tin, lead and bismuth). One example was 'Cornish Reducing Flux' (sometimes called 'black reducing flux') which consisted of tartar (crude potassium tartrate) 60 per cent, saltpetre 21 per cent and borax 19 per cent.[38]

Glass

In keeping with the beliefs about borax's ancient lineage, it is often assumed that its well-known uses today in glass, enamels and ceramic glazes were also known to the older civilizations which developed these arts. By 1500 glass-making was widely practised in Northern Europe, and in the great centre of Venice several writers – for example Biringuccio (1540) and Agricola (1556) – alike describe both the raw materials in use and the preparation of the alkaline salt constituents (today known as soda-ash and potash) from natural vegetation.[39] However, there are no references to the use of borax.

Moving to the seventeenth century, a certain Antonio Neri, a priest from Florence who had experience of glass-making and lived near Venice, wrote the first textbook on the subject, *L'Arte Verraria* (1614), which was translated into English in 1662 by Christopher Merrett, a Fellow of the Royal Society and a distinguished London physician, who also added his own observations and commentary.[40] The only mention of borax is in Merrett's description of the making of spherical mirrors from solid metal, with borax being used to remove the last traces of impurities at the molten state. The earliest European mention of borax in glass occurs in a German work by Johann Kunckel in 1679,[41] which includes borax in the descriptions of several glass compositions used for making artificial precious stones.

There is a passing reference to the use of borax in glass in Peter Shaw's 'Chemical Lectures (1734)', and there are numerous formulae involving the use of borax as a flux in counterfeiting precious stones and in the preparation of 'fine crystal' glass in a work called *The Laboratory or School of Arts* (1738), translated from the German by G. Smith.[42] Crystal and white flint glass were terms used for clear, colourless glass. Johann Cramer (1739) recommends for crystal glass:[43] '3 parts of prepared flints [silica], 1 part of purest alkaline salt [potash], and 1 part of burnt borace [borax]'. He also describes how the burnt borax should be prepared: 'Vitrifications by gentle burning, whereby a small quantity swells to a prodigious, moist, light, and spongy mass of a very white colour. If you neglect this preparation, the mixture to which it is added will boil over, the vessels though ever so large.'

In the currency of 1750 borax was exorbitantly expensive, selling in London at over £700 a ton, and Robert Dossie (1758) puts the role of borax in glass-making in perspective at that time when he says:[44] 'Borax is the most powerful flux of all the salts or indeed of any known substance whatever, but on account of its great price can only be admitted into the composition of glass for looking-glasses as plates or other purposes where a considerable value can be set on the product or where the quantity wanted is very small.' Dossie says the best composition for looking-glass plates is one containing white sand [silica] 56 per cent, pearl ashes [potash] 23½ per cent, saltpetre 14 per cent, borax 6½ per cent; and he also notes that borax helps glass to receive certain colours.

In trying to fix the first use of borax in glass it needs to be remembered that, prior to the nineteenth century, many of the accounts of glass and allied technology were written by scholars and observers who were not themselves involved in the art. The secrets were passed on by word of mouth and practical instruction, and those who knew most were not given to writing for the benefit of others. How long any important method may have been in use before it was reported in print remains a matter of conjecture. Dossie mentions the great secrecy about fluxes which was maintained in centres like Venice and Dresden, but it seems unlikely that borax was in use in the glass industry in Europe (except for making artificial gems) until well into the eighteenth century, and then in a very limited way.

The only earlier reference to borax in glass comes from China, and is an intriguing one. Chau Ju-kua, who wrote 'On the Chinese and Arab Trade in the 12th and 13th Centuries', describes a glass (*Lui-li*) which came to China from the Ta-shi Arab countries (Syria).[45] He says this special *Lui-li* was made by adding 'southern borax' to the same basic ingredients used in China – namely, lead oxide, nitrate of potash and gypsum – which 'causes the glass to be elastic without being brittle and indifferent to temperature'. Gypsum would not be used in glass, and this is probably a translator's mistake for calcium carbonate. However, in the case of borax the question arises as to whether this is an early description of borosilicate glass, which has been widely developed in the twentieth century for heat-resistant purposes, as in ovenware and laboratory glass.

The writer goes on to say that Chinese *Lui-li* was fragile and broke at once if hot wine were poured into it, while the Arab glass did not. There is no suggestion that the Arab glass resisted fire, and a better-quality glass could have resisted hot wine and yet not have contained borax. In view of the wide use of the term borax by the Arabs, it is difficult to conclude that 'southern borax' was sodium borate, as the

Arabs have left no description of a glass containing borax, nor has its presence been detected in Arab glasses by chemical analysis. More needs to be known before we can attribute the discovery of heat-resistant borosilicate glass to this era.

Enamel

Closely allied to glass-making is the art of enamelling. Enamel is a vitreous coating fused to a metallic base – gold, silver, copper, bronze or iron. In certain periods decorative enamels were also fused to a glass base. The coating material is referred to as a 'frit'. The art of enamelling began to take definite form in the early Byzantine era, but in spite of the important role which borax played later it seems it was not used in enamel frits applied to metals until the middle of the eighteenth century. The early frits were coloured ground glass, and were used almost entirely for decorative purposes, and then in small quantities. The merits of borax as an enamel flux are first elucidated by Dossie (1758), and none of the earlier writers on enamels such as Neri, Kunckel or De Blancourt mention it.[46] In enamel-work, Dossie says,[47] borax promotes 'the fusion of any glass when vitrified in a greater degree than any substance known', and that this has the 'greatest consequence in forming fluxes for enamel. Its use is not much known in common practice'. He describes borax as the most kindly and powerful flux for vitrifying zaffer (the impure cobalt oxide used for blues) and for many other colours, and he quotes its use in quantities varying from 4 to 18 per cent of the composition of the frit. But the main increase in the use of borax in enamel-work did not come about until the enamelling of iron created a new industry in the nineteenth century.

Glazes

Allied to enamelling is the glazing and decoration of porcelain, pottery and ceramics generally, also a process of coating with glass. The history of glazes, starting with the ancient Egyptian, Chinese, Babylonian and Greek civilizations, is lengthy and complex. The earliest reference that has been found to borax as a glaze is in an account by Kenzan I, the Japanese artist and designer, of his discussion with the master potter Ninzei in the year 1699.[48] From 1648 to 1703 Ninzei was active in the development of *raku* (a low-fired faience pottery), and he developed the use of vitrified enamels as an 'overglaze'; for gold he used a mixture of one part of fine gold dust and one-fifth part borax, and for red a mixture of iron sulphate and borax. For a transparent raku glaze he used a mixture of white lead, boracic frit and quartz. It is possible that the use of borax as a glaze was introduced into China at about the same time, but there is no analytical evidence to confirm this, and earlier the Chinese had relied on lime and potash for good-quality glazes.

An anonymous tract printed in Berlin in 1750 describes porcelain-manufacture in China and Saxony;[49] and a similar account appeared in Dossie's *Handmaid of the Arts* (1753).[50] Neither author gives the source of his information (which shows certain similarities), but both state that the glaze and flux for Chinese porcelain was said to be made from a paste of a secret mineral called petuntze★ mixed with a residue obtained from a mixture of calcined lime, ashes from burnt vegetation

---

★For petuntze see Glossary, p.292.

(potash), and borax (about 1 per cent). The glazes used in Saxony were secret, and were not described. Against this, Pierre Macquer, professor of chemistry in the Jardin du Roi at Paris, who worked on the development of true porcelain at Sèvres, makes no mention of borax in his account of porcelain and the glazing process written in 1778.[51] At the high temperature at which true porcelain is fired there would be little reason for using it.

In preparing colours for painting on chinaware, Dossie (1758) mentions blue zaffer (cobalt oxide) vitrified with calcined borax, and in describing glazes for earthenware he says:[52] 'When it can be afforded, glazing is improved by adding 1 to 2½ per cent of borax and reducing the pearl ash [potash] content by 6 parts for each part of borax.' As with glass and enamel, cost was a main reason why little borax was used in glazing ceramics in Europe before the nineteenth century. In about 1780 the price of borax in Amsterdam was about a hundred times that of materials like alum and saltpetre.[53]

Chemical analysis

In the middle of the eighteenth century borax featured in the new methods which were developed in Sweden for identifying minerals and chemicals, and which played an important role in analysis for nearly two hundred years. The use of the blowpipe – sometimes referred to as the chemist's stethoscope – combined with certain reagents such as borax, soda and microcosmic salt (sodium ammonium phosphate), became the basis of a new technique developed by Cronstedt and von Engestrom in the mineral field[54] and by Bergmann, Gahn, and others in the chemical laboratory.[55] Almost all substances were found to dissolve when fused with borax and many metals could be identified by the colour which a salt or mineral imparted to a melted lens of borax held in a small loop of platinum wire. This became known universally as the borax bead test.

The nineteenth century and the Industrial Revolution

From the middle of the nineteenth century, developments in enamelware and pottery and china saw a rapid increase in the demand for borax. This coincided with its increasing availability from Italy and a substantial reduction in price, which enabled it to be used in entirely new roles. In 1815 borax sold in London at about £400 per ton compared to £750 in 1750; but by the 1850s, when Italian borax had captured almost the whole of the European market, the price had fallen below £100 per ton.

Glass

The glass industry continued to use borax sparingly. In 1824 Michael Faraday was commissioned by the Royal Society to carry out the experimental part of a programme initiated to improve optical glasses.[56] Little of significance resulted from several years' work, but he did develop a lead borate glass of remarkably high refractive index containing 18 per cent of boric oxide and known as Faraday's heavy glass, but it was not durable enough to have any practical use in optical work. However, many years later in September 1845 its presence in Faraday's laboratory was a decisive factor in his momentous discovery that a magnetic field could cause the plane of polarized light to rotate.

Faraday's search to find a link between the mysterious forces related to electricity, magnetism and light had produced only negative results. His electromagnets had no effect on polarized light when passed through a number of

Borax-refining in the sixteenth century. Illustration from Agricola's De Re Metallica (1556); mixing vat (A), dissolving cauldron (B), crystallization tub (C) with copper wires (D) and pestle and mortar (E)

Early alternative alchemical symbols for borax

Above: Ulexite (sodium calcium borate), an exotic form called 'Bunny tail'
Above left: Tincal (sodium borate)
Left: Colemanite (calcium borate)
Below: Kernite (sodium borate)

Clear Lake, Lake County, California. First discovery of borax in North America

Death Valley, California

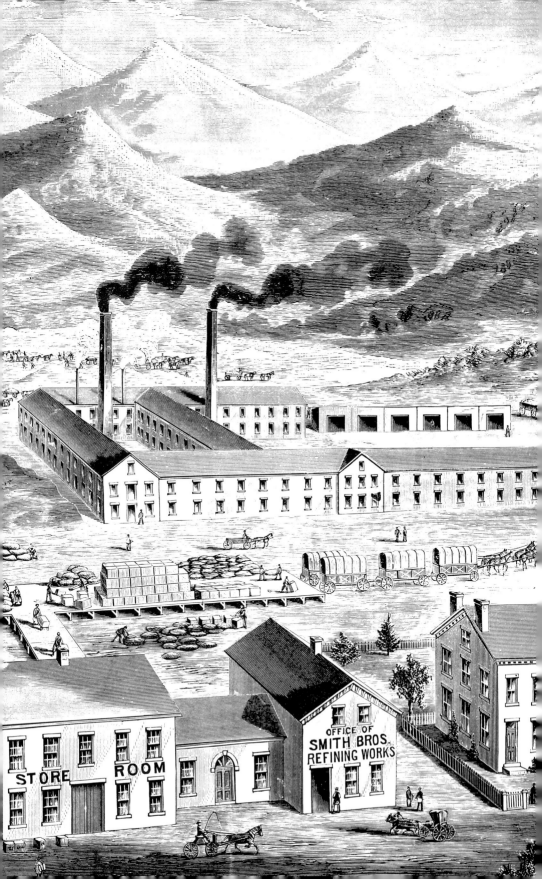

materials such as flint glass, rock crystal, or calcareous spar. However, using 'a piece of heavy glass . . . which was 2 inches by 1·8 inches and 0·5 of an inch thick being of a silico borate of lead, and polished on the two shortest edges . . . there was an effect on the polarised light ray, and thus magnetic force and light were proved to have relation to each other'. (Faraday's Diary, Entry 7504.) Now known as the Faraday Effect, this was the beginning of a new era carried further by Faraday's research on diamagnetism and later by Clerk Maxwell and others, who established the mathematics and laws governing the field theory of electromagnetism, at the core of which was Faraday's work. The application of this made possible the engineering of the most remarkable series of technological changes in industry and domestic life that man in the course of civilization had yet experienced.

An understanding of the way in which boric oxide could enhance the quality and performance of glass commenced to evolve when Otto Schott persuaded Ernst Abbe from the University of Jena to join him and Karl Zeiss in forming the Jena glassworks of Schott and Sons in Germany in 1884. Schott and Abbe investigated the introduction of a wide range of chemical elements into glass compositions: boric oxide was found to increase the dispersion of the long light-waves and to decrease that of short waves, making the dispersion more uniform, so that it became a valuable constituent of optical glasses.

Schott and Abbe succeeded where others had failed in improving the optical properties of a glass, and at the same time increasing its resistance to water and chemical attack. They also discovered that glasses containing boric oxide (called borosilicates) could be formulated which would withstand sudden changes in temperature, and from 1892 onward the manufacture of many articles such as lamps, thermometers and laboratory equipment based on borosilicate glasses was developed in Germany, Bohemia (now Czechoslovakia) and Russia. The price at which borax and boric acid was now available enabled revolutionary improvements in glass to take place, and its thermal endurance was extended beyond limits which had hitherto been regarded as final. The dangers and disasters experienced on North American railroads, where the glass in oil lamps frequently shattered in the extreme temperatures of winter (thus causing vital signals to be extinguished) were overcome by the production of a special borosilicate glass in the United States. With the outbreak of war in 1914 supplies of laboratory ware from Germany were no longer available, and Corning in the United States became a new centre for the production of borosilicate glasses. Laboratory ware was followed by battery jars and the now familiar heat-resistant oven-ware for domestic use. These heat and chemical resistance borosilicate glasses usually contain significant amounts of boric oxide, varying between 5 and 20 per cent in the end-product.

Enamel

The first enamelling of iron took place in the early part of the nineteenth century. Cast-iron shapes such as cooking pots were heated in furnaces, and enamel frit was dusted on to the metal as a dry powder which melted and stuck to the iron. The article was then returned to the furnace and the enamel melted to a smooth glaze, other coats of enamel being added later. The enamel had to be easily fusible, and borax became an important ingredient.

Enamel was first applied to sheet iron and steel in Austria and Germany about 1850. By the end of the century a worldwide trade had developed in all kinds of household goods such as dishes, bowls, buckets, bathtubs, as well as durable advertisement displays, street names and signs of all kinds. A highly technical frit industry, working closely with iron and steel fabricators, evolved, and the improved quality of borax, soda ash and other chemicals enabled rapid progress to be made. At the start of the twentieth century enamel frit was the largest single use of borax.

China glaze

In this same period of the Industrial Revolution the production of chinaware in the potteries of Europe and America saw another rapid expansion. Before this time even in the better homes, wooden bowls and common earthenware glazed with salt and magnesia had been in general use, and washbasins and water-jugs made in chinaware were known only to a few. Soon, like enamelware, these new ceramic products became part of worldwide trade, and by the 1850s a typical glaze on Staffordshire earthenware or soft porcelain often contained between 12 and 25 per cent of borax. Glazes containing borax (as boric acid) were developed in the nineteenth century to produce leadless earthenware glazes in order to remove the scourge of lead poisoning in the industry.

Medicine

Seventeenth- and eighteenth-century works on pharmacy show that the earlier medical interest in borax had virtually disappeared; however, boric acid became prominent overnight when in 1875 Joseph Lister, FRS (later Lord Lister), who had introduced the use of antiseptics in surgery, published papers in *The Lancet* entitled 'Recent Improvements in the Details of Antiseptic Surgery'.[57] He announced a new antiseptic – 'both highly efficient and much less irritating than carbolic acid' – and described his work, which had started three years previously:

> Boracic [boric] acid was then little more than a chemical curiosity. But I succeeded in obtaining in Edinburgh a sufficient quantity to enable me to test its properties. A striking instance of its antiseptic efficiency as well as its therapeutic value was at once presented by a case of 'pruritus ani' of upwards of ten years standing. . . The result was immediate relief . . .

Lister describes the preparation of boracic lint and its use, and the many areas of surgical application in which he had applied it, such as treatment of ulcers and skin-grafting. This started the worldwide use of boracic acid, in solution and in lint, both in hospitals and in the home. This was continued for over seventy years until superseded by other products considered medically more effective, and perhaps less open to misuse.

Boric acid was used extensively in foot powders and, particularly in the First World War, in the harsh conditions of the trenches.

Food-preservation

Another important discovery by Professor Theophilus Redwood in London in about 1886 was the use of boric acid as a food-preservative,[58] particularly in the case of margarine, butter and bacon. This was extremely valuable prior to the availability of industrial and domestic refrigeration, and it was practised extensively in Britain for specific applications until the middle of the twentieth century.

Production and prices

By the end of the nineteenth century world production of borates was shared between the five main areas.[59] Average production in metric tons in these areas for the years 1895–1905 were: United States 25,000 (calcium borates), Chile 10,000 (calcium borates), Peru 5,000 (calcium borates), Turkey 9,000 (calcium borates), Italy 2,700 (crude boric acid). By 1900 the tincal trade from Tibet was reduced to a few hundred tons a year, and this seems to have ceased altogether in the early years of the twentieth century. Production in Italy and Turkey remained at a fairly constant level. By 1910 the United States and Chile were the main producing areas, averaging about 40,000 tons and 30,000 tons respectively, and provided 85 per cent of world production.

The changed pattern of supply and demand had a substantial effect on the level of borax prices. In England these declined from £100 per ton before the 1850s to £35 per ton by 1880, but demand began to outstrip the ability to supply, so that by 1884 the price was up to £60 per ton. However, with increasing production in California supply again soon exceeded demand, and by 1890 the price fell to £30 per ton, and by 1900 it had declined further to about £16 (3½ US cents per lb). Such fluctuations were, of course, a great hazard to the many companies endeavouring to produce and create successful business enterprises, and set a problem not of mining or marketing but of survival. As the century neared its end there emerged the men capable of facing this problem, and the remainder of this book is in effect their story.

REFERENCES

1. Chaptal, J. A.: *Chimie Appliquée aux Arts* (1807), Vol. IV, p.249 (English translation 1807, Vol. IV, p.256).
2. 'Preparations Book' kept by Luke Howard 1799–1803: Laporte Industries archives; also article in *Chemist & Druggist* 1914.
3. Shaw, P.: *A New Method of Chemistry*, 'translated from the original Latin of Dr Boerhaave's *Elementa Chemiae*, to which are added Notes'. 2nd Edition (London, 1741) – footnote to Pt I, p.109.
4. Scott, D. F. S.: *Luke Howard* (Wm. Sessions, York 1976).
5. Demachy, M.: *L'Art du Distillateur d'Eaux-Fortes* (Paris, 1773), p.132.
6. Muspratt, S.: *Chemistry, Theoretical, Practical and Analytical* (1860), Vol. I, p.347.
7. Robottom, A.: *Travels in Search of New Trade Products* (London, 1893), p.32.
8. Anon: *Historical Notes*, c. 1903 (Archives – Borax House, London).
9. Tavernier, J. B.: *Travels in India* (English translation by V. Ball). (London, 1889), Vol. II, p.16 (1st French Edition 1676).
10. Mandelso, J.: *Voyages célèbres et remarkables fait de Perse aux Indes oriental.* New Edition (2 parts), Amsterdam 1727, p.202 (1st Edn. Leyden 1718).
11. Hill, J.: *Theophrastus's History of Stones* English version (London, 1746) p.70 (footnote).
12. Model, J. G.: *Chymische Nebenstuden* (St Petersburg, 1762), p.192.
13. Demachy, M., *op. cit.*, pp.132 *et seq.*
14. Kirwan, R.: *Elements of Mineralogy* (London, 1784), p.206.
15. *Phil. Trans.* London, 17 May 1787 (1) Blane – 'Some Particulars relative to the production of borax', Vol. 77, pp.297–303. (2) Rovato: Letter 'Concerning some observations relative to borax', Vol. 77, pp.471–3.
16. Saunders. R.: (i) *Phil. Trans.*, Vol. LXXIX (1789), p.79; also (ii) 'Observations sur l'origine du tincal au borax' in *Annales de Chimie* (1789), Vol. 2, pp.299–301.
17. Turner, Capt. R.: *Account of an Embassy to the Court of Teshoo Lama in Tibet* (London, 1800), p.281.

18. Cibot, P. M.: see Ch. I, *n*. 22.
19. Montgomerie, T. G.: *Journal of the Royal Geographic Society* (1868), Vol. 36, pp.129–219. 'Report on the Route Survey made by Pundit – from Nepal to Lhasa' and 'Extracts from a Diary kept by Pundit – during his journey from Nepal to Lhasa' – ref. Borax, p.161. (Pundit Nian Singh's name was withheld for security reasons.)
20. Anon: *A History of the Borax Industry*, *c*. 1903 (Archives – Borax House, London).
21. Robottom, A., *op. cit.*, pp.29–30.
22. Cunningham, Major Alex.: *Ladok* (London, 1854).
23. von Schlagentweit, H.: 'Ueber das Auftreten von Bor-Verbindungen in Tibet', in *Sitzungberichte der Math-phys. (Cl.d.k.b.) Akademie des Wissenschaften* (Munich, 1878), Vol. 8, pp.509–38.
24. Burrard, S. G.: *A Sketch of the Geography and Geology of the Himalayan Mountains and Tibet* (Calcutta, 1907), p.265.
25. Watt, G.: *A Dictionary of Economic Products of India* (Calcutta, 1889).
26. Chaptal, J. A.: *Chimie Appliquée aux Arts* (1807), Vol. IV, p.247.
27. Chardin, J.: *Voyages du Chevalier Chardin en Perse* (Amsterdam, 1735).
28. Bergman, T.: *Manuel de Mineralogiste ou Sciographie* (Paris, 1784), p.59 and Chaptal, J., *op. cit.* Vol. IV, p.248, *n*. 25.
29. Borax Consolidated, Ltd: 'Report on a Visit to Iran by K. R. Greenleaves' (October 1960).
30. Tancred, T.: *On the collection of Boracic Acid from the Lagoni of Tuscany* (The Ashmolean Society, Oxford 1837), p.11.
31. Robottom, A.: *op cit.*, pp.36–87.
32. Mazzoni, A.: *The Steam Vents of Tuscany and the Lardarello Plant* (Bologna, 1954).
33. Chaptal, J. A.: *Elemens de Chimie* (Montpelier, 1970) Vol. 1, p.242.
34. Robottom, A.: *op. cit.*, pp.159–63; and Anon: *A History of the Borax Industry* (*c*. 1903), Borax House archives.
35. (i) Veatch, J. A.: *California Acad. Sci. Proc.*, Vol. 2, pp.7–8, 1863 (read before the Society 17 Jan. 1859) gives Jan. 1856 for the discovery of borate of soda in the mineral spring in Tehama County. (ii) Browne, J. R.: *Mineral Resources of the States west of the Rocky Mountains* pp.178–87 (1867) quotes Veatch's earliest report as June 1857, and describes the process of borax-production at Borax Lake, Lake County, Cal.
36. (i) Biringuccio, V.: *Pyrotechnica* (Venice, 1540).
    (ii) Agricola, G. – see Ch. 1, *n*. 13.
    (iii) Ercker, L. *Beschreibung allerfürmeisten Mineralischen Ertzt vund Bergwercks arten* (Prague, 1574).
    English Translation by Sir John Pettus *Fleta Minor* (London, 1683), pp.116, 213.
37. Cramer, J.: *Elementa Artis Documasticae* (Leyden, 1739). English translation *Elements of the Art of assaying metals* (London, 1741), p.41.
38. Pryce, W.: *Mineralogia Cornubiensis* (London, 1778), p.249; and Richardson, W.: *The Chemical Principles of the Metallic Arts* (Birmingham, 1790), p.60.
39. Biringuccio, V.: *op. cit.*, Book II, Chapter 14; Agricola: *De Re Metallica* (Basel, 1556), Book XII.
40. Neri, A.: *The Art of Glass* (London, 1662); 1st English Edition (translated by Christopher Merrett), p.340.
41. Kunckel, J.: *Ars Vitraria Experimentalis* (Frankfurt and Leipzig, 1679), Part I, p.206; Part II, pp.57–9.
42. London 1738, pp.70–2, 77–8, 82. (The date and author of the original German text is obscure.)
43. Cramer, J.: See above, English translation 1741, p.440, *n*. 33.
44. Dossie, R.: *Handmaid of the Arts* (London, 1796), New Edition, Vol. II, pp.171 and 188–9 *op. cit.*
45. Hirth and Roskill: 'Chau Ju-Kua on the Chinese and Arab Trade in the 12th and 13th Centuries' (1911), Imperial Academy of Sciences, St Petersburg.
46. (i) Neri, Merrett, Kunckel: *Art de la Verrerie*, French translation (Paris, 1752), 6th Book, pp.203–18. (ii) de Blancourt, M. H. *De l'Art de la Verrerie* (Paris, 1697), pp.329–411.
47. Dossie, R.: *Handmaid of the Arts*. New Edn. (London 1796), Vol. I, pp.196, 202.
48. Leach, Bernard: *Kenzan and His Tradition* (London, 1966).
49. Anon: Tract entitled 'Secret des Vraies Porcelaines de la Chine et Saxe' (Berlin, 1750), included in above ref. 46 Neri etc. pp.606–16.
50. Dossie, R.: *Handmaid of the Arts*, 2nd Edn. (1764), Vol II, p.360.
51. Macquer, P. J.: *Dictionnaire de Chimie* (Paris, 1766). Article: 'Porcelaine'.
52. Dossie, R.: *Handmaid of the Arts* (1764) 2nd Edn., Vol. II, p.367.
53. Ricard, S.: *Traité général du commerce* (Amsterdam, 1781), Vol. 1, pp.83–105.

54. Cronstedt, A. F.: *An Essay towards a System of Mineralogy* (1st Edition, Swedish 1759). Trans. by Gustav von Engestrom, to which is added a Treatise on the Pocket-Laboratory written by the Translator – Revised and Corrected by E. M. da Costa, London 1770.
55. Bergman, T.: *De Tubo Ferruminatorio* (Vienna, 1779). English version included in *Physical & Chemical Essays*, trans. by Dr Cullen, London (1788), Vol. II, Essay 25. (Bergman attributes first use of the blowpipe technique – used with a candle-flame – to Swab, a Swedish mining metallurgist, in about 1733.)
56. Faraday, M.: *Experimental Researches in Chemistry and Physics*, (London, 1859), pp.231, 256.
57. Lister, J., FRS: 'On Recent Improvements in the Details of Antiseptic Surgery', in *The Lancet*, 1 May 1875, pp.603–5, and 5 June 1875, pp.787–9.
58. Redwood, Prof. T.: Manuscript Report, dated 14 August 1887.
59. Vale, C. G. and Gale, H. S.: *The Production of Borax in 1911* US Geological Survey (Washington, 1912), p.6. Production Statistics.

3

Where North American borax began

The currently accepted date for the earliest discovery of borax in the United States of America is 8 January 1856, when Dr John A. Veatch evaporated some water from a mineral spring situated a few miles east of Red Bluff in Tehama County, California.[1] These springs were already being used by the health resort of Lick Springs, and the Doctor was trying to ascertain more precisely the medicinal properties of the water. His test produced crystals which he recognized as 'biborate of soda',* and being an American doctor in an age of colonial expansion in the Far West, his mind at once turned from medicine to the commercial possibilities opened up by his discovery.

The USA had already been importing borax for some years when this discovery was made, and imports are estimated to have reached $200,000 in total annual value, with the retail price in drug stores running at some 50 cents a pound. The lateness of the discovery of extensive indigenous deposits of borax can possibly be explained by the fact that their locations, the states of California and Nevada, were provinces of Mexico until 1848.

Needless to say, this discovery – made at a time when the market price for imported borax was so high – started a borax rush which went on most of the time until the end of the century. It was to have consequences for the market price which the first fortune-seekers had not paused to consider.

Dr Veatch started exploring the country around Lick Springs with Dr William Ayers, and found other springs containing borates. Local rumours about hills and flats white with some efflorescent substance were also followed up, and eventually a small muddy lake, distinguished by the overpowering smell of 'rotten eggs' (sulphuretted hydrogen), was found in the vicinity of Clear Lake (Lake County). This aroma was a known feature of the fumaroles of Tuscany and of the Tibetan borax lakes, and bottles of the unattractive liquid from the Clear Lake area were found, on analysis, to contain around 18 per cent of borax.

Obviously, the water of this small lake and the mud beneath it were

---

*A chemical name frequently given to borax at this time.

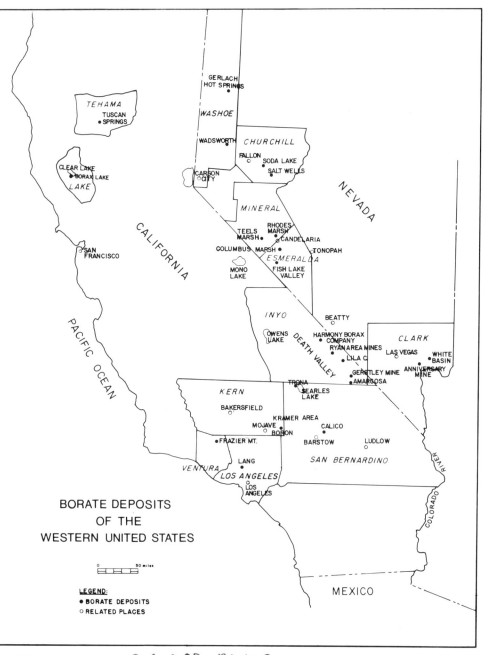

Map III

impregnated with borax, and after the two doctors had obtained a location permit and had driven a pipe down through the mud they found perfect crystals of pure borax weighing up to a pound each.[2] Large heaps of these were recovered from the depths of the lake, and it seemed to them that, calculating at the current market price of borax, large fortunes lay just round the corner. However, the doctors had no capital of their own, and while they were trying to raise it in San Francisco the American Civil War broke out, with the result that the 'Borax Company of California' did not start operating until 1864. In the first year only twelve tons were produced from 'Borax Lake', as this small site of about fifty to two hundred acres (depending on seasonal rainfall) came to be called, and annual production never exceeded a total of 220 tons in the five hopeful years of its existence.[3] However, expectations about the project ran into astronomically high forecasts of output, including ten million tons in a Government statement, so that in anticipation of a glut the American retail price of borax dropped rapidly from fifty to thirty cents a pound. Unfortunately, by 1868 the crystals in the mud at the bottom of the lake were already beginning to show signs of rapid depletion, and the whole project suddenly collapsed with the unexpected dilution of the waters of the lake. This was variously ascribed at the time to an artesian well sunk into its bed, becoming an uncontrollable 'gusher', and to excessive rain. Whatever the reason, hopes of further profitable recovery of borax ceased.

This was a major setback for the two doctors, as Lake Hachinhama – which was a smaller edition of Borax Lake, and was discovered in the same district – contained no crystals.[4] However, borax recovered from its waters by evaporation enabled production to be restarted in 1872. Although only 140 tons were produced in that year it sold at thirty-two cents a pound.

In 1860, following his discoveries at Clear Lake, Dr Veatch had also tested the waters of Mono Lake near the California–Nevada state boundary, had found traces of borax in them, and had mentioned this to a William Troup of Virginia City, who had started prospecting on his own. Eleven years later, in 1871, Troup found 'cottonballs'* (i.e., ulexite – sodium calcium borate) both on the desert a few miles from Columbus Marsh, Nevada, and also at Salt Wells, forty-five miles south-east of Ragtown, near Dayton, Nevada.[5] Folk-tales have it that Troup took his Salt Wells sample to Ragtown, borrowed the wash-boiler of a Mrs Kenyon, boiled up the cottonball with some water and carbonate of soda, and found he had produced the first borax ever made in Nevada.

Others, however, were on his trail. A Nevada teamster had brought back another sample of cottonball from near Wadsworth in 1869, and it is said to have got into the hands of a chemist in San Francisco who put a pack of businessmen on the scent. In fact, once this easily identifiable stuff had been recognized as containing borax there was bound to be others who would make finds, and soon the whole area around Columbus Marsh was bristling with claims, processing plants and new borax companies. By 1872 the Pacific Borax Company had started operations at Columbus Marsh and another group, the American Borax Company, moved in on the property at Salt Wells (where Troup had found the

---

*'Thus named from the silky felted or interlaced crystals which the globular masses show when broken. They were from the size of peas up to twelve inches in diameter' – Undated (probably about 1887) Report of the State Mineralogist.

cottonball which he cooked in the wash-boiler). Troup himself thereupon disappears from history, following Dr Veatch and other pioneers into oblivion.

1872, however, is remembered in borax circles for quite a different reason. In that year Francis Marion Smith, the thrusting youngest son of a small Wisconsin farmer (who had been wandering round the mining camps of the West since 1867, and rapidly changing jobs from dishwashing to mine prospecting, to running a restaurant tent, and then to picking up 'wood-ranches' and supplying timber and fuel to the silver camps in Esmeralda County), decided to move his wood-selling pitch and his pack-train to where the action was at Columbus Marsh.[6] Needless to say, he thought of joining in himself, but saw little hope of quick profit with so many competitors about – until one day his attention was attracted to a similar but smaller alkali flat some twenty miles north-west of Columbus, and hitherto lying absolutely deserted in a barren, featureless landscape.

Smith hopefully set himself the impossible objective of trying to 'corner' the 12–15 square miles of this area, 'Teels Marsh' – provided, of course, that the glittering white surface in the distance really turned out to be cottonball. It was easy enough to visit it with two of his wood-choppers to take samples, but the problem was to get them analysed without starting desert rumours buzzing. The only assayer for hundreds of miles was the one at Columbus Marsh, and inevitably when Smith returned to his temporary camp on the marsh (after collecting a certificate of assay, which showed his samples to be extremely fine specimens of borate of soda) a friend of the assayer was waiting for him, and asking to be directed to Teels Marsh. Smith sent him off in the wrong direction with one of his choppers, and proceeded at once to stake out the whole Marsh in locations – although he left one for the assayer who turned up a few days later. This did not help Smith much, however, for when the United States Commissioner of Mines, Drummond, was besieged by the losers in the race for Teels Marsh he ruled that borax lands came under a new Act of 1872, which limited each mineral claim to 20 acres instead of the 160 acres allowed under the saline laws of the state of Nevada. This was to prove only the first of several occasions on which the question of whether borax is by definition a mineral or a salt – and therefore which particular piece of legislation governs it – has upset the best-laid plans of expanding borax companies.

In this instance the decision caused Smith several complicated and difficult months. He had now to admit legitimate claimants to areas that he could no longer retain officially in his own name, or in those of his friends. Also he had frequently to run off claim-jumpers, by appeal to the courts when they arrived in dangerous-looking gangs or at the point of his own gun when he judged this as likely to be sufficient to put them to flight. Soon, however, the laws of economics changed in his favour. The price of borax started falling with increased production, while the scarcity and cost of wood, labour, and transport remained burdensome, so that the realization started dawning on the small men in the game that processing and moving borax from that remote area was far more difficult than they had imagined. Smith then moved in to start buying them up.

The wood-contractor, however, needed a great deal more capital if he were to succeed in this objective, and one of the less well-documented aspects of this part of the career of Francis Marion Smith is the sources from which he drew the initial wherewithal to build up his own empire – mainly by buying up other people's.

In the case of this first venture into borax, however, the explanation is clear. He went into partnership with his elder brother Julius, who was in the farm implement and machinery business, and managed to interest the Chicago firm of Storey Brothers in the venture. Since this was a time when output of borax from the established plants at Columbus Marsh and Little Borax Lake (Hachinhama) had already depressed the market, and the price had fallen from thirty to ten cents a pound, Smith brothers managed to buy out Storey Brothers' interest within a few months, and to go on to buy out many other locators on Teels Marsh as well. Long before the process was over Teels Marsh had become an official Mining District, with scores of registered locators. By May 1874 several of these, including Smith Brothers, had set up their own small mills and sported chimney-stacks. By the end of 1874, however, Smith Brothers and then Blanco Vale Borax Company – in which Smith had bought out the seven other original members – were the sole producers on the Marsh.

An explanation of how Smith managed to achieve all this lies in the connection he had managed to establish with William T. Coleman and Co., 'a large and representative commission house in San Francisco' to whom Smith shipped his borax for sale. A document, dated 25 November 1874,[7] shows that Coleman lent Smith Brothers 'ten thousand seven hundred and fifty dollars gold coin' to acquire borax properties in the Teels Marsh area and there is also reference to an earlier agreement dated 18 April 1874 made for the same purpose. From that time Smith and Coleman established a working understanding, while remaining independent in ownership and control of their respective companies. It was a successful relationship that covered a period of some fifteen years.

William T. Coleman had risen from nothing, like almost all the other great figures in the industrial history of the United States in this period of expansion after the Civil War; but he was one of the few at this time to forsake the California gold-rush of 1849 and to see the steadier advantages to be gained from setting up as a middleman on the fringe of this and various other get-rich-quick ventures in the deserts of the hinterland. He established his commission house in San Francisco in 1850 and became a most respected citizen through acting as chairman of a new Committee of Vigilance which successfully cleaned up an appalling level of crime in the town in 1856.[8] His business spread from minerals to the export of California dried fruits, to fish-canneries, and to financial support for most of the pioneer industries in the state, while branches sprang up in Los Angeles, Chicago, New York, and London.

By the time that commercial production of borax first began at Borax Lake in 1864 W. T. Coleman & Co were one of the most solidly established businesses on the Pacific Coast, and their unhesitating pursuit of the commercial advantages in disposing of the borax on behalf of its producers gave the latter a novel air of respectability. Smith Brothers started using Coleman's services as selling agents in 1873, a year when Government action reinforced the effect of a steadily falling price in shaking out of the industry all except the toughest firms. A duty of five cents on the import of crude boric acid (from Tuscany) and crude borates had existed since 1867, but both articles were placed on the free list in 1872, and the tariff barrier remained down until 1883. On top of this the legislature of Nevada began taxing the proceeds of borax and soda mines in 1873, and the overall result of these various impositions was that by February 1875 only five 'marsh borax' companies remained in existence[9] – the Smith Brothers on Teels Marsh, the Pacific Borax Company on Columbus Marsh (later acquired by F. M. Smith), Griffing & Wyman at Fish Lake fifteen miles south of Columbus, Rhodes &

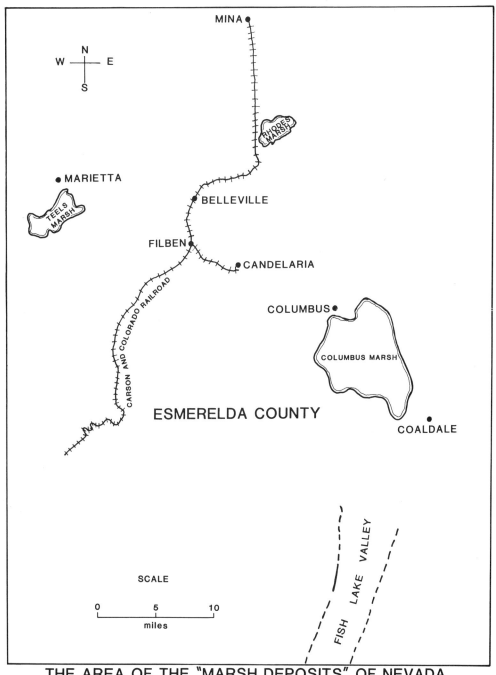

THE AREA OF THE "MARSH DEPOSITS" OF NEVADA
worked from 1871 until about 1890

Map IV

Wasson at Rhodes Marsh, and Searles & Co. near Panamint, at what later came to be named Searles Lake.

As mentioned earlier, Smith was allied with Coleman, and the only rival they feared was Searles & Co. John and Dennis Searles had been gold-hunters in the Mojave region of California, and Dennis had visited Death Valley in 1860, long before borax was discovered there. About 1864 they heard about the money being made out of borax from Borax Lake many miles to the north and this recalled to their minds the great white expanse of dry lake that they had once crossed while prospecting near the Slate Range, south-west of Death Valley. They revisited it to take samples for testing at San Francisco, but at first they received little encouragement, so forgot about it for nearly nine years. They then heard about the borax stampede precipitated by Smith's discovery of borax in precisely similar surroundings at Teels Marsh, and at once rushed off to 'discover' Searles Lake officially on 14 February 1873. As the later chapters in this book will show, this was an event destined to be of great significance for the future, but the immediate sequel was almost identical with the story of Teels Marsh – a host of placer* claims, the disappointing limitation by law to twenty acres a claim, and the gradual disappearance of all rivals, with the Searles brothers being left in sole possession. They started production in the spring of 1874, and took on as sales agents W. L. Babcock & Co of San Francisco. Smith and Coleman, however, soon found that they had little to fear from the rivalry of Searles and Babcock. An analysis of Searles Lake surface material which had been obtained by stealth showed it to be of decidedly lower grade than that at Teels Marsh.

Pacific Borax Company had been started at Columbus Marsh in January 1872 by Joseph Mosheimer and Emile Stevenot, who earlier had first refined borax in San Francisco;[10] however, with declining borax prices it did not prosper, and it closed in August 1876, owing money to Coleman among others. In anticipation of the arrival of the narrow-gauge Carson and Colorado railroad in 1881, which linked the Candelaria area with main-line routes, Smith acquired the assets of the bankrupt Pacific Borax Company, and erected a new borax-refinery at the north end of Fish Lake. This started production in mid-1882, and with the Teels Marsh refinery now also close to rail transport, he had the foremost position in the area, or indeed anywhere in North America.

By 1878 the price of refined borax declined disastrously to just under 9 cents a pound, but sales remained unresponsive, so that the six tons or so a day produced by the little group of 'marsh borax' producers still sufficed to meet existing demand. Clearly the market must be expanded, and it was at this point, it is said, that Julius Smith conceived the idea of popularizing borax in package form in the

---

*On 16 May 1919, while discussing a current case, Frank R. Wehe, a lawyer advising Pacific Coast Borax Co., summarized part of the judgment in Duffield v. San Francisco Chemical Co. (205 Fed. Rep. p. 480) as follows: 'The Court then discussed the definitions of placer and lode deposits, which hold that all rocks in place or deposit whether solid or broken, between walls are lode deposits and that placer deposits are superficial deposits, and [by this] is meant ground within defined boundaries which contains mineral within its earth, sand or gravel, ground that includes valuable deposits not in place – that is, not fixed in rock but which are in a loose state. . . .' Making a claim to a new discovery under the correct definition of 'placer' or 'lode' has usually been of vital importance to subsequent ownership assertions.

Eastern States, and he took a store with an office at 155 Water Street, New York. However, the pulverized substance which first arrived from the West was not suitable for household use, so a small plant for turning it into powder was set up at Jersey City. F. M. Smith then initiated a promotion campaign to push sales of borax in the households of New York. He never did things by halves, and the literature used contained as exaggerated a 'puff' as has ever appeared before or since. Of the small deposit and modest plant at Teels Marsh it was written: 'The brothers have perfected an immense establishment and are producing an enormous quantity of a chemically pure article of borax which stands first and is in demand in every household, to whom it is supplied by grocers and druggists throughout the country.'

Each package of borax bore a view of 'Smith Brothers' Mine and works in Nevada' which was actually entirely a figment of the artist's imagination. The selling agents for Smith Brothers set about educating the public in the uses of the product 'which have hitherto been imperfectly apprehended', and bringing to it a better appreciation of the properties and peculiar characteristics of borax.

Some of the 'characteristics' they claimed for it were indeed peculiar. The use of borax in washing-water, for example, would prevent diphtheria, which 'chemical investigation' had traced to the poisons in soap fat. Lung fever and kidney trouble had their origin in this same 'morbid source'. Borax in the water would counteract the baneful effects of these poisons, and at the same time make the skin beautifully clear and smooth. If you suffered from nothing more noxious than a nervous headache borax would relieve that too. The directions were simple: a shampoo with a strong solution of borax, followed by a rinse in clear water. Then

> let the person thus suffering remain in a quiet, well ventilated room until the hair is nearly or quite dry and, if possible, indulge in a short sleep, and there will remain scarcely a trace of the headache. If clergymen, teachers and others who have an undue amount of brain work for the kind and quality of physical exercise usually taken, would shampoo the head in this manner but once a week, and then undertake no more brain work until the following morning, they would be surprized to find how clear and strong the faculties had become, and there is reason to hope there would be much less premature decay of the mental faculties.

Certainly the Smith brothers themselves never appear to have shown any sign of premature decay of the mental faculties, but very little evidence has been left to show how this effort to increase business succeeded over the next few years, even with the aid of a declining borax price. This price trend was reversed from late 1878, the California borax-producers raising prices by 25 per cent by mutual agreement, and prices continued to improve in spite of severe competition from borax made from imported Italian crude boric acid. Then in 1883 the five cents import duty was reimposed on boric acid, with a similar duty on imported refined borax and a three cents duty on imported borate ores, which had not hitherto been protected. This general renewal of protection had paradoxically the effect of once more reducing borax prices in the American market, as it encouraged increased home production from abundant sources. There were also large stocks of crude boric acid overhanging the market[11] which had been imported by Pfizer and Co. and others just before reimposition of the tariff.

It was in this unfavourable climate that Smith began a series of further moves that laid the foundations of his personal fortune. As a first step, his partnership with his brother came to an end in 1884. Although no evidence remains of whether this parting was friendly or otherwise, it was probably unfriendly, since F. M. Smith (having intentionally allowed himself to be bought out in return for payment in borax costed at market price over a long period of time) promptly started to develop other borax deposits that he had acquired at Columbus Marsh. Julius Smith, apparently, did not foresee that with the new tariff of 1883 overproduction in the USA might again depress prices, and so found himself with a ruinous and profitless obligation to continue paying in devalued borax to his brother. F. M. Smith, therefore, did not have to wait long to buy back his own interest in Smith Brothers and in turn buy out his brother, who retired to a vineyard in California.

F. M. Smith thus became sole owner of Teels Marsh, and incidentally of a valuable interest comprising 16,000 acres and a store on some town lots in the town of Columbus, commonly known as the Pacific Borax Property.

The events that followed show Smith in his prime, combining his hard-won experience of the mining game in the unfriendly desert with a flair for sales promotion and all the instincts of a company promoter bent on impressing others with the soundness of his financial position. The next step was the incorporation under California law on the 29th of March 1886 of the Pacifix Borax, Salt and Soda Company,[12] with an authorized capital of half a million dollars consisting of 100,000 shares of five dollars each, of which 38,000 dollars had already been subscribed for 7,600 shares. Smith now had an office in San Francisco at 54/55 Merchants Exchange, and the first meeting was held there on 20 April. Smith with 5,000 shares was elected President, and J. W. Mather, a commission agent who had been involved in the borax trade since its very beginning in California, was elected Vice President with 1,000 shares. The remainder of the 1,600 issued shares were held by the other three directors, Alton Clough, F. B. Pritchard and J. H. Maynard. A week later the Board met again and agreed to the acquisition from Smith of the Pacific Borax property referred to above in exchange for the balance of the (92,400) unissued shares to be placed in the name of 'Smith himself or any such persons desired by him'. Teels Marsh remained in Smith's ownership.

An important factor in Smith's future was the energetic and resourceful team that he had now got together, and which history shows played a large part in his success. In addition to J. W. Mather, there was John Ryan, a young Irish immigrant who had joined Smith in Teels Marsh back in 1873. He was to become a tower of strength, first to Smith and later to the Pacific Coast Borax Company, and apart from a short spell prospecting in South Africa he stayed with them for the rest of his life.

About the same time another young man, Chris Zabriskie[13] from the little town of Candelaria in Nevada, was appointed Superintendent of the Pacific Borax, Salt & Soda Company. Zabriskie was reputed a smart lad locally and a Jack-of-all-trades, knowing something of mining, engineering, telegraphy and bookkeeping. Early on he was in partnership with Mr Albright, the local undertaker, whose place of business carried the inconspicuous sign:

<pre>
          ALBRIGHT AND ZABRISKIE
          A to Z —   We Get Them All
               You Kick the Bucket and
                   We Do the Rest
</pre>

He joined Smith at the age of twenty-two, and also stayed with the borax business for the rest of his life.

The third participant from the same area was Fred Corkill, a mining engineer from the Isle of Man, who had worked previously in a Nevada silver-mine. Corkill was Smith's most experienced mining man, and for nearly thirty years he was in overall charge of Smith's mining operations and the all-important field of mineral exploration and acquisition possibilities.

Over-production among American producers had by 1887 resulted in a disastrous situation, and the price of refined borax had reached a record low of 5¾ cents per pound. In its first year Smith's new company had paid one 10 cents dividend, but by mid-1887 there was little prospect of further dividends or profit for any of those marsh borax producers that still survived.

There is no evidence that Smith had any formal partnership or joint company link with Coleman, other than the loans which Smith received in 1874 to develop his production at Teels Marsh, and which may well have been repaid with refined borax. As will be described later, Coleman himself became involved in borax-production early in the 1880s. Many of his early mining claims in the Furnace Creek area were filed in several names because of the 20-acre limitation imposed on individuals.[14] Some do show the names of both Coleman and Smith, which has probably given rise to the suggestion made by several writers[15] that they were partners in the borax business, specifically in Coleman's Harmony Borax Works. Records show that by the end of 1882 Coleman owned outright most if not all of his claims,[16] and there is nothing in the records of Smith's various companies or in his own account that supports this idea of a formal link with Coleman.

At the beginning Coleman had been Smith's sole link with the market, but to Coleman borax was a small part of his merchanting business, while to Smith it was his whole life, an important part of which became the understanding and the influencing of the market. However, he was shrewd enough not to clash with Coleman, and at the end of 1883, when Coleman started to produce borax himself,[17] the friendship and understanding that Smith and Coleman had developed was an important factor in keeping this new industry viable under conditions which steadily declining prices had made very difficult. Thus, in the gloom of 1887 it would have been Smith and Coleman who put their heads together and organized the 'San Francisco Borax Board', the constitution of which was agreed on 15 October 1887,[18] by the following five persons and the interests they represented:

(1) F. M. Smith – The Pacific Borax, Salt and Soda Company (Apr. 1886) and the Teels Marsh Borax Company (May 1887).
(2) W. T. Coleman – The Harmony Borax Company (May 1884) and the Meridian Borax Company (November 1884).
(3) H. L. Coye – Nevada Salt and Borax Company (January 1882).
(4) E. C. Calm – The Columbus Borax Company.
(5) J. W. Searles – The San Bernardino Borax Mining Company (June 1878).

At its first meeting the Board agreed that the voting strength of each of the parties and the allocation of annual production of refined borax should be:

Smith (28 votes) 1,400 tons; Coleman (24 votes) 1,200 tons; Coye (9 votes) 450 tons; Calm (5 votes) 250 tons; Searles (10 votes) 500 tons.

They also agreed to dispose of the production at agreed prices. The Board met monthly over a period of about eighteen months. The effect on the finances of Smith and the Pacific Borax Salt and Soda Company was instantaneous, and at the end of November the Company declared a dividend of one dollar per share on a restructured capital of 5,000 shares of $100 each, and in the next three years continued to declare a dividend of this amount at approximately monthly intervals. Thus Smith and his nominees began receiving dividend income of about $58,000 a year. The decline in the price halted, and borax moved up to 6½ cents by the year's end and to 7½ cents per pound during 1888, where it remained for some years.

In February 1888 Smith made a further significant move. He resigned from the Board of Pacific Borax, both as President and as Director, and at a meeting held three days later the Board approved the first of many offers from an 'independent' Smith to purchase borax for resale at an agreed price.[19] For the next eighteen months almost every Pacific Borax board meeting records the acceptance of a formal offer from Smith by letter in the following somewhat quaint terms:

> Knowing your company to have in storage and en route to San Francisco, borax amounting to somewhere in the neighbourhood of 145 tons, I herewith submit you the following proposition: Will pay six and one half cents per pound less five per cent discount, for the whole quantity, delivered San Francisco. In payment for above borax, I will give my note, payable one day after date, and bearing interest at the rate of six per cent annum. An early answer is desirable.

This scheme was maturing nicely when suddenly on 8 May 1888 an unexpected event occurred of great significance to Smith's future plans. The rumours of the previous day crystallized into the definite news that Coleman had crashed, with liabilities that he could not meet, to the extent of some twenty million dollars. It was not losses on his recently acquired borax-mining activities which had ruined him. It was the old and, alas, ever new story of a commercial empire built on credit, and of a temporary lapse of confidence preventing the right finance coming forward to plug the gap left by an expired loan at precisely the right time. Ironically enough, the immediate cause of his fall was, it is said, that he was about to sell his borax interests – in which he had invested about a million dollars – to a group of foreign capitalists (unnamed) for twice this sum to raise money. Then came news of the proposed Mills Tariff Bill, which threatened to put borax back on the free import list again, and the deal collapsed. The Mills Bill never became law, but by 8 May 1888 Coleman's properties, including his borax interests, were in the hands of a receiver.

Shortly after this event Smith further strengthened his position by becoming 'Sole Agent of the Borax Producers of the Pacific Coast'[20] and handling officially all the production of the members of the San Francisco Board. This Board continued to meet until May 1889, by which time Smith was already complaining that he was having to carry the whole thing on his back. In the meantime, the proceeds of the sales of Coleman's borax were being paid to his estate and he continued to attend the meetings of the Board in spite of his financial troubles. By December most of his creditors had agreed to settle[21] for 40 cents in the dollar; however, Coleman was a man of integrity, and before he died in 1893 he had paid off all his debts in full, although he was not legally bound to do so.[22]

The newspapers, especially the *Examiner*, were now starting to work up an outcry against trusts and protective tariffs in general, with the particular example of borax as a sitting duck, but by 2 July 1890, when the famous Sherman

Left: An imaginative advertisement drawn by a New York artist of Smith Bros works at Columbus Marsh, Nevada 1876

Mule Team transport at Harmony Borax Works about 1888. (This picture shows only eighteen animals.)

Smith's letter-heading after setting up as an independent agent of 'The San Francisco Board'

OFFICE OF F. M. SMITH
SOLE AGENT OF THE BORAX PRODUCERS OF THE PACIFIC COAST
230 MONTGOMERY STREET

SAN FRANCISCO, Nov. 9th 1888.

Messrs. J. W. Mather.

Dear Sirs:

BORAX: We are now prepared to supply the Trade with all grades of Borax, viz: Concentrated, Refined and Powdered.

We quote you prices, net cash, f. o. b. at San Francisco, for the current month as follows:

Concentrated Borax, in Sacks 6 3/4 cents per pound. ⎫
Refined Borax, in Barrels   7 1/2   ”   ”   ⎬ Carload Lots.
Powdered Borax, in Barrels   7 1/2   ”   ”   ⎭

We respectfully solicit your orders.

Yours truly, F. M. Smith.

Professor Theophilus Redwood, pioneer of the Pharmaceutical Society

Professor Redwood's opinion on the merit of 'Preservitas' as a food-preservative

Dr REDWOODS,
ANALYTICAL DEPARTMENT.

17, Bloomsbury Square,
London, 14th August 1887
W.C.

I have made numerous experiments with the Saline Compound termed Preservitas, for the purpose of testing its properties as an antiseptic.

Its effect in counteracting or retarding the changes to which many kinds of food are subject is of a very marked character. It appears to check the tendency to decay in organic substances used as food, by increasing the stability without altering the nutritive qualities of their constituents. Being itself like common salt, both neutral and harmless, its addition to food in the small quantities in which alone it is required, is free from any objection. In my opinion founded on experiments, it is the most efficacious and best of the preparations used for the preservation of food.

T. Redwood, Ph.D., F.I.C., F.C.S.
Professor of Chemistry and Pharmacy

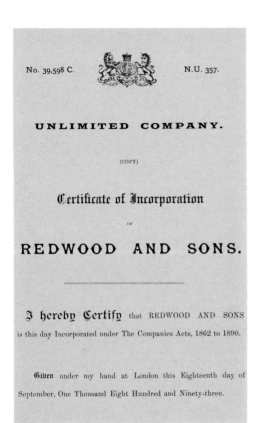

Early trademarks registered by R. C. Baker in the name of the Preservitas Company and used for Redwood's products

'Redwoods Unlimited', the company formed to produce food-preservatives containing borax and boric acid

The Founding Fathers of Pacific Borax and Redwood's Chemical Works Ltd, later to become Borax Consolidated Ltd. From left to right, R. C. Baker, F. M. Smith and J. Gerstley

A selection of early 'Borax' letterheads

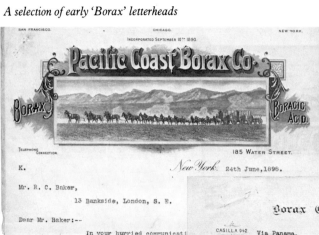

Anti-Trust Bill became law, the San Francisco Borax Board had been out of existence for over a year. Meanwhile Smith, with the assistance of the able Mather (who had been shopping around the banks in New York), managed to get together the funds to acquire in March 1890 all Coleman's borax properties from the trustees for $580,000, payable by instalments.[23] Then, in October of the same year Smith floated the Pacific Coast Borax Company, which acquired all the assets of the Pacific Borax, Salt and Soda Company along with Smith's own Teels Marsh Borax Company and the all-important Coleman properties. Thus, fortified with dividends, sales commissions and cash from the sale of borax bought from his own company with the 6 per cent notes which he issued, Smith was well on his way to becoming the Borax King of the Western World.

The story of how these Coleman borax interests were built up still remains a fascinating one, while Death Valley holds its place in the romance and history not only of California, but also of the great nineteenth-century epic of America – the conquest of the West. Today when thousands of '49ers annually commemorate the early history of California they choose to meet in the National Monument of Death Valley. The beauty and colour of the area remains unchanged and unspoilt, but today the tale of man's effort and ingenuity in overcoming the severity of the environment, in the days when the West was really wild, is largely told in the history of the borax industry. Wisely, the remaining traces of that industry's presence have now been carefully preserved.

Death Valley proper is a long, narrow fault basin, and lies in California near the Nevada state line. It is about 130 miles long and from 6 to 15 miles wide, and it is walled in by towering mountain ranges, the Panamint Range on the west and the Funeral and Black mountains on the east. It has the distinction of being the lowest, hottest, driest spot on the American continent, and is at one point called Badwater 279·4 feet below sea-level.

Current geological opinion puts a date of around ten million years ago on the thrust faulting which produced the two mountain boundaries of today's Death Valley,[24] although other geological features of the area – known as the Titus Canyon, the Artists Drive, and the Furnace Creek formations – are products of various stages in the preceding thirty million years. During these incomprehensibly vast ages rain and snow undoubtedly fell from time to time, bringing mineral salts down from the higher levels into land-locked lakes, and producing temporary spells of vegetable and animal life on the valley floors. Animal footprints have been found in the Furnace Creek formation dating from perhaps ten million years ago, and actual animal remains have been collected from the gravels and debris of the last glacial epoch (the Wisconsin) of the American Pleistocene ice age (c. 1½–1 million years B.P. to 10,000 years B.P. on current European dating). Smith's down-to-earth mine superintendent, John Ryan, wrote in March 1900:[25] 'The remains of three mastodons were found in those mud deposits back of the house at Amargosa. I saw part of a tusk and an immense thigh bone in the office. . . . It shows that there was lots of vegetation in this country once.'

Mastodons, which possibly resembled shaggy elephants, are believed to have predated the Pleistocene and become extinct only some ten thousand years ago.[26] For about the last ten thousand years of their existence they were probably contemporaneous with prehistoric Indian (Palaeoindian) settlements. These were people who, now bearing the modern label of 'the Desert Culture peoples of the Great Basin', occupied caves in Nevada from at least 7,000 B.C.,[27] and wandered far afield in their perpetual search for food. It seems more than likely that, in the arid conditions after the Ice Age, they roamed as far as the dried-up lakes of Death Valley millennia before gold, silver and borax became desirable objects.

The Paiute were one of the four tribes encountered by white explorers in the Great Basin in the early nineteenth century,[28] and are alleged to have given the name 'Tomesha' or 'Ground Afire' to the area later known as Death Valley. Some decades later other white men arrived there in a truly desperate plight. In 1849 the gold-rush was on in California, and a party of emigrants and their families set out from Salt Lake City rather too late in the year to chance a crossing of the Sierra Nevada to reach the location direct. So they hired a Mormon leader, Captain Hunt, to take them by a more southerly route via the Old Spanish Trail to Los Angeles. Some of them got impatient along the route and broke away to take an uncharted short-cut due west across the desert. They stumbled into the burning salt valley bottom of Tomesha, and, before the survivors got out of it one had looked back from the summit of the Panamint Range to which he had struggled, and is alleged to have said, 'Goodbye, Death Valley.' Coming as he did from Mormon and Bible-reading Salt Lake City, it is possible that if he had enough voice left to say anything at that moment he would have used the phrase 'Valley of Death'; but whether the name arose from such an incident or was invented by the newspapers when news of this disastrous expedition spread to the world at large, the name 'Death Valley' has stuck to this desolate spot from that day to this.*

All this seems a long way from the practical history of borax, and so it probably is – except for the story that one of the survivors staggered back along his tracks to look for a lost gunsight and picked up a glittering hunk of what looked like silver ore.[29] Thus, it is said, the legend of the gunsight silver lode grew up in the mining camps of California, so that in May 1860, when the prospects for gold in the California river-beds were starting to look decidedly thin, a Dr Darwin French organized a party to search Death Valley for its rumoured silver lode. Dennis Searles, brother of John Searles, went with him, and must have tramped over acres of borax without realizing it in his abortive search for non-existent silver. A Dr George followed with a party in the autumn of that year, and found nothing except an antimony lode. Further expeditions in 1861 also found nothing except a thin vein of gold near Saratoga Springs at the southern end of the valley. Interest waned, but the many myths about Death Valley grew. Jacob Breyfogle, a blacksmith from Austin, Nevada, is said to have stumbled out of the Valley, nearly dead from heat, thirst and loss of blood and mentally crazed, but with his pockets full of fabulously rich gold ore which he claimed that he found somewhere between the Panamints and the Funeral Range. He thus added a gold bait to the existing silver one. The Governor of Nevada therefore took a party to the Valley in 1866, army engineers crossed it in 1867, and in 1871 a large government expedition, consisting of seventy-five officers and men, was sent to survey it and provide gazetteer-type information. But they produced no samples of 'cottonball', which were there, plain as a pikestaff, for anyone who had impartial eyes to see it, nor did they make any scientific reports of importance.

Finally, in 1875, Death Valley once more justified its post-1849 description in the toll that it took of the last of these transient expeditions before the discovery

---

*Early maps show Death Valley as Dry Valley, and the US General Land Office unsuccessfully resisted the more morbid name for several years.

of borax there ended its isolation for ever. There had been a silver rush in the Panamints Range, Death Valley's western boundary, in the early 1870s, but by 1880 a Frenchman, Isadore Daunet, and six of his friends had had enough of disappointments there, and decided to push off towards new hopes in Arizona by the shortest possible route.[30] This lay across Death Valley, and it being then midsummer, three of the party died from the furnace-like heat of the former Tomesha – Ground Afire – while the remaining four were rescued by descendants of the Indians who had given the valley that name. One of those saved was Isadore Daunet, who was to return later after the chain of events about to be related put Death Valley on the map for all time. Surprisingly, in spite of heat, death and disappointment, 'desert rats' had started to drift into Death Valley in the early 1870s and to settle there and do a little cultivation, or even keep a head or two of cattle around the rare springs and watercourses. They rejoiced in names like Bellerin' Teck Bennet, and Cub and Phi' Lee, two of a quartet of brothers christened Leander, Philander, Meander and Salamander Lee respectively. It seems unlikely that these various characters were serious agriculturists. It is far more likely that they were the flotsam and jetsam from earlier gold and silver rushes in the west and keeping an eye open for the chance of a lucky strike in Death Valley.

One of these was Aaron Winters, who lived in an isolated spot called Ash Meadows just east of Death Valley, where there was bunch-grass enough to feed a few cattle, mesquite trees to supply beans for flour and a spring of good water. He must have a story all his own, as in spite of its improbabilities the account of how he discovered borax is one of the most popular folk-tales of Death Valley.

Sooner or later someone was bound to notice the resemblance between the material which was being scraped off the floor of Teels and Columbus Marshes many miles farther north-west and that lying all over acres of the floor of Death Valley, particularly as parts of the Valley were obviously becoming a highway for casual migrating miners. One of these is alleged to have stopped off for the night at Aaron Winters's home[31] – 'half hewn out of the rock, half canvas lean-to' – where he lived, surprisingly enough, with an attractive-looking young Spanish wife called Rosie. The traveller (Harry Spiller by name) was on his way south. He talked about the borax deposits up north, and the fortune that could await the man who found more. Winters listened, and casually asked how you could tell whether it was the real thing. He explained to Winters that if a sample were treated with a mixture of alcohol and sulphuric acid and then set alight, it would burn green if borax were present.

As soon as Spiller had ridden away the next day Aaron and Rosie somehow equipped themselves with alcohol, sulphuric acid and matches and set out on the forty-mile journey to Furnace Creek and the floor of Death Valley. After collecting some of the stuff that looked like cottonball Aaron performed the experiment, the flame burnt the right colour, and he is alleged to have shouted the famous words, 'She burns green! Rosie, by God, we're rich.'

This is the story as Winters told it, and it makes a good one, although the details may well have been embroidered at a later date. However, without detracting from Winters's 'discovery' role in these events, it is interesting to note that a report in the *Inyo Independent* as early as 10 May 1873 refers to borates in Death Valley and, coming from a source that is unknown, the article says: 'Extensive deposits (of borate) have recently been found and located at two or three different points in Death Valley. There, it is said to be, on the ground from one inch to a foot in depth but containing a large percentage of borate of lime and other impurities. . . . One of these deposits is in the immediate vicinity where those '49 immigrants abandoned their wagons.'

It is of course not surprising that this knowledge failed to arouse interest, considering the unfavourable climate and transport problems; the difficulties and economics of manufacturing in Death Valley at that time compared poorly with the more 'comfortable' situation at the Nevada Marsh deposits.

Winters was probably another of the wanderers who had left the California goldfields for brighter hopes elsewhere and not simply a poor farmer/squatter ignorant of the mining game. Certainly he behaved in a shrewd manner as soon as he had made his discovery. He did not attempt to stake claims and set up a business for himself, but sent samples of his material to William Coleman in San Francisco. Then he quietly slipped off and filed claims on the water rights at Furnace Creek, knowing apparently that a borax plant cannot run without water.

All this occurred in 1881, and the Coleman representative arrived at this outlandish spot within a few weeks, but before disclosing to him where the cottonball deposit lay Winters haggled until he had secured the promise of a cheque for $20,000 for the discovery rights, to be paid just as soon as he had shown Coleman's man the deposit. The Coleman man may have been a little disconcerted when he found that the deposit was miles from anywhere, at the bottom of Death Valley, but he handed over the cheque and started staking his claims. Then he discovered that Winters had pulled a fast one on the water rights, as described above, and had no option but to hand over another cheque (for $2500 this time) to secure these rights also. Winters then withdrew from this particular borax field and bought a ranch at Pahrump thirty miles away. He was to make a further incursion into the history of borax the following year before disappearing from the borax scene for good.

Naturally, as in the case of the discovery of Columbus Marsh, of Teels Marsh, and of practically every other mining-field in nineteenth-century history, there was something of a rush by other speculators just as soon as the news of Winters's discovery spread around. But this time, owing to the poor price of borax in the early 1880s and the remoteness and harsh reputation of Death Valley, Coleman was left with only one serious rival on the spot.

The Frenchman Isadore Daunet, who had narrowly escaped death when he crossed the Valley in 1875, had now returned with four companions, located some 320 acres of borax land near the well named after Bellerin' Teck Bennet, and erected the first borax factory in Death Valley, under the name of The Eagle Borax Company,[32] before Coleman had started erecting his. A huge iron boiling pan ($5' \times 20' \times 3'$ deep) and a dozen 1,000 gallon crystallizers were hauled in over the mountains and fifty men were hired to gather cottonball from the valley bed.[33] Production started in autumn 1882, the purity of the end-product started to improve in the second season and Daunet's company began to make a little money, but costs were against him. Daunet had recently married a French-Canadian, but by 1884 the business was nearing bankruptcy and the marriage went wrong. After following his wife to San Francisco and failing to effect a reconciliation Daunet committed suicide and the factory passed eventually to Coleman, who let it fall into ruin.

Coleman had completed his Harmony Borax Works towards the end of 1883,[34] some twenty-two miles north of the Eagle Borax site and near the marsh sold to his representative by Aaron Winters. It consisted principally of a giant boiler housed in an adobe building while the crystallizing vats were out in the open. The process was a simple one. The cottonball was scooped from the surface of the marsh and shovelled into wagons by some thirty or forty Chinese, who were hired in San Francisco at $1.25 a day for a seventy-hour week. Few ever came back for a second year to earn a wage increase of 25 cents, as the social amenities of the camp and locality were insufficient to satisfy even them. The cottonball was hauled to the plant and dumped into the giant tank to be mixed with water pumped from a remote spring. Crude carbonate of soda, possibly found in the locality, but in abundance in the form of trona at Owens Lake, was added in order to remove the calcium in the ulexite as insoluble calcium carbonate. Mesquite, which flourishes in the deserts of both North and South America, was used as fuel to heat the tank but the main problem (which had earlier hit the Eagle Borax Works also) was to find some method of insulating the crystallizing vats against the tremendous heat of a Death Valley summer. No solution was ever found in the few brief years during which these operations mattered, and it became accepted that the factory would have to close during the summer months.[35] Fortunately, Coleman was able to minimize the commercial loss by using a second works based on a smaller deposit in the 'cooler' (110°F average compared with about 130°F at the Harmony site) atmosphere east of Death Valley at a point overlooking the bed of the normally dry Amargosa river and the Amargosa works were completed and started a few months before Harmony. This enterprise had had its origin in two men, by name Parks and Ellis, picking up samples at Amargosa and taking them to our old friend, Winters, for his 'expert' opinion.[36] He once more found that 'she burned green', and offered to find them a buyer for their deposit. They accepted and he contacted Coleman's representative and finished up with another cheque for $5,000 for himself, as well as cheques for similar amounts for Parks and Ellis. Parks went off east, and Ellis was killed by a cook in a row over a poker game at Amargosa the following year.

Coleman was thus able to use a minor works at Amargosa for production during June, July and August and to use the Harmony Works for the rest of the year. To formalize the position two separate companies had been incorporated in 1884, under the names of the Harmony Borax Mining Company and the Meridian Borax Company, but it had not been until the autumn of that year that Harmony Works had got going in a big way, and by May 1888 Coleman was bankrupt[37] and a new epoch was beginning, both in borax-production and in the kind of personality behind it.

These few brief years of activity by Coleman's companies created a landmark in the history of transport, which has subsequently stolen the show in the majority of histories of Death Valley and borax-mining.

This event was the famous Mule Team trek which later formed the inspiration of the now famous '20 Mule Team' trade mark used first by Pacific Coast Borax and thereafter world-wide by all parts of the Borax Consolidated, Limited's organization and by its successor companies. Treated in context it does not loom quite so large in history as it does in public relations; but for all that it remains a wonderful feat of ingenuity in the face of nature at its harshest. It was also a story of great physical endurance by the unfortunate mules, a factor which usually receives less notice than the undoubted skill of the driver and 'swamper' in handling them over such difficult terrain.

The fact that it should be found necessary to build special wagons and train a team of as many as twenty animals to transport borax from Death Valley 165 miles over a roughly hacked out road, and sometimes through difficult mountainous territory, to the railhead at Mojave underlines a vitally important point in the history of the borax industry. As has been suggested in Chapter 1, borax was scarce in earlier ages in Europe not so much because, by comparison with most other minerals and chemicals it is hard to mine or refine but because of the remote and inhospitable areas in which it is often found. The desert areas of Nevada and California (see map III on page 39) were certainly nearer to a civilized market than were the lakes of Tibet, but the various journeys required still presented formidable problems and the ports on the west coast of California, before the Panama Canal opened in 1914, were remote from the markets of eastern USA and Europe. The Tuscan and Turkish sources had certain obvious advantages where the European market was concerned, while the South American sources were in general just as inconveniently placed as the North American ones, with the added disadvantage of high altitude.

Transport problems, therefore, feature largely in any history of the borax industry, even though Doctors Veatch and Ayers – who as already related made the first discoveries in the USA at Borax Lake – probably had few of these, as they were near a trunk railroad and the coast. However, these short-lived deposits produced only a few hundred tons. When Columbus Marsh and Teels Marsh went into production in the early 1870s the problems were much greater. The borax was hauled by mule team from the marshes to Wadsworth railroad station, some 125 miles north of Columbus town and on the recently completed Central Pacific Railway. This was only thirty-nine miles shorter than the distance covered later by the famous twenty-mule teams from Death Valley; but the terrain covered was not so difficult, so no-one has written a rival saga. The operation must nevertheless have been expensive, as hay at $60 a ton and grain at $140 a ton had to be brought twenty-five miles from Columbus along a rough toll road to feed the mules. However, from 1882 the Carson and Colorado Railroad transformed the transport situation for the Nevada deposits. Death Valley, on the other hand, had no railroad anywhere near, and the problem was to plan suitable routes, and to devise some practical and economic method of getting the borax out in quantity over terrain that was inevitably difficult whichever routes were chosen.

The word 'routes' is used because it is sometimes overlooked in accounts of the Twenty Mule Team episode that there were two of these – the famous 165-mile route from the Harmony borax-works to Mojave and, for about three months of the year, the 135-mile route from the Amargosa borax-works down the desert and around the southern end of Death Valley to the town of Daggett, also on the Santa Fe railroad. Recalling the event some fifty years later,[38] Ed Stiles the teamster – who had joined Coleman after the Eagle Borax Works closed – said that the first trials with the Twenty Mule Team took place at Amargosa, when the twelve mules which he drove were hitched together with a team of eight, which belonged to Bennett, the Amargosa haulage contractor, and Stiles demonstrated that he and the swamper could manage the lot.

To adapt this equipage so that it could draw much heavier loads over the far rougher route to Mojave called for a piece of engineering and planning remarkable for the Wild West in the 1880s. First of all, special wagons had to be built which would hold the tremendous loads such teams were capable of hauling. Each wagon was to carry ten tons – about half the capacity of a modern railway freight car – but it had to be tough enough to grind through sand and gravel,

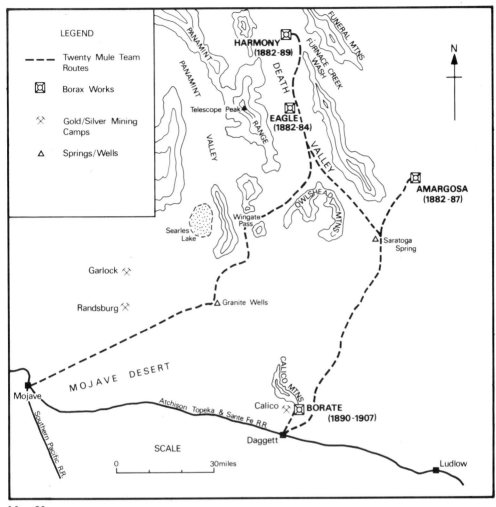

Map V

lurching over boulders and pitching in and out of chuck-holes, and to tackle steep mountain grades both ways.

When the blueprints had been worked out it was found that this called for wagons with rear wheels 7 feet high and front wheels 5 feet high, each with steel tyres 8 inches wide and 1 inch thick.[39] The hubs were 18 inches in diameter and 22 inches in length. The spokes, of split oak, measured 5½ inches wide at the hub and 4 inches wide at the point. The axle-trees were made of solid steel bars, 3 inches square. The wagon beds were 16 feet long, 4 feet wide and 6 feet deep. Each wagon, empty, weighed 7,800 pounds. Loaded with borax it would weigh 31,800 pounds. Two such loaded wagons, plus the weight of a separate water wagon – which held 1,200 gallons and weighed 9,600 pounds – made a total of 73,200 pounds, or 36½ (short) tons. The cost was about $900 a wagon. They were designed mainly by J. W. S. Perry, a pharmacist by training but taken on by Coleman to manage shipping and transport, and they were built by Delameter the blacksmith at Mojave. The excellence of the design and construction is demonstrated by the fact that during the seven years of constant operation there is no report of mechanical failure or the need for change. Since then they have toured the United States, appeared at World Fairs and State Expositions and rolled along numerous city streets in parades, while in 1937, fifty-two years after they were built, the wagon train travelled the original route from Death Valley to Mojave behind a twenty-mule team – again without a breakdown. In 1981 after almost a hundred years these splendidly engineered wagons, drawn by twenty mules, appeared once again at Bishop, California, in a Memorial Day parade.

While the wagons were being built, other men tackled the problem of the routes themselves. To reach the western side of Death Valley and the pass over the Panamints the teams would have to cross the salt bed known today as the Devil's Golf Course, its surface being a mass of jagged pinnacles, anywhere from 6 inches to 3½ feet in height, hard as flint and sharp as knives. A crew of Chinese was put to work, hacking off these pinnacles of rock salt until they had beaten a wagon road, six feet wide and eight miles long.

The main route led past the site of the old Eagle Borax Works along the edges of the Panamints and on up the long, steep grade over Wingate Pass (known to teamsters as Windy Gap). Once out of Death Valley and over the summit of the mountains, there remained a hundred miles of desert before reaching Mojave – a hundred miles in which there was no human habitation, and only three springs of water. The loaded team could travel on average from fifteen to eighteen miles a day. At that rate it required ten days to make the 165-mile journey, which meant ten overnight stops with at least half of them dry camps. In these camps water tanks were provided, which were towed by the teams from the springs to the dry camps and back again to be refilled. The conception and execution of this bold plan was undoubtedly the work of many, but in the main credit for its success must be coupled with the names of Perry, Delameter, Stiles and Coleman.

The twenty-mule teams were controlled by a cord jerk-line attached to the bridle of the leading left mule.[40] Experience proved that it was sometimes preferable to have a pair of fine, strong draught horses in the shafts as 'wheelers', but opinions about this differed: horses required more food and care than the resilient and more resourceful mules. There were special problems when rounding a sharp curve. By the time six or seven pairs of animals had turned, tremendous power was exerted at an angle, which could easily pull the wagons off the road. To counteract this the pointers and sixes (the second and third pairs) were trained to jump over the chain and pull furiously at an angle to the rest of the team until the wagons had safely reached the point of the turn – a very difficult manœuvre.

It was the teamster's job to harness the mules each morning, inspect each piece of harness, and hook up the outfit. He rode on the high wheeler and handled the brake of the lead wagon. The swamper had many duties. He rode on the rear wagon and managed its brake on the downgrades. He kept the teamster supplied with rocks with which to 'encourage' the mules on the upgrades. When they stopped to camp he gathered the fuel, cooked, and washed the dishes. There was little drinking on the twenty-day round trip, but at Mojave the men generally made up for it. Between mid-afternoon (when they reached the little railroad settlement) and the following morning (when they started for Death Valley again) many a teamster blew his entire pay on liquor and faro.

Much has been written about the romance of the twenty-mule teams (whose contemporary name was 'the big teams') but to the men who drove them they were simply a practical means of hauling some two and a half million pounds of borax from Death Valley each year. Nothing survives which gives much of a hint as to how J. W. S. Perry organized the running schedule. From the number of wagons constructed (ten) and reference to the numbering of the teams, there were eventually five wagon-trains in operation, but no illustration, real or imaginary, survives of two teams crossing along the route. Another point is that when J. R. Spears in his *Illustrated Sketches of Death Valley* (1892) first brought life there to the attention of the American public he fixed for ever the size of these big mule teams as twenty. However, a picture of the mules at Harmony borax works in 1883 shows a team of eighteen animals, and whether these big teams were actually eighteen or twenty is still subject to conflicting evidence. It seems probable that both were in use.

There are, of course, a good many stories and legends surviving of how the theories outlined above worked out in actual practice – endless plodding monotony for the teams and drivers, but occasionally and inevitably accidents, madness, sudden death and even murder – usually when the teamster got on the swamper's nerves.

It is perhaps the occasional surviving letter that illumines the realities of Death Valley far more effectively than the books written subsequently with publicity in mind. For instance, J. A. Delameter, the blacksmith who originally built the wagons, wrote in 1917 to C. R. Dudley, in the office at Oakland, applying to be driver of the mock-up twenty-mule team due to be sent round the principal cities of the United States for publicity purposes. His letter turned into reminiscences of old times, and he finished by writing about 'Wm. Shadley, who died near an old salt well 44 miles south of Furnace Creek, and it was 28 days before we found him, when he was mummified and only weighed 100 lbs and stood up on his feet when placed in that position and there was no odor from his body. There are many old time events that happened in the Valley.'

But, 'Without pride of ancestry, without hope of posterity', and with the woefully unjust label of obstinate stupidity given to it by man, the mule deserves the last word in this epic. In truth no animal has exceeded the endurance and bravery of the mule in the service of man. On the other side of the world to Death Valley, the mule was a key to the military superiority which for more than a hundred years enabled the British to secure the North-West Frontier of India against marauding tribesmen. The Indian mountain artillery regiments were designed around their mules, who carried the 'screw guns' to the craggy heights overlooking the Khyber and other strategic passes from the north, where they were assembled for action. Only the mule could have scaled these mountain ledges and remained calm in battle. Kipling's poem 'The Screw Gun' (the 2·5" Rifled Muzzle Loading Howitzer) tells the story.

REFERENCES

1. Veatch, J. A.: (read before the Society 17 Jan. 1859). *Proceedings of California Acd. of Science* Vol. 2 (1863), pp.7–8. First publication of discovery of borates in Tehama County in 1856.
2. Browne, J. R.: *Mineral Resources of the States and Territories West of the Rocky Mountains* (1867), pp.118–87. Describes production at Borax Lake, Lake County (Washington).
3. Bailey, G. E.: 'The Saline Deposits of California'. *California State Mining Bureau Bull. 24 Part II*, Borates, pp.33–90 (1902).
   Anon: '*Borax*', US Geol. Survey Mineral Resources 1882, pp.566–77 (Washington, 1883).
   These are summaries of the historic development of the borax industry in the United States up to that time.
4. Ayers, W. O.: 'Borax in America', in *Popular Science Monthly*, Vol. 21, pp.350–61, July 1882. A description of personal experience at Borax Lake and Hachinhama.
5. Hanks, Henry G.: 'Report on the Borax deposits of California and Nevada' in *Cal. State Mining Bureau Third Annual Report Part 2* (Sacramento, J. J. Ayres 1883) p.45 (ref. American Borax Co.); p.53 (Pacific Borax Co.)
6. Spears, J. R.: *Illustrated Sketches of Death Valley; and other Borax Deserts of the Pacific Coast* (Chicago; Rand, McNally & Co. 1892), pp.178–86. (A reprinted interview with F. M. Smith.)
7. Agreement. Frank M. Smith and Julius P. Smith with Wm. T. Coleman dated 25 Nov. 1874 (Borax House Archives).
8. 'Constitution and Address Committee of Vigilance of San Francisco', in *San Francisco Mining Globe Print*, 1856.
9. *The Borax Miner*, Feb. and March issues 1875.
10. (a) Certificate of Incorporation of the Pacific Coast-Borax Company d. 31.1.1872.
    (b) Stevenot, A. D.: 'Cottonball Refinery in California' in *The Pioneer* Jan. 1967 (U.S. Borax house journal).
11. Bailey, G. E., *op. cit.*, p.42.
12. Minute Book. Pacific Borax, Salt and Soda Co. (U.S. Borax Archives).
13. 'A Profile of Christian Beevoort Zabriskie by his friend Horace M. Albright', in *The Pioneer*, Vol. 4, No. 97, Sept. 1963.
14. Ledger. Death Valley Mining District 1881–1886, pp.2–112 (Borax Museum, Furnace Creek Ranch, Death Valley, Cal.).
15. (i) Stiles, Edward: 'Saga of the Twenty-Mule Team', in *Westways*, Vol. 31, No. 12 (Dec. 1939), p.8. This states that Smith was Coleman's partner.
    (ii) Spears, *op. cit.*, p.26, says Smith and Coleman were partners in Harmony Works.
16. Property Records, Death Valley Hotel Co. (files in U.S. Borax archives).
17. Hanks, *op. cit.*, pp.36–7; and *Inyo Independent* 20 Jan 1883.
18. Minute Book, The San Francisco Borax Board (U.S. Borax archives).
19. Minute Book, Pacific Borax, Salt and Soda Co. (U.S. Borax archives).
20. F. M. Smith's letter-heading.
21. *The Engineering and Mining Journal*, 28 July and 8 Dec. 1888.
22. Scherer, J. A. B.: '*Lion of the Vigilantes*'; *William T. Coleman and the Life of Old San Francisco* (Indianapolis, c. 1939), pp.309–10.
23. Minute Book, Pacific Borax, Salt and Soda Co.; Letter by F. M. Smith to J. W. Mather, 6 Nov. 1890.
24. *Death Valley – Origin and Scenery* by John H. Maxson, Fellow of the Geological Society of America. Published in 1963 by the Death Valley Natural History Association.
25. Ryan, J. to F. M. Smith, 10. 3. 1900.
26. *Encyclopedia Britannica*, 15th Edn.

27. Clark, G.: *World Prehistory* (Cambridge University Press, 1961).
    Clark, G. and Piggott, S.: *Prehistoric Societies* (Hutchinson, 1965).
28. *Encyclopedia Britannica*.
29. Spears, *op. cit.*, pp.20–22; p.47.
30. Spears, *op. cit.*, pp.51–4.
31. Spears, *op. cit.*, pp.56–9.
32. Inyo County Tax Assessment Rolls, 1883, p.40 – Inyo Court House, Inyo, Cal.
33. Hanks, *op. cit.*, pp.36–7; *San Francisco Mining and Scientific Press*, 30 June 1883.
34. *Engineering and Mining Journal*, 10 Sept 1892; also *Fourth Annual Report of the State Mineralogist*, 15 May 1884 (J. Ayres, Sacramento).
35. Hanks, *op. cit.*, pp.36–7; also *Inyo Independent* 20 Jan 1883.
36. Spears, *op. cit.*, p.61.
37. *San Francisco Mining and Scientific Press*, 2 June 1888.
38. Stiles, E.: 'Saga of the Twenty-Mule Team', in *Westways*, Vol. 30, No. 12 (Dec. 1939), pp.8–9.
39. Spears, *op. cit.*, pp.86–90.
40. (i) Wills, Dr Irving: 'The Jerk Line Team', pp.2–5.
    (ii) 'Notices' Vol. v., No. 1, 1959. *Bull. of Santa Barbara Hist. Soc.*

4

Colemanite and consolidation in the USA

Smith's acquisition of the Harmony and Meridian Borax Companies from Coleman's assignees on 12 March 1890 marked a decisive step towards a far wider amalgamation of the borax industry, which was to follow in 1896 and 1899 on a world-wide scale. Smith – obviously a man of volatile temperament – probably had not formed any such intention as early as 1890, but when he began to sort out his existing borax properties and his new acquisitions the way ahead became clearer. Among the new assets were two which Coleman had acquired in the 1880s, the Calico Mine and a factory at Alameda Point on San Francisco Bay, for which his plans had not materialized.

The discoveries which led to a change from one type of raw material to another which these places symbolize began back in October 1882. At that time Coleman, convinced that so far they had literally scratched only the surface of the borax deposits in and around Death Valley, had sent some of his men to prospect in the area. One of these, R. Neuschwander, while working his way up a small canyon on the east side of Death Valley, found a quartz-like mineral outcropping on the walls of the canyon.[1] Assays proved it to be a calcium borate, quite unlike the familiar cottonball (which was ulexite, a sodium calcium borate) and of a significantly higher boric acid content. It was at once named 'colemanite', for obvious reasons. Other ledges of this material were found around Death Valley, but no plans were made to exploit them. The material required a different and more expensive refining process, at a time when the existing market and current price-levels offered no encouragement to expand. However, the following year a prospector called Hugh Stevens, while prospecting for silver in the Calico Mountains about a hundred miles south of Death Valley, near Daggett, came upon a white ore which he sent to Thomas Price of San Francisco.[2] Price in 1871 had analysed a piece of calcium borate ore discovered in Oregon which had been named priceite, and in March 1883 he was now first to analyse samples of colemanite from Death Valley, which contained calcium borate in a new and different form to that in priceite.[3]

So Price passed word of the new discovery to Coleman in this same town of San Francisco, and Coleman at once sent his son-in-law, Robertson, to acquire the

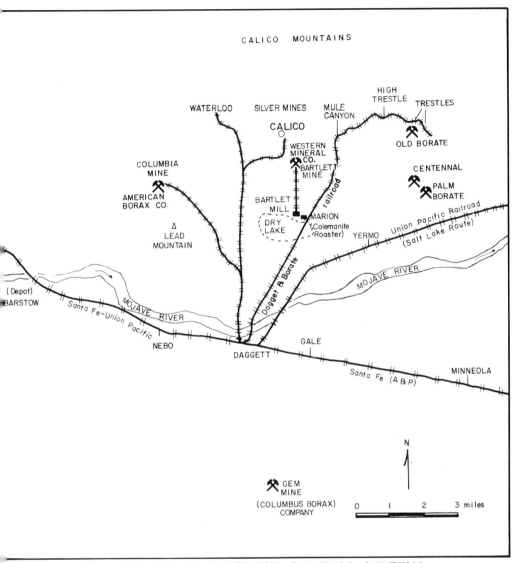

BORATE DEPOSITS, MINES AND RAILROAD SYSTEMS IN THE CALICO AREA. AS DEVELOPED BETWEEN 1899 AND 1907.

Map VI

Calico deposit. At the same time he acquired a further set of claims in the Calico mountains from Marion Neill, another prospector. But these were left unexploited, and the same thing happened with yet further colemanite discoveries located by Coleman's men in the Death Valley region. These latter included well-known names in later colemanite-mining like Monte Blanco, which was a 1,000-foot hill near Furnace Wash, the Biddy McCarthy and Played Out lodes discovered in 1882; the Widow and Lizzie V. Oakley lodes found early in 1884; and the Lila C. lode, thirty miles east of Death Valley, and named after Coleman's daughter Lila. Obviously Coleman had intended to do something about exploiting these widely scattered claims, as his Harmony Borax Company had bought a factory at Alameda suitable for a refinery, which had belonged formerly to the Royal Soap Company.

The discovery of these extensive vein deposits of colemanite, particularly those in the Calico area (see map VI, p.61), changed the whole basis of the borax industry in the United States[4] and by the early 1890s the surface 'marsh' and dry lake deposits from which cottonball (ulexite) had been harvested were exhausted or were dwindling rapidly. Smith's company now had a high proportion of all the known colemanite under his control, and it was therefore logical for him to concentrate on working the Calico colemanite deposit in the Mojave Desert, situated only about ten miles from a main-line railroad, thus eliminating the need for a 165-mile haul from Death Valley. His new mine, christened Borate, started production in 1888, and soon became the principal source of borax in the United States.

The start of the new mine at Borate was followed by the incorporation of a new company on 5 September 1890, named Pacific Coast Borax Company.[5] No prospectus survives, and the Company's copies may have been lost in the San Francisco fire of 1906, which destroyed most of Smith's earlier records. However, a letter which Smith wrote in confident mood in October 1890 provided his own prospectus for the new company:[6]

> As you are aware I am the owner and control a number of borax properties being the Pacific Borax Salt & Soda Company, the Teel's Marsh Company, the Oregon property and the former Wm. T. Coleman interests. I have recently organized a new company called the Pacific Coast Borax Company, and it is proposed that the company purchase all the assets of whatever kind and nature, of the Teel's Marsh Company, and the Pacific Borax (Salt & Soda) Company. The new company will make an issue of 15,000 shares of the par value of $100 each. The proposition is to purchase off the Pacific Borax Company all their property and issue therefor, upon transfer of such property, 5,000 shares of the Capital Stock of the Pacific Borax Company. You know that part of the property is situated in Death Valley and is very rich in Borax. This property, however, we are not working at present. The property we *are* working is located at Calico, about 8 miles from Daggett, a station of the A. & P. [Atlantic & Pacific – later Sante Fe] Road. We have a ledge there of nearly two miles in length, capable of producing a vast amount of crude material from which we make our borax, in fact this property alone could produce more borax than the market could take in the next 100 years. Our refinery at Alameda at present produces 150 tons per month and it can produce about 170 tons with present capacity. I desire to sell about 1,000 shares in the new company. I hold this stock at $70 per share. Do you know anybody who wants any of the stock? It is a good investment for men with surplus capital. The company will pay dividends of $1.00 per share which is certainly a good rate of interest on the investment.

Smith thus acquired an undisputed majority in what had become the largest borax-producing concern in the world. The dividends of $1.00 per share to which he referred had for the last three years been paid monthly by his old company Pacific Borax, which was indeed a good return.

The first shaft at Borate was sunk at High Point, where the finest ore lay, and was at an angle of 45 degrees, eventually reaching a depth of 350 feet. The ground here was extremely heavy, necessitating a tremendous amount of timbering. Load after load of 10×12 inch Douglas fir timber was hauled to the mine by the mule teams and taken underground, but by the end of the year most of it had to be replaced. The terrific weight of the ground had flattened the caps (overhead cross-pieces) to a mere two inches. Not only did the ground settle from the top; it also swelled from below. Old photographs show the narrow car-tracks on the surface humped like a gentle roller-coaster.

Gas engines hoisted the ore up the shaft in a 50-gallon whisky barrel. The men below were supposed to break the rock into small chunks before loading it into the barrel, but often it was easier to dump in a large boulder intact – which was hard on the man at the top, whose job it was to empty the contents of the barrel into a waiting ore-car. The barrel had a ring on the bottom to which he held on as he tipped it all over to dump out the ore; but a large boulder in the barrel would often carry him unexpectedly head over heels into the ore-car along with the rock.

The eleven-mile run from the mine to the Daggett railroad station was again provided by mule teams. The route was extremely tortuous and steep but these teams have not competed for the glamour bestowed on the Death Valley lot, although the steepest part has been named Mule Team Canyon in their memory.

They outlasted Smith's remarkable experiment of 1894, when he purchased a traction engine, named *Old Dinah*, and two steel wagons, and tried them out on the Daggett Borate run, which very quickly knocked and choked *Old Dinah* to a mechanical standstill. She must have been an extraordinary sight, careering along at 3½ miles an hour with her two rear wheels making heavy weather of the gradient and her one front wheel frequently rearing up in the air out of control, while the mule teams snorted and backed with fear and fury as she passed. But these more traditional alternatives to *Old Dinah* were not to last for long, as in 1898 work began on the narrow-gauge Borate and Daggett Railroad to span the eleven miles to the mine, with some sharp bends, phenomenal gradients, and a flimsy high trestle bridge to cross at the finish. Two Heisler narrow-gauge tank locomotives supplied the motive power, and were named *Francis* and *Marion* in honour of Smith.

As regards the mining camp at Borate, it was a typical framehouse affair in a canyon, with bunkhouse accommodation for the all-male working force, three meals a day cooked and served by pig-tailed Chinese, and with the catering arranged by the mine superintendent at a personal profit. This was the first underground venture in the mining of colemanite in the USA, and Wash Cahill, the company storekeeper, recalled its atmosphere perfectly:[7]

> There was only one recreational centre in the camp, a small 'reading-room', and everyone worked seven days a week. That was standard practice. Then one day, like a bolt from the blue, came an order from headquarters – no Sunday work. Was this greeted with cheers and huzzas? It was not. No Sunday work meant a cut of $3.50 a week in a miner's pay. and what was there to do in a mining camp like Borate on a Sunday but work? If the Company thought that 120 men were going to spend the day in the Reading room . . . ! They expressed their sentiments to Fred Corkill, the Superintendent at Borate, but there was nothing he could do about it. He was only acting on orders from Mr Smith who, it seemed, had been persuaded by somebody that it was

un-Christian for an employer to have his men work on Sunday. The men decided, in that case, they would not work at all. The quit their jobs en masse, went down to Daggett, and proceeded to get gloriously drunk at Ma Preston's [the local boarding-house bar]. The strike lasted for several days. Then word came from San Francisco that the 'No Sunday Work' order had been rescinded. The men could work every damn day of the week if they wanted. The mines started operating again the very next day.

Cahill's recollections also covered life in Daggett, including Ma Preston's bathing habits. She used a large whisky barrel which stood out in the yard, and one night in trying to climb out of it the barrel tipped over and she broke both wrists. After that she just stood out in front of the house in her birthday clothes and let her husband Old Tom hose her off. It is said that passers-by would tip their hats politely in recognition and be cordially greeted. But woe to him who presumed to turn around and look back. He was no gentleman, and Ma Preston let him know it.

With the arrival of the railroad a second shaft was sunk, and from 1899 the annual production at Borate reached about 22,000 tons, which made it the largest borate mine in the world at that time.[8] In retracing the railroad's track today, one can see what a remarkable feat of engineering was involved. Steep gradients wind up through eroded hills and rocks reminiscent of Death Valley in its contours and the yellow and purple colours, which gave the Calico mountains their name. The only signs of the railroad which remain are the crumbled foundations of the high trestle bridge near the end of the line, while Borate itself is a ghost mine hidden in the hills just a few miles from the freeways and a busy world below.

Mining continued at Borate until 1907, when Pacific Coast Borax moved its mining operations to the Lila C. mine in Death Valley.[9]

As regards the processing into borax of the raw material from the deposits and mines under Smith's control, the development of the Alameda Works in this period marked a considerable step forward from the situation in the autumn of 1872 when Julius Smith brought boilers, tanks, crystallizers and other plant from Chicago to F. M. Smith's new location at Teels Marsh. The *Report of the California State Mineralogist* of 1883[10] stated that the method employed by Smith Brothers for the production of borax from the Marsh crude material was by solution, separation of mechanical impurities by settling, and crystallization, the result being concentrated borax. When this was recrystallized it was known as 'refined borax'. Smith Brothers also built a refinery at Fish Lake, about twelve miles south of Columbus Marsh,[11] and were the only company operating on the Nevada marshes who actually refined their product there. Among the others the larger companies merely concentrated the cottonball there in order to economize on the rail freight into San Francisco, where it was dealt with in other people's refineries; but the smaller companies, while they survived, probably sent their cottonball to San Francisco in its raw state. When Borate mine (Calico) started up around 1888 new methods had to be found for upgrading colemanite. A calcining plant was installed four miles north of Daggett on the edge of the plain at Marion, to concentrate the ore. The colemanite when heated lost water and conveniently disintegrated into a fine flour that passed through screens and was separated from the shale and other impurities. It was then sent to Alameda, where it was converted to a hot solution of sodium borate by treatment with carbonate of soda (soda-ash), and then separated from the impurities by filtration and crystallization. Later the company began shipping colemanite to England. The ore was run over 'grisslies' at the mine to separate high-grade 'English' ore (not less than

40 per cent boric oxide) from the Alameda ore. The 'English' ore was then crushed to the size of a walnut at Marion and railed to Los Angeles for shipment.

By 1890 Smith had established almost unrivalled supremacy in the borax industry. Whether he achieved the status of millionaire at this time or a little later is difficult to judge. It was of course an age when Rockefeller, Carnegie and other prophets of materialism were rising from complete obscurity to become multi-millionaires through exploitation of the apparently boundless resources of the New World. However, in the years before 1890 the borax industry of California was still relatively small and borax alone can hardly explain how he was enabled to live in style from about 1892 onward. By then he had built for himself a large mansion, Arbor Villa, at Oakland, and acquired another, Presdeleau, on Shelter Island, New York, together with a yacht or two, moose and duck shooting concessions, and other conventional trappings of the rich at the time. Also about this time he started playing the big boss and spending freely, partly on grandiose schemes which had no connection at all with borax. The explanation for the apparent speed of this transition appears more clearly in his dealings with his vice-president and principal financial adviser, J. W. Mather. Much of Smith's financial success at this time was undoubtedly due to the support and intercession of Mather, a man twenty-six years older than Smith and possessed of long commercial experience. More important still, Mather had a good reputation among the banks and commission houses of the Eastern States at a time when Smith was still an unknown quantity. Mather had joined Smith when the Pacific Borax, Salt and Soda Coy. was formed in 1886,[12] and when Coleman crashed in May 1888 Mather already was on the east coast, lobbying in favour of a higher duty on borax and against the Mills Bill, which threatened to remove the duty entirely. This threat had helped to topple Coleman, but its subsequent disappearance helped to establish Smith. Following the death of his younger son, Mather had arranged to move to New York, and it took only a little encouragement to persuade Smith to open an office at 48 Wall Street, New York, on 1 January 1889. This was called a sales centre, and obviously Mather was the man to take charge of it; but it is equally obvious that his main duty in these crucial months of 1889 was to find the finance for Smith to take over the Coleman properties.

Probably Mather had personal motives for following the rising star in the world that he knew best, borax, and for putting at Smith's disposal in California $50,000 of his own personal funds and adding a further $20,000 in New York. Presumably he had a commission agent's eye for a likely profit, and whereas some months earlier Mather's efforts to get the stock of the Pacific Borax, Salt and Soda Company listed on the New York Stock Exchange had failed – which made it unlikely to be acceptable as security for bank loans – the outlook had suddenly changed. Now, with the emergence of Smith as a potential successor to Coleman, Mather was able to use his own reputation and business record to secure extensive credit to enable Smith to establish an expanding borax empire. As he wrote to Smith in October 1889, 'Our financial standing is A1 with the Bank of New York as well as with other banks and business men in general'. This was certainly the end of the beginning for Smith, but perhaps it was also the beginning of the end. A headstrong operator of his type was likely to look on bank credit once granted as an endless source of funds, and from this point it seems clear that Smith's intentions and Mather's began parting company. While until 1896 it can be said that the Pacific Coast Borax Company prospered a good deal through the financial and sales experience of J. W. Mather, and through the genius for pushing sales shown by his son Steve, this was in spite of the problems created by the financial antics of Smith, which the elder Mather endeavoured to restrain. In May 1891, for example, J. W. Mather wrote to Smith:

> I had a talk with Mr Fry, president of the Bank of New York, about the future in money matters, and he wished me to state to you as coming from him that you would do well to get your business in as snug shape as possible, as the outlook in financial matters is not encouraging. His expression was that "Gold is going out from us very much as though it went into a rat hole", by which I understand him to mean it was not likely to come back very soon.

Apparently Smith resented this advice, and Mather had then to assure him that the bank president's advice was meant 'in a purely friendly way'. A few months later their temperaments again clashed. There was a disagreement as to whether the admittedly handsome profits of $26,000 in April 1891 justified raising the stock dividend. Mather wanted something put to reserve. Smith wanted cash.

Mather claimed that he carried Smith through the nation-wide financial panic of 1893, and that he himself raised from his friends over half the loans of $150,000 standing in Smith's name in that year. Whether this was true is hard to say, as by that time his opinion of Smith had been soured by the latter's high-handed and questionable interventions on the sales side, and above all by his ill-advised impulse in appointing the Coffin Redington Company as his New York sales agents in place of Mather in 1891. However, the old man (he was by now in his seventies) hung on for another five years, hoping that the whole business could be sold to or merged with an American chemical company: he had Pfizer in mind. Such a move could not only have brought financial gain but also ended the erratic financial behaviour of Smith, which was not to Mather's taste.

Nevertheless, the 1890s were a period of great prosperity for the Pacific Coast Borax Company. In the very first year of Mather Senior's administration of the New York sales office sales quadrupled. The position was still further improved by the advent as part-time adviser in 1891 of his son Steve, who brought new and original thinking to the problem of increasing borax sales. Steve Mather was working as a reporter for the *New York Sun*, and it was he who first hit on the idea of making commercial capital out of the story of the borax operations of Coleman's companies in Death Valley from 1883 to 1888, and in particular out of the journeys of the big mule-teams. A fellow-reporter at the *Sun*, called J. R. Spears, agreed to write a book to help to give borax glamour in the eyes of the general public, and by February 1892 *Illustrated Sketches of Death Valley and Other Borax Deserts of the Pacific Coast* had been so well written that the Twenty Mule Team story has become enshrined to this day in American history and the commercial literature of borax. Mather's proposal to use the Twenty Mule Team as a symbol was not at first well received. Smith's comment was[13] 'No, I cannot say I like the idea of the "mule team" brand of borax. My name and that of the company should be in the foreground.' But Smith eventually had to agree that his own name, Francis Marion Smith (which he blazoned widely wherever opportunity offered), should yield place to an illustration of the Twenty Mule Team as the company's principal trademark.

Work expanded so much that the services of Coffin Redington were dispensed with and Steve Mather left journalism and became full-time promotions and advertising manager in the New York office of Pacific Coast Borax on 1 January 1893 at the age of twenty-six. He at once started a vigorous campaign to familiarize the public with borax and its uses. This would be the obvious duty of any advertising manager, but surprisingly enough he found himself hampered initially by Smith, who was opposed in these early days to spending money on advertising space. As an alternative to direct advertising, Mather recommended that they should offer to those who wrote for the news syndicates so much a word for anything accepted which stressed borax. Smith agreed:[14] 'I say the newspapers should not receive a dollar of our money; [we should] deal with the contributors only.'

It was brother Julius who had been the real enthusiast for pushing the sale of packaged borax, and Julius had retained the rights to the name 'Smith Brothers' when the partnership broke up in 1885, and had even come out of his vineyard again to use this brand-name in selling packaged borax, made from materials obtained from sources other than his brother's company. Presumably this competition drove Smith to follow Mather's advice to him to buy out Julius; and he did so, in 1893, for $6,000.

At this time Steve Mather was getting round Smith's parsimony over taking up advertising space by writing letters about borax to the leading household magazines for publication, and pretending that they came from housewives. He managed to get some fifteen of them published, and also used his contacts in journalism to plant syndicated news items on borax in the newspapers. Next Mather introduced a new scheme whereby each package of borax contained a coupon offering a one-dollar prize for any letter to the editor that was published on the merits of borax. This produced quite a harvest, and as a result he assembled a booklet entitled *Borax: From the Desert, Through the Press, Into the Home: 200 Best Borax Recipes from More than 800 Issues of 250 Different Publications in 33 States of the Union.* The introduction reads as follows:

> Borax is not a proprietary article, the popular demand for which has been created by any system of advertising. On the contrary, it is a staple mineral product found in numerous deposits and can be mined and manufactured by whoever may chance or choose.
>
> The United States consumption of Borax has increased more than 1,500 per cent since its first production in California thirty years ago. This fact attests not only its value as an article of general use, but also the existence of some agency whereby its intrinsic merits have been made known. That agency is the Press, largely the domestic columns of the daily and weekly family journals which have most liberally set forth the many uses of BORAX in the household.
>
> The Pacific Coast Borax Company – the largest single producer of Borax – has, during the past year or two, carefully collected such of those newspaper items as have been easily procurable, and this little pamphlet is the partial result. While it is by no means complete, it presents a fair specimen of the 'Opinions of the Press', which have been voluntarily offered by those most competent to instruct the public.

This succeeded in putting a box of borax into a great many homes in the United States, and there is no doubt that many of Mather's ideas were highly original, and are of interest in the study of the early history of advertising and sales promotion. His flair for communication and public relations – learnt first in journalism and later in the promotion of borax – was, as will be seen, a rare talent, and later it took him far afield in another career. Eventually Smith was won over by Steve Mather to pay for proper commercial advertising, and became an enthusiastic convert. He opened a further sales office at 253 Kinsie

Street, Chicago, late in 1894, with Steve Mather in charge.

With increasing profits Smith was starting to get the impression that by now the sky was the limit, but in 1894 the old bogey of competition from borax made from ore imported through East Coast ports suddenly raised its head again. This was stimulated greatly by the Wilson (Tariff) Act (1894), pushed through by the Democrat administration, which had brought about among other things a reduction in the duty on refined borax from 5 to 3 cents a pound, and on the raw material, borate of lime, from 3 to 1½ cents. As a result, imports of borax and boric acid for the year 1895 showed a substantial increase compared with the previous years.[15] The trend resulting from the tariff cut also appeared in the fall in the price of borax from 7½ cents a pound in 1893 to 5 cents a pound in 1896, the lowest price hitherto recorded on the New York market.

But by this time also Smith had undoubtedly become a Republican of some consequence and the unpredictable effect of politics on tariff protection, and with it the security of the American borax industry, was probably one of the concerns uppermost in his mind. This must have strengthened his conviction, now that he had absorbed so many of his competitors in the USA, that it was time to investigate the situation overseas.

REFERENCES

1. Jackson, A. W.: 'on the Morphology of Colemanite', in *California Academy Sci. Bull.*, No. 2 (1885), p.3. (Names Neuschwander, and dates discovery.)
2. Anon, 'Borax', in the *US Geological Survey – Mineral Resources 1883 – 84*, pp.859–63, and 1885. (Contains announcement of the discovery of the new deposits of borates in the eastern part of the Calico District, San Bernardino County, Cal.); also W. W. Cahill to C. B. Zabriskie, 7.8.31.
3. Jackson, A. W.: 'On Colemanite, a new borate of lime', in *Am. Jour. Sci.* (1884), 3rd series, Vol. 28, p.447.
   Bailey, G. E.: 'The Saline Deposits of California', in *Cal. State Mining Bureau Bull.* 24 (1902), pp.88–90.
4. Campbell, Marius R.: 'Reconnaissance of the Borax Deposits of Death Valley and Mojave Desert' (*Bulletin of the US Geological Survey* No. 200, Washington (1902), pp.7–8.
5. Minute Book of the Pacific Coast-Borax, Salt and Soda Coy.: Meeting 28.10.90.
6. Letter from F. M. Smith dated San Francisco, 23 October 1890.
7. Description of life and work at Calico Mine at Borate given to Ruth Woodman by W. W. Cahill, formerly storekeeper at Borate Mine (d. about 1948).
8. Schedule attached to BCL prospectus 1899, and F. M. Smith – Annual Report to Directors of Borax Consolidated 1900/01.
9. Smith, F. M.: Annual Report to the Board of Borax Consolidated, Ltd 1906/07.
10. Hanks, Henry J.: *Third Annual Report of the California State Mineralogist* (1883), pp.45–7.
11. Anon, in the *US Geological Survey – Mineral Resources 1883 – 84*, p.861 (Washington, 1885).
12. Minute Book of Pacific Borax, Salt and Soda Co.; also Robert Shankland, *Steve Mather of the National Parks* (New York, 1970), p.20.
13. Letter Smith to Mather 22 Sept. 1893 (Bancroft Library, California).
14. Letters Smith to Mather 20–25 June 1892 (Bancroft Library, California).
15. Yale, Charles G.: 'The Production of Borax in 1903', in the *Dept of Interior – US Geological Survey* (Washington, 1904): Imports 1867–1903, p.9.

5

Redwood and Sons, Unlimited

Whereas the start of the company which was to become Borax Consolidated, Limited can be identified in the USA with Smith's presence at Teels Marsh in 1872, that of the English counterpart in the alliance is not so easy to fix precisely. Redwood and Sons was founded by Professor Theophilus Redwood, Ph.D, FIC, FCS, whose research had provided an answer to a serious problem created by the sharp rise in the urban population in the nineteenth century. The scramble for empire overseas had helped to supply the extra food needed for the increasing number of mouths in England; but in the days before refrigeration and food-canning the problem was how to preserve the more perishable foods in the stage between production and consumption, particularly when they had to be transported long distances. An advance on the centuries-old salting technique was urgently needed.

Writing about the beginning of Redwood's business, Gerstley says[1] that the leading Dutch manufacturers of margarine had asked Professor Redwood to investigate the possibility of an effective preservative, which would not affect the flavour. Margarine was a product in its infancy, and a satisfactory solution to this problem was much needed. The Professor discovered that boric acid, sometimes mixed with borax and used at levels between ½ and 1 per cent, produced excellent results, and that this could be extended to butter and a wide range of foodstuffs.

Redwood was a hard-working and versatile Victorian with a fine record of achievement. Although research on borax was only a small part of his life's work, he did provide an essential link in a vital chain of events for the developing borax industry. He was born in March 1806 at Boverton, Glamorgan, and died there in March 1892. Son of a village schoolteacher, he was apprenticed at the age of fourteen to an apothecary in Cardiff, and two years later, thanks to the assistance of a Quaker lady, joined the historic London apothecary 'John Bell of Oxford Street'. This led to a most fruitful relationship with Jacob Bell that extended far beyond the frontier of their immediate business. Redwood was self-educated, worked long hours and later attended lectures at the Royal Institution, where he must have listened to another self-educated scientist of the Victorian era, Michael Faraday.

From about 1840 Redwood, together with Bell, embarked on a programme which was to change the whole background of the apothecary's trade and to raise it to a profession with the higher educational standards associated with science and medicine. Bell took on the lion's share of public and political work, while Redwood became the active promoter of the educational and scientific side. The Pharmaceutical Society was formed. *The Pharmaceutical Journal and Transactions* was published. In 1842 Redwood became first professor of pharmacy at the newly established school, and in 1846 became professor of chemistry and pharmacy, a position he held for the next forty years. A school with lectures, laboratories, museum and library was created, with a professor who maintained his own research programme, and who had a record of scientific achievement.

To quote from Redwood's obituary in *The Pharmaceutical Journal and Transactions*, March 1892:

> It must not be forgotten that fifty years ago the chemists and druggists possessed the most elementary knowledge of many subjects which are now familiar to the average student. Professor Redwood was unrivalled in his exposition of chemical and pharmaceutical processes which from time to time demanded explanation. He had a share in promoting an honourable understanding between Medicine and Pharmacy and in paving the way for that recognition of the mutual position of the physician and the pharmacist which of late has been so conspicuously advanced.

In 1865 the Medical Council appointed Redwood to edit a new *British Pharmacopoeia*, and he was still the editor of a revised edition in his eightieth year. He was a vice-president of the Chemical Society from 1869 to 1872. The only honour he received was a Ph.D from the University of Giessen. This was at the request of the celebrated German chemist Professor Liebig on his retirement in 1853, for he wished to perform a graceful act on leaving an institution where chemistry and pharmacy had been taught together successfully, and in the manner which Redwood had pioneered in Britain.

Professor Redwood's eldest son, Boverton, born in 1846, studied chemistry and became a Fellow of the Chemical Society before the age of twenty. Like his father, he dispensed with a university education, but gained a worldwide reputation as a consultant in the petroleum and mineral oil industry, and for many years was an honorary technical adviser to the British Government. He was knighted in 1905 and created a baronet in 1913, holding important posts in the Ministry of Munitions in the First World War. However, it was two of Theophilus Redwood's other sons, Theophilus Horne and a younger brother, George, who together with R. C. Baker were to feature in borax history.

Mr R. C. Baker was a frequent visitor to the Redwoods' home in London. At some time before 1887 a partnership called Redwood and Sons was formed to produce food-preservatives based on the Professor's research work. R. C. Baker, who was less than thirty, was invited to join the business in which the two Redwood sons were the active members. Redwood senior would have been about eighty at this time, but he was still very active; however, it seems that for professional reasons he had not been visibly connected with 'trade' since he left the apothecary's shop fifty years earlier.

A great deal will appear about R. C. Baker later in this book, particularly as he became a dominating figure in the worldwide borax industry from 1899 until just before his death in 1937. According to Gerstley,[2] he had some experience in accountancy and banking, and before the age of thirty-five he was living at King's

Mill House, Nutfield, Surrey, which would be taken as a sign of worldly prosperity even at the present day. From personal recollection of company pensioners, he was a man of lofty and imposing stature and deep voice, and from his portrait might well be mistaken for one of the late Victorian Forsytes of recent television fame. The toothbrush moustache, the immaculate dark suit and white stiff collar, and the slightly unwelcoming mien, bespeak the business-man of authority that he obviously was. There are the stories of his love of cricket, his competing in penny-farthing bicycle races, and his love of sport generally, riding, shooting and fishing. He collected antique violins, and so maybe like Sherlock Holmes he played a little. He also held charm for the opposite sex, and enjoyed exercising it. These things, and the meticulous attention that he paid to what would now be described as personnel matters, all betray the more human side of his nature.

Before meeting the Redwoods, Baker had established his own trading business, Burton, Baker & Co. His colleague James Gerstley felt that Baker's uncle was the famous Sir Richard Burton, scholar, athlete, rake of many parts and partner with Speke in the exploration of the White Nile. Sir Richard Burton's mother was one of three daughters of a Bond Street merchant of substantial means called Richard Baker, living at Barham House, Herts. (Barham is now called Boreham Wood.) Richard Baker had an only son by an earlier marriage, and as Burton wrote,[3] 'My mother had a wild half-brother, Richard Baker Jr., a barrister-at-law, who refused a judgeship in Australia and died a soap-boiler.' Burton himself used the pseudonym Baker in some of his writings; however, nothing is known of any surviving family belonging to the wild soap-boiler. Richard Charles Baker of borax fame seems to have been reticent about his background and early life, and it has proved difficult to trace any connection with these Bakers or the Burton family. Perhaps he thought that Burton, Baker & Co. sounded a good name for the business which he later operated with the assistance of James Gerstley, who in turn brought with him considerable business experience and knowledge of European languages, and who seems to be the only close friend Baker is ever known to have had.

Perhaps in order to keep the Redwood name out of the public eye, Burton, Baker & Co. assumed the role of managing agents, performing both the buying and selling function for Redwood and Sons, and as a result Baker was able to acquire a knowledge of the borax industry as a whole. A small factory was leased near Kennington Oval and was known as the Cornwall Works, the landlord being the Duchy of Cornwall. Theophilus Horne Redwood, an analytical chemist about ten years older than Baker, and George Redwood as works superintendent were responsible for technical and production matters.

Referring to the site, Gerstley noted: 'This location being possibly chosen due to its vicinity to the famous M.C.C. ground, both the Redwoods and R. C. Baker being ardent cricketers.' It is a nice thought to picture the London birthplace of today's great enterprise resounding throughout the summer to Cockney cheers as their heroes Bobby Abel and Tom Hayward opened for Surrey and attacked the enemy of the day; but the Surrey crowd would not have appreciated the association of their beloved Oval with the MCC, a *faux pas* which is equivalent to ascribing the home of the Brooklyn Dodgers to the New York Yankees.

The nearest which any records get to the start of business is a professional opinion written by Professor Redwood on 14 August 1887 from the Pharmaceutical Society laboratory at 17 Bloomsbury Square, London, advocating a product called 'Preservitas' for food-preservation;[4] and an application to register the name as a trademark which was filed on 24 August by R. C. Baker. A later similar trademark application (30 September 1891) refers to Mr Baker trading as The Preservitas Company. The success of these products resulted in this soon becoming Baker's main activity; although Redwood and Sons and Burton, Baker & Co. remained different entities, it seems that their financial interests in the food-preservative products were 'merged' into the accounts of Redwood and Sons.

In September 1893, eighteen months after Professor Redwood died, Redwood and Sons was incorporated as an unlimited company, and among the shareholders were the two Redwood sons, two unmarried daughters of the Professor, and R. C. Baker and his wife. James Gerstley joined Redwood and Sons in 1894,[5] and with his assistance Baker started expanding business by leaps and bounds, furnishing important quantities to Australia and New Zealand in connection with the rapidly growing butter industry; by travelling all over Western Europe they succeeded in opening connections with many of the most important manufacturers of perishable food commodities in Holland, Belgium, France, the Scandinavian countries and elsewhere.

The volume of production required to meet this growing trade was plainly becoming too heavy for the small Kennington works. Production in the early days was based on crude boric acid imported from Italy, which was converted to refined borax and boric acid. As the business developed they purchased refined borax and boric acid in increasing quantities, furnished by a consortium of refiners who were virtually controlled by the owners of the boracite mines in Asia Minor, and who instead of encouraging consumption by reducing prices as the demand increased were constantly raising them. Accordingly Redwood's resolved at the first favourable opportunity to establish their own borax-refinery, and with this object in view they acquired in 1895 the freehold of a larger factory, situated on the Thames at Belvedere in Kent,[6] which brought with it the use of cheap water transport.

This decision would have been made by Redwood and Sons in the light of the considerable knowledge Baker had already acquired about the European borax business. In order to understand the moves which he and Gerstley made later, it is necessary at this point to identify some of the principal forces at work. In the last quarter of the nineteenth century the availability of raw material for the manufacture of borax from the different competitive sources in Italy, Turkey, South America and California, together with a new level of demand for borax for industrial uses, produced a sharp increase in the number of refiners who set up in many industrial centres. During that time it is possible to trace the following number of companies who were engaged in borax-refining in the Western world: sixteen in England, fourteen in Germany, seven in the USA, three in Austria, three in France, two in Russia.

The severe competition which resulted, and the speculative nature of the raw-material market, produced some serious losses for refiners, and for most survival in the borax business became difficult and risky. In an effort to improve the situation a group of five leading German refiners formed in November 1886 what was known as the 'German Union', whose purpose was to regulate the manufacture and sale of borax and boric acid.*

---

*These were: Chemische Fabrik in Billewärder (formerly Hell and Stammer) of Hamburg; Kunheim & Co. of Berlin; Ertel Bieber & Co. of Hamburg; Morgenstern & Bigot & Co. of Hamburg; and Julius Grossman of Hamburg.

In March 1889 a further important development took place which united the German Union with two other substantial companies[7] – The Borax Company with its refinery at Maisons Lafitte near Paris and extensive boracite deposits in Turkey, and H. Coghill & Son with a large refinery at Liverpool. These companies made an agreement called the International Union Contract. Later additions to membership were the refiners Joseph Townsend, Ltd of Glasgow in January 1890, while the refiners Mear and Green of Staffordshire and the Société Lyonnaise des Mines et Usines de Borax – also with a Turkish source of raw material, and with refineries Lyons and Vienna – both joined in 1892.

In April 1894, in order to maintain co-operation and to provide improved stability of market conditions, these companies formed the International Borax Union[8] and worked out a tripartite agreement between the two companies producing raw material, which were referred to as 'the companies' (i.e., The Borax Company and Société Lyonnaise), 'The English Union' (the three British refiners) and 'The German Union' (the five German refiners). This agreement provided for the supply and restriction of raw material to these refiners by 'the companies', except in the case of Russia and the United States, where raw material could be supplied at prices high enough to exclude any competition in refined products. It covered selling prices of refined borates and quotas of production. The agreement was a complex document covering accounting procedures and many practical aspects of doing business.

The International Borax Union included no party concerned with Italian, South American or United States raw-material sources, and although it apparently worked well for all concerned for some years, it was not surprising that with increasing production in Italy and South America making its impact in Europe prices became depressed, and that the members of the International Borax Union found it impossible to work their system. In May 1897 the Union was dissolved, and as a result market conditions became somewhat chaotic.

These events would all be known to Baker and Gerstley, in making their plan to enter the borax-refining business in 1895, and they would have been giving some thought to the politics of the Union and the market forces external to it. How an unexpected event provided Redwood and Sons with the opportunity to enter the borax-refining business with a position and a voice that commanded attention from the start is the subject of the next chapter.

REFERENCES

1. *History of Borax Consolidated Limited* by James Gerstley, written about 1945 (unpublished).
2. *Ibid.*
3. Quoted in F. M. Brodie, *The Devil Drives: a life of Sir Richard Burton* (Eyre & Spottiswoode, 1967).
4. Manuscript letter by Prof. Redwood, 14.7.87 (U.S. Borax archives).
5. Gerstley, *op. cit.*
6. The conveyance of Belvedere Works to Redwood and Sons is missing along with the rest of their records, but other deeds at the present Borax House show that a factory was built on this site in 1877–8 by one Eli Earl Collins, who acquired the site in 1877. No evidence is traceable of the purpose for which this factory was used.
7. The International Borax Union – Copy No. 4 of an Agreement made 19 April 1894 – consists of 14 pages and 55 articles. Reference is made to the International Borax Union Contract of 22 March 1889 (Borax House Archives).
8. *Ibid.*

6

Smith goes to London

It would be interesting to know what 'Borax' Smith (as the American Press now dubbed him) had in mind when he embarked at New York on his first visit to Europe at the end of 1895. As mentioned in the last chapter, he was becoming increasingly concerned about competitive imports to the USA, and as a dynamic expansionist he was always looking for new markets for his own products, while never far from his thoughts was the possibility of raising additional capital. It would fit Smith's personality if the trip were based on a hunch that it was time for a 'hunting expedition', to meet and size up the other half of the borax world which dealt with sources coming from South America, Turkey and Italy – all of which from his point of view were becoming increasingly bad news.

It was doubtful whether Smith had ever heard of Redwood and Sons when he landed in England. In a brief history of Borax Consolidated which he circulated privately in 1946, James Gerstley describes the first meeting between Baker and Smith as a 'fortunate coincidence'. Gerstley's German uncle, Lewis Gerstle, who was at that time Vice-President of the Alaska Commercial Company, came over from San Francisco on a visit to his brother, James Gerstley's father. When he inquired how his nephew was getting on he learned that he was interested in borax, and was looking for alternative sources of raw material. He mentioned that he had happened to have come over on the same ship with the 'Borax King', F. M. Smith, who was accompanied by his lawyer, a Mr Chickering, with whom Lewis Gerstle's son Mark was working. He added that they were staying at his hotel, and that he would be delighted to introduce them. Smith, Baker and Gerstley met, and the latter reports:

> We learnt that Mr. Smith, owner of the Pacific Coast Borax Company, had enjoyed the lion's share of the American market, possessed mines of practically inexhaustible supplies, but recently, owing to a reduction in the tariff on the crude material, had suffered from Asia Minor competition. His object in coming over was to increase his volume of trade by securing a share of the European Market.

In fact, any arrangement with Smith was not likely to solve Redwood's immediate problem of an improved and cheap source of borates for their new Belvedere

Works; however, they would have had a good deal to talk about in the process of getting to know each other.

One is left with the impression that the momentous plan to merge Pacific Coast Borax with Redwood's, and most of the major details, were agreed in an evening's discussion at an hotel and in a subsequent weekend at Baker's house at Nutfield, and that it largely depended upon two people, Baker and Smith, finding that sympathetic rapport which sometimes develops between opposite but complementary personalities. As a result the two men decided to do business on the basis of a joint future. No trace exists of any previous contact between them, but the circumstances of the time indicate distinct advantages to both in joining forces.

Redwood's was one of several small chemical companies using borax in Britain, and its authorized capital in 1896 was still only £25,000.[1] They were already committed to the Belvedere Works and to refining borax, and one can be sure that Baker and Gerstley had plans, with or without Smith, to secure a raw-material position in South America.

A connection with a much larger American company with a strong raw-material position in the same line of business must have seemed a stroke of good fortune for Redwood's at a time when they needed a voice in the European affairs of borax. It would be surprising if Baker had not discussed with Smith the problem of survival of the International Borax Union, and also the scope existing for a 'group' that could command raw-material resources in South America and the United States. Such a group could play an effective part in stabilizing borax prices, and also in strengthening the United States borax industry against foreign price-competition.

In 1896 there was probably no sign of the forthcoming government opposition to the use of borax and boric acid in food-preservatives, and if this had come a few years earlier Redwood's could hardly have survived, and certainly would have been unable to make a deal of this magnitude. Smith's position has already been mentioned: he needed to continue generating money out of borax for his other ventures, and it seems possible that if it had not been for a failure in the tariff-protection that he expected between 1894 and 1896 he would never have been worried enough about the borax price to start considering ventures outside the USA. Indeed, Smith must later have been greatly encouraged by the electoral victory of McKinley in 1896, after which the Republicans restored the import duty on refined borax to five cents.

The empire that Smith and Baker were to create between 1896 and 1900 was massive in comparison with their previous efforts, or with any other borax group existing at the time; and although reasons for the acquisition of particular parts of it can be advanced, it is still possible to wonder whether any overall 'grand design' was intended. It was never to include Lardarello and the other Tuscan producers, while the wild-goose chases in South America did not succeed in netting all the borax prospects there. In the USA Pfizer, Stauffer and some other tough opposition appeared and stayed outside, and their attitudes hardened as the years went by. Nevertheless, the overall effect was to produce a combine which would exercise a considerable influence on the stability of this all too easily produced substance, and this was probably the common purpose of Smith and Baker in putting together a group on an international scale.

The details and mechanics of the mergers that eventually produced Borax Consolidated, Limited in 1899 are uncomplicated to relate. The first stage in 1896 was the merger between Smith's borax interests – which had been concentrated since 1890 in the Pacific Coast Borax Company – and Redwood and Sons into a company with the unwieldy name of 'The Pacific Borax and Redwood's Chemical Works Limited'. Burton, Baker & Co. was not included, and Baker, strangely enough, became the only Redwood and Sons director to be given a place on the board of the new company.[2] The brothers Theophilus and George Redwood seem to have left the scene at this point; Theophilus was a pioneer in landscape photography before the days of films and dry plates, a Fellow of the Royal Photographic Society, and worked as a chemical consultant. He was afflicted by poor health, and died in 1909 at the age of sixty. However, a relation, Iltyd Redwood, became technical chief of the new company in Britain, with the title of Works Superintendent, and held the post until his death in 1910. The suggestion of sending him to the States to advise John Ryan, the Production Superintendent, on milling techniques sparked off possibly the rudest exchange of letters to be found in the company's files, and another fifty years were to elapse before any serious attempt at a transatlantic exchange on technical matters emerged.

Incorporation of the company took place on 19 June 1896, and its assets included all the borax properties of Pacific Coast Borax. These included: 'The Neel Consolidated Borate Mine' (alias 'Borate' or 'Calico') and 'The Stevens Consolidated Borate Mine' (unexploited) in the Calico Mountains – which have already been referred to collectively as the Daggett deposit – along with the Alameda refinery and the Oregon, Columbus and Teels Marsh properties, and a whole host of properties and placer claims still held in Death Valley. The prospectus included a report on Daggett by Thomas Price & Son of San Francisco, stating that 'There can be no question but that this deposit of Borate is positively unique among deposits of this nature', and that 'there certainly seems to exist here an almost inexhaustible deposit of Calcium Borate.'

From the Redwood side only the newly acquired Belvedere Mills were included, as the works at Kennington were on the point of being closed. The only comment in the prospectus about Redwood and Sons stated: '[It] has shown increasing annual profits from its commencement. It is intended to extend the manufacturing operations, at present carried on by this firm, to the United States, where it is believed a large and profitable business in addition to that now existing can be done.'

No sales or separate profit record for Redwood's was given. The combined profits of the two companies were stated thus – £81,036 in 1893/94 and £43,481 in 1894/95, the fall in profits being due to 'the reduction in the price of borax by the American company in the USA'. Prospective shareholders were not told a great deal in those days.

The eventual price paid to the two founder companies for these assets and for the accompanying businesses and goodwill was set out in an agreement of 21 August 1896. 'The Pacific Borax and Redwood's Chemical Works Limited' had been incorporated with authorized capital of £545,000, divided into 22,500 6 per cent. Cumulative Preference shares of £10 each and 32,000 Ordinary shares, also of £10 each, and certain very obvious points are noticeable. The Redwood interest was assessed at only something just over a quarter of the value of the new combine – i.e., it was to all intents and purposes an American takeover of Redwood's in spite of the British registration.[3] On the American side, however, Smith very carefully kept the name of The Pacific Coast Borax Company out of it, and took approximately three-quarters of the shares and cash in his own name, presumably as trustee for himself and the other Pacific shareholders. Surprisingly also the minute book of the new Anglo-American merged company shows that it had to send a loan of £30,000 to the USA without delay, shortly after its formation, to enable Pacific Coast Borax to pay off its liabilities, and thus to stop its creditors stepping in to attach assets earmarked for transfer to the new company. This perhaps provides an insight into Smith's enthusiasm to bring the deal to a rapid conclusion.

The Pacific Coast Borax Company continued in being, and the world in general was given little encouragement to find out that its assets were actually now owned by a British-based company. Presumably it would have been politically unwise, under a Democrat regime in 1896, to have publicized the fact that the greater part of the American borax industry was now owned, officially, overseas: and in any case it may seem surprising that London and not somewhere in the USA was taken as the centre of the group. The probable reason was that in the USA Mather's plan for the future – namely, to link Pacific Coast Borax with the larger established chemical company Pfizer – would not have suited Smith's plans for creating a financial empire of his own, and that throughout his visit to London Smith may well have seen a new opportunity to rid himself of the strictures of Mather and escape into a new world of financial respectability based on London, with freedom to go his own way in the United States.

Smith's move, made without consultation with his staff in the United States, could have caused considerable resentment, but only Mather senior was really in a position to react to Smith's strategy. Mather was deeply upset, and his resignation followed.[4] He assumed that the new British partners would have the same financial outlook as Smith, and he referred to them as 'those foreign sharps', and the depth of his frustration comes out in writing to a friend – 'I have tried Smith's style of business, if business it can be called. His methods have been devious, and his opinions liable to change as a woman's. . . . My old fashioned ideas and notions about business do not seem to be in order any more however, and I think it best to retire.' The rest of Smith's team were either unquestioning loyalist mining-men in the Far West or young commercial men in the sales offices whose dependence on Smith he took for granted.

The agreement of 1896 made it necessary to provide underwriters for the issue of Preference shares to the public and a reputable institution to act as trustees of the company's assets on behalf of the holders of the £100,000 worth of Debenture Bonds, which were to provide working capital. Thus the Indian and General Investment Trust Limited was called in and it also provided the new company with its first and only chairman in Sir Alexander Wilson, one of the Trust's directors, who stayed on to become the first Chairman of Borax Consolidated, Limited in 1899. Sir Alexander Wilson had been a well-known figure in the Calcutta of his day. He was Chairman of the Mercantile Bank of India, a director of the leading firm of Jardine Skinner, and a member of the Indian Legislative Council.

The two joint managing directors were of course Smith and Baker, although Smith's appearances at the Board were rare. He is shown as turning up on four successive occasions between 14 and 25 May 1897,[5] clearly in order to push through amendments to the Articles which would excuse him from attendance while resident in the USA and which would make all directors, except the managing directors, liable to resign and seek re-election each year instead of by rotation. This accorded with the views that he expressed to Baker:[6]

I want to know or feel that the Board will do as the majority of the stockholders direct, and that they will act in sympathy with your management there and my management here, leaving the exploiting, the connubiating, and the manipulating to yourself and myself, and back us in what we believe to be the best interests of the stockholders, and not cater to the small minority of preference shareholders. What I was thinking of more particularly a year ago was the apparent disposition of those members of the Board who had no investments with us to override the wishes and policy of those who held the principal investments.

Smith did not appear at Board meetings again until once late in 1898 when matters were already far gone towards the formation of a wider merger and the formation of the company which was to be known as Borax Consolidated, Limited.

The first move towards expansion came within a few months of the start of the new company (1896), when Mear & Green Limited of Staffordshire offered their business for sale. The two principal shareholders in this small company were Stephen Mear, a timber merchant of Longton, Staffordshire, and Thomas G. Green, an 'earthenware manufacturer' (still famous) of Burton-on-Trent.[7] They had three factories for production of borax and boric acid, situated respectively at Connah's Quay in Flintshire and at Tunstall and Kidsgrove in Staffordshire, and when Baker took a chemist friend of his, W. B. Giles, FIC, to see these factories he reported favourably on the state of the plants, and that the sites were well placed in relation to one of the largest markets for borax, 'The Potteries' of Midland Britain. He thought that the amount at which they stood in the account books – £35,000 – was a fair one, and accordingly the sale went ahead very rapidly on a basis of £60,000.

Although Mear & Green stated that their net profits had averaged £8,000 over the previous five years, events soon proved Giles's favourable opinion of the operations to have been mistaken. For many years Baker relied on Giles and his small laboratory at his house at Leytonstone for scientific advice and most of the analytical work required: and Giles's many letters, now in the archives of Borax House, London, are masterpieces of clarity and precision. However, he was never again asked to make an assessment of the commercial worth of a factory.

Another miscalculation – this time about people – was more immediately and financially embarrassing. Horsfall, who was left managing the three small factories, kicked over the traces as soon as he was free of Mear & Green, and during Baker's absence in the USA for a few weeks he purchased nearly two thousand tons more of South American borate of lime than the stock of 288 tons which Baker had authorized before his departure. This tied up all the company's funds and left the three brokers concerned uncovered for a handsome sum. Horsfall appears to have given no explanation of his conduct to the meticulous Baker, and when asked not to make any more purchases without the latter's approval he promptly resigned and started selling crude borate in New York, right within Smith's sacred preserve of the USA and in breach of Mear & Green's undertaking to the new owners. Not unnaturally, Stephen Mear, who had been invited to join the board of Pacific Borax and Redwood's in December 1896 (before his erstwhile employee Horsfall had started creating trouble) disappeared from it again in mid-1898.

All this time – in theory at least – the board was meant to be responsible for the affairs of Pacific Coast Borax in the USA as well as of Redwood's in England. But it is clear that with Smith the headstrong and forthright major shareholder in charge in the USA the British board was likely in practice to be confined to matters outside that vast country. Certainly Smith's first major step after the 1896 link-up of Pacific Coast Borax with Redwood's was never reported to the new London board. For a number of years Smith had had his eyes on acquiring their interest in Searles Lake from the San Bernardino Borax Mining Company: and the borax price slump of 1894–7 at last dropped it into his lap. The Searles brothers had had to close the works early in 1895, and in January 1897 Smith was able to acquire 70 per cent of the stock of San Bernardino, which he put in the name of the still existing Pacific Coast Borax Company.[8]

Smith's second major move after 1896 had, however, the London board's full approval. When the incoming Republican administration once more raised tariffs it was considered that the time had come to consolidate the advantage that this gave to the borax business in the eastern states against the growing menace of competition from imports. It was decided to build a factory near New York,[9] and after one or two unsuccessful attempts a 'suitable' site was reported as secured for the proposed factory in December 1897. This site was located at Bayonne, New Jersey, on the Kill von Kull, almost directly across from Sailors Snug Harbour on Staten Island, New York. It seemed a dubious choice, as it was entirely surrounded by Standard Oil Company property, except for one roadway and the waterfront, and the only access by rail was over the Standard Oil Company's private yard line, so that in these cramping circumstances most of the raw material seems to have reached the place eventually by canal from the docks. Nevertheless, on this site a building measuring about 125 by 250 feet was built by a firm of contractors called Ransome and Smith (actually none other than F. M. Smith). It had the distinction of being the first reinforced concrete building to be built in the eastern states, and it followed Ransome's successful effort in building a concrete extension to the works at Alameda, which was claimed as another 'first' – the first industrial building in concrete in the USA.

The methods used to counter competition from material imported into the eastern states soon went further than this. Many of the borax-refiners in Britain were trying, as long as their independence lasted, to buy up cheap South American material, refine it and export it to the growing American market in the face of Smith's opposition. Some of these had connections with those who claimed to 'own' or operate certain deposits in South America, and it was with one of these that Pacific Borax and Redwood's made its first (and, for several years at least) most important contact in its efforts to buy up or neutralize potential South American competition at its source. On 2 February 1898 Baker reported to the board the result of the negotiations for the acquisition of the borax deposits of the Empresa d'Ascotan Company in Chile, and the terms on which these could be acquired.

Baker, in spite of many a good intention, never managed to find time to visit South America before 1916, and there is no indication of where and when he met Norman A. Walker, who according to Gerstley, had come to England with the vague intention of disposing of 'Ascotan', and had promised Pacific Borax and Redwood's first option if he decided to sell. But no sooner had Walker made his airy-fairy promise than he seems to have gone off and sold the property to the syndicate of German refiners – who had hitherto been its customers for the raw material – through the Deutsche Bank in Berlin. Negotiations therefore had to begin all over again, this time with the Deutsche Bank. Gerstley was sent over to Berlin to negotiate,[10] although it might seem that Pacific Borax and Redwood's had little to offer other than money in order to retrieve the situation. However, the time was in their favour, and gave them the first chance to exert the strength of the new combine. The 'German Union' of refiners had, since the dissolution of the International Borax Union in the previous year, seen a chaotic situation develop with regard to raw material and refined-borax prices. They probably had no real wish to become joint mine-owners in South America, but saw no way to secure some stability except to take up the offer of the Ascotan deposit. Gerstley no doubt presented the picture of increasing competition from the United States and other properties in South America which Pacific Coast and Redwood's could acquire, and the 'German Union', remembering the better days of the International Union, agreed to sell to Pacific Coast and Redwood's, who in return offered a new agreement to the 'German Union'. The Ascotan property was purchased for £125,000, and the so-called 'Ascotan Agreement' became part of the deal.[11]

Under this agreement the German Union – to which were added the names of Schering of Berlin and Balzer of Gruneau, making a total of seven – agreed to buy from Pacific Borax and Redwood's specified quantities on specified terms for fifteen years, with provision for withdrawing from the agreement at the end of 1903 or 1908.[12] The maximum to be supplied from the Ascotan mine was fixed at 8,000 tons a year, but the amount would vary according to a formula relating it to 'world demand'. The agreement controlled the whole operation right through to the prices to be charged for the end-products, refined borax and boric acid, and the refiners were debarred from obtaining crude material from other sources as long as the company maintained sufficient stocks at Hamburg and Berlin. Perhaps the most important clauses in the agreement were the undertakings of the refiners not knowingly to sell borax or boric acid for export to the USA or to resell crude material from Ascotan except among themselves, and the undertaking of the company not to export any refined products from the USA and not knowingly to sell any crude or refined products to any concern in the USA where there was any risk of them being exported.

This was a tough agreement, which no doubt pleased Smith and which set the tone of commercial policy towards Europe for a considerable number of years, and it is amazing that the constraints that it imposed worked as well as they did. Inevitably the agreement gave rise to never-ending anxiety among the parties that one or the other of them might be stealing a march on the rest by committing an undetected breach of the rules, but for all this the agreement seemed the only way of preventing a disastrous price-cutting war in the limited European market of the time. Meanwhile Pacific Borax and Redwood's had pressed on further with their apparent aim of controlling the sources of South American borate of lime. Messrs F. Rosenstern & Co had been appointed the company's agents in Hamburg,[13] and in March 1898 Friedrich Lesser, a cousin of Rosenstern, was seconded to the company in order that he might use his knowledge of South America and of Spanish to add further mining prospects to the Ascotan concession. Lesser (whose Christian name became transmuted from Friedrich to Federico in South America) was soon on his way, although he was not appointed officially 'General Manager on the Pacific Coast' until a year later, while the remarkable increase in the number of South American negotiations recorded during 1898 indicates that he was soon a very active agent.

The first borate of lime property to come up after Ascotan was an option on a property called Pintados, which was acquired in two parts in 1898–9,[14] while

Carcote – an area of some 6,900 acres near Ascotan, and estimated at the time to contain about 300,000 tons of borate of lime – had been purchased with the minimum of delay from the Sociedad Boratera of that name for £15,000 in July 1898.[15] (See map IX, page 239.)

By this time the idea of the forthcoming Borax Consolidated merger was starting to take shape, and Lafayette Hoyt de Friese, a member of a firm of international lawyers, was called in, made a director and asked to act as nominal holder of the South American properties that Lesser continued to acquire, so that they could be transferred direct to the new creation when it emerged.[16] Under this arrangement considerable sums of money were laid out, within less than six months in 1898, on acquiring further sizable borate of lime properties.[17] There was Chillicolpa (120 acres), then in Chile but now in Peru; Cosapilla, in Chile (343 acres of good-grade material); and finally in Peru the property of the Compania Boratera de Arequipa, consisting of about 1,280 acres with depots, warehouses, dwelling-houses for 500 men, and calcining plant at Arequipa and Salinas, together with carts, mules and donkeys for transport and estimated reserves of a million tons of borate.

More properties were acquired in the following decade, and Part II of this book has been devoted to illuminating strange names and far-off places with stories of the realities found there by company employees, and to trying to unravel the directors' intent in acquiring, and occasionally operating, so many scattered possessions.

Meanwhile, what was perhaps the most important segment of the world's borax industry at the time had to be tamed by negotiation if this far-reaching attempt to bring about stability in world borax prices were to succeed. In the event the main Turkish mining companies appear to have been taken over without any serious opposition, and as no document has survived from the actual negotiations, it can only be assumed from the facts recounted that the financial state of these companies and the venal atmosphere in which the directors operated made purchase for a handsome price a sufficient inducement.

It has already been recounted in Chapter 2 (p.28) how a French engineer working in Paris, and called Desmazures, discovered that two statuettes brought from Turkey and given to him were made almost entirely of calcium borate: also how he traced the source and obtained a 'permis' in 1865 from the Sultan's government to excavate the mineral. The 'permis' was exchanged for a full Imperial Firman of concession for fifty years on 21 June 1887, and Desmazures and his partner Groppler are alleged by Robottom to have made fortunes for themselves by exporting the calcium borate mineral called pandermite to France and to England, where it was used at the Belvedere works.

Desmazures had also started a borax-refinery at Maisons Lafitte, near Paris, and the Société Desmazures et Compagnie was acquired later by a new company called The Borax Company, registered in London on 1 December 1887.[18] By this time Desmazures et Cie had alongside their concession in Turkey another one granted (also in 1887) to Charles Hanson & Co.,[19] bankers of Constantinople, and to Frederick Giove, but by 1898 both the Desmazures and the Hanson concessions had been amalgamated and came under the control of The Borax Company, of which the Chairman was the Hon. H. A. Lawrence. Robottom says that both Groppler and Desmazures were still associated with the company in 1898, and died shortly afterwards.

The prospectus describes the properties which The Borax Company made over to the proposed new Borax Consolidated as the boracite mines and works near Sultan Tchair, in the Vilayet of Karassi, Asia Minor. (See map X, page 259.) This was near the port of Panderma, on the Sea of Marmora, with an area of 12,000 acres, and included with it was all buildings, works, plant, machinery and stock. These mines were fully developed and opened out, and had produced an average of nearly 8,000 tons per annum during the past three years. The boracite in sight was estimated at 250,000 tons, and the supply appeared practically inexhaustible. The Borax Company also owned the land, factory and refinery at Maisons Lafitte near Paris already mentioned, which comprised an area of about 8¾ acres with all buildings, modern plant and machinery complete, and capable of producing about 3,000 tons per annum of borax and boric acid, exclusive of other products. The price paid to The Borax Company was £320,000, in a mixture of shares, debentures and cash.[20]

On the same day (29 November 1898) the property and assets of the second European company operating in north-west Anatolia passed to de Friese, acting as stakeholder for the proposed new Borax Consolidated. This was 'La Société Lyonnaise des Mines et Usines de Borax', whose head office was at Lyons. There is no evidence of the year in which it was founded, but it was based on a concession granted by the Sultan's government in 1889, originally to His Excellency Fuad Pasha, 'my general aide-de-camp' (i.e., one of the hangers-on at the Palace), in return for payment of 2,440 piastres a year to the Department of Mines.[21] The Pasha sold the concession to M. Auguste Falcouz, a banker of Lyons, on 15 April 1891. In the previous year 'La Compagnie France-Autrichienne de Produits Chimiques' had leased from one J. Sperber, a manufacturer, the borax works at No. 256 Brigittaplatz, Vienna, and two years later Auguste Falcouz, on behalf of Société Lyonnaise, leased the borax works situated at No. 43 Scaronne Road in the Mouche district of Lyons from Francis Vial, a chemical-manufacturer. These two works and the mine in Turkey passed to de Friese, the Borax Consolidated representative, as the assets of the Société Lyonnaise des Mines et Usines de Borax for the equivalent of £240,000, which was £80,000 less than the price of The Borax Company.

From their description the mines of the Société Lyonnaise at Karassi were a few miles south of The Borax Company's mines and very much smaller in extent, consisting of only just over 600 acres compared with 12,000 acres. However, their output compared remarkably well. They had produced from about 7,000 to 7,500 tons per annum, and there was an estimated quantity of 150,000 tons of mineral in sight. The refinery at Lyons was described as fitted with modern machinery and plant, but no evidence survives to show how La Société Lyonnaise had prospered before it joined Borax Consolidated.

According to letters from Smith to Baker, it was the threat to these French companies' European markets for refined material caused by Pacific Borax and Redwood's acquisition of a cheap source of raw material at Ascotan which was probably the main factor that drove them to give up their independence and join the new combination.[22] The collapse of the International Borax Union in 1897, and the chaotic situation of the borax market thereafter, would also have made a serious impact on profits and outlook for the future.

The incorporation of Borax Consolidated, Limited followed on 11 January 1899. A valuation of the new company given in the prospectus – which added stocks, cash, and working capital to the book value of fixed assets, and to fifteen years' purchase of the mines and deposits at the figure for profits calculated by the auditors – produced a figure of £3,148,412, which was a considerable sum for 1899. Of the stock available to them, the public took up less than a fifth of the voteless preference shares offered, but more than three-quarters of the much safer debentures. They were obviously not very impressed with the profit forecast in the prospectus which showed that, although the assets looked valuable, the combined profits of Pacific Borax and Redwood's, Mear & Green and the two French companies for five years up to March 1898 had declined from £175,711 to £119,457.

To the Ordinary shareholders the average annual profit forecast on the £600,000 worth of Ordinary shares was just over £82,000 before deductions for directors' fees, depreciation and reserves – say something just over 10 per cent. However, if good management, rationalization and dynamism were introduced into this mixed bag of assets, and if the market for borax continued to expand, those who had acquired the Ordinary shares had good prospects ahead.

The prospectus shyly admitted that 'Messrs F. M. Smith and R. C. Baker were interested in the sale to the company and would receive part of the profit to be derived from it'. De Friese had in fact been carrying on his job of collecting all these assets by means of money supplied principally by these two. So when the list of Ordinary shareholders settled down after a few weeks, Smith appeared as far and away the largest shareholder, with Baker and the two French companies as very modest runners-up, and with no one else with any outsize holdings at all. This remained the position right up to Smith's financial crisis in 1913; but because the company was nominally centred in London – from which all the assets except those in the USA were managed – and because Smith was habitually absent from Britain in these fifteen years, few seem to have noticed the real position at the time. The reports of annual general meetings in particular read as though the company was composed of a wide spread of shareholders from the general public. This illusion had its advantages, as will be seen from the next chapter.

From this point Smith appears to have been content to leave detail and many of the major decisions connected with the Company's extensive borax interests outside the USA to those in London, and there is no evidence that he visited Britain at all after January 1899.

An important outcome of the formation of Borax Consolidated, Limited was the extension of the Ascotan Agreement to include the companies which controlled all the Turkish borate mines, the majority of the refining business of France and Austria and further properties in South America. Later six British companies were included in the arrangement as 'associated refiners'. These were Coghill & Sons (Liverpool); Joseph Townsend (Glasgow); Sir John Cuthbertson (Glasgow); Boron Products (Stroud); John Jones (Stoke-on-Trent); Howard's (Stratford, London). This represented a formidable array of interests in the world of borax, and was a considerable achievement for Baker and Smith, particularly the former, taking into account the size and position of Redwood and Sons only four years before.

Finally, it is intriguing to reflect for a moment on the contrasting personalities of the partnership formed by Smith and Baker. Smith was an inspired adventurer from what was still the Wild West – he had successfully emerged from the world of mining camps, fortune-hunters and prospectors who roved the scorching deserts of California and Nevada on mules, carrying their rock specimens in their saddlebags along with the rest of their worldly goods. Smith's impatience with detail and paper-work and his driving urge for risk-taking and expansion made a splendid contrast with Baker and his new-found partner Gerstley – both experienced City business-men, ignorant of the world of mining but with a shrewd understanding of the financial realities and pitfalls of commercial life.

No detailed comparisons are needed after a glance at the portraits of Smith and Baker, except perhaps the remark that one could never imagine Baker running a man off a claim with a gun or building a mansion like Arbor Villa in Oakland. It is no wonder that they never got down to Christian names, but remained 'Mr Smith' and 'Mr Baker' in their numerous communications for a period of over eighteen years as business colleagues. However, a voluminous correspondence between these two men scarcely contains a single discordant note, and reflects a mutual respect for each other's often differing views on business matters. The relationship proved highly successful in dealing with the difficulties of communication and decision-making at a distance of over six thousand miles, between two parts of the world which could be accurately described as poles apart. Business always came first, but many letters – more by Smith than by Baker – include passing references to such matters as grouse-shooting, deer-stalking, duck-hunting or yachting, indicating a common bond in the world of sport; but neither in their leisure moments seems to have sought the company of the other.

REFERENCES

1. Memorandum and Articles of Redwood and Sons.
2. Minute Book of Pacific Borax and Redwood's Chemical Works Ltd.
3. Prospectus of the Pacific Borax and Redwood's Chemical Works Ltd.
4. Shankland, R.: *Steve Mather of the National Parks* (New York, 1951), p.35.
5. Minute Book.
6. Letter F.M.S. to R.C.B., 15.9.98.
7. The Minute Books of Mear & Green and of Pacific Borax and Redwood's contain together an unusual amount of detail which forms the basis of this account.
8. Teeple, John E.: 'Industrial Development of Searles Lake Brines', *Am. Chem. Soc. Monograph* 49 (New York, 1929).
9. Minute Book.
10, 11, 12. Minute Book entries of 2 February, 13 May, and 25 March 1898 respectively.
13. Minute Book and Gerstley's *History*.
14. Minute Book.
15. *Ibid.*
16. *Ibid.*, and Smith's letters.
17. *Ibid.*, and deeds now at Borax House, London.
18. Deeds now at Borax House.
19. Attachment to a letter from T. Browne, Vice-Consul to F. M. Smith, dated 10 February 1898.
20. Deed of 29 November 1898.
21. Translation of Imperial Firman.
22. F.M.S. to R.C.B., 22.11.98.

7

The effects of the Borax Consolidated merger

There was a special feature about the activities and intrigues which produced Borax Consolidated, Limited early in 1899 that many a chairman of a modern merger might envy. Instead of the drive and purpose being diluted by the prearranged admission of the heads of the other companies concerned to the new board, the board of Pacific Borax and Redwood admitted only one 'outsider' when it transformed itself into the board of Borax Consolidated, Limited in 1899.

That this one outsider, the Hon. H. A. Lawrence, chairman of The Borax Company, was quite sufficient for the time being is apparent from F. M. Smith's comment to Baker in June 1899:[1] 'I was much surprised to learn that Lawrence was capable of taking advantage of his position with our company as director, to work out his own personal schemes. It would appear as though he was not in the least interested in the success of the Borax Consolidated. . . . It is no longer strange that 'The Borax Company' was a failure.' Lawrence, however, was not destined to trouble the Smith-Baker axis for long. He died in 1902.

The directors of the other French company in the merger, the Société Lyonnaise, Ms Vial, Pradel and Jacquier, were safely fobbed off for the time being with membership of an 'advisory committee' set up in France.[2] They were joined also by Ms Monchicourt and Geisenheimer, Paris agents of The Borax Company.

The merger was given an official reason in the prospectus of the new company, Borax Consolidated, Limited, published in London, which, after listing the companies and properties comprised in it proclaimed:[3]

> The said properties are believed to comprise all the important mines and sources of production of raw material from which nearly the whole supply of the world has hitherto been obtained. The acquisition and consolidation of these properties will enable the Company to put an end to the competition which has hitherto existed between them, and the economies which will be effected by the various properties and concerns being under one administration should substantially increase the profits hitherto made, without any undue inflation of prices.

The first sentence of this statement, although drafted as a belief designed to encourage the public to buy shares and debentures, was hardly true at the time, and was perhaps a plan for the future. It paid no regard to the Italian producers' considerable influence in the European market, to the continuing trickle of Tibetan tincal over the passes into India and thence to Europe, or to formidable competition that already existed in the USA and South America. The second sentence was a declaration of policy which might have been more tactfully phrased in view of the existence in the USA from 1890 of the Sherman Anti-Trust Law. This law, in the first of its eight sections, laid down that 'every contract, combination in the form of trust or otherwise, or conspiracy, in restraint of trade or commerce among the several states, or with foreign nations, is . . . illegal' and followed up with enlargement of this main theme in the remaining seven sections.

There was, however, no real reason in 1899 for alleging that the new company had been formed in breach of the Sherman Act, whatever the Democrat Press in the USA might allege at the time. It had fallen far short of establishing a monopoly over all the world's important sources of supply, and this fact was to become all too painfully obvious to the board as the years went on. Moreover, the borax price in the USA showed no reaction at all to the formation of the world's largest borax merger up to that time. It was seven cents a pound in 1898, and it did not rise at all until 1902, when it rose by only one-quarter of a cent, and remained at that level until a gradual decline started in 1905, which was accentuated by a substantial cut in the American protective tariff in 1909.[4]

Baker expressed the general thinking behind the merger most aptly in a letter written to Señor Ezcurra, Borax Consolidated's representative in the Argentine, on 9 July 1907,[5] when advising him how to reply to an attack on 'the borate industry' which had appeared in a pamphlet by a Dr Reichert. He suggested a response pointing out that the borate industry was not of such importance as was attributed to it, that the total consumption of the world was a very small figure, and that the borate deposits (being the mines in the USA and in Asia Minor and those which were then being worked in Chile, Peru and the Argentine) were much more than was necessary to supply the limited world demand, which he calculated as only about 30,000 tons at that time. It would be impossible for any deposits which were not favourably situated to compete, and any effort in that direction could only lead (as it had done in the past) to a loss of money for those who wished to interest themselves in the business. There was no monopoly in borates but only the harmony of interests between those producers who had for years past supplied the trade, and it was only this which had prevented the utter collapse of the business. Any one of the favourably situated properties which were then being worked could produce for many years to come the whole requirements of the world at that time. The 'harmony of interests' between existing producers was in fact more the sort of harmony produced among enemies by the grim fact that they all possess the same devastating weapon.

However, it was to preserve this harmony, and to protect the variegated collection of assets thrown together by the merger from the effects of any breach of it, that the main efforts of the board were directed for the first five years or so of Borax Consolidated's existence. Protection, of course, inevitably involved an element of internal strengthening, through the need to turn into some sort of homogeneous group assets and employees spread over four continents. That some sort of effective rationalizaion was achieved in so few years was due almost entirely to the rare administrative skill of Baker and to the shrewd way in which he chose and controlled his immediate subordinates in order to carry out his purpose. In reading through volume after volume of his correspondence one feels that one has at last found the man with the 'infinite capacity for taking pains' which is one of the popular definitions of genius. He obviously enjoyed the excitement of taking over other companies and making international arrangements as much as any other tycoon; but he possessed also the much rarer gift for the duller job of tidying up afterwards and for administering the new acquisitions with all the care that he had given previously to the small Redwood's chemical works at Kennington Green. Inevitably on occasions he carried this meticulous attention to detail too far, and sorely tried the patience of otherwise loyal colleagues – notably in his endless nagging attempts to bring the vagaries of costs and employees in remote dust-covered places in South America and Turkey under the same sort of scrutiny as those of Belvedere. But financially his sphere of responsibility in the new group – everything outside the USA – prospered in these years largely as a result of his eagle eye in management.

This was to prove the salvation of Borax Consolidated in the long run, even though the American side, under the preponderating shareholder, F. M. Smith, started with the confident assumption that these many interests in three other continents had been taken over with the main object of protecting its home market in the USA and providing it with expanding outlets overseas.

Meanwhile the major problem of the first decade soon became the question of how deeply the new company must involve itself in South American mining prospects in order to protect its interests elsewhere.

South America

The colourful story of the Borax Consolidated efforts to secure some sort of control over the most important of the numerous sources of supply in that vast (and at that time only partly explored) continent has already been mentioned, and is told in its human aspects in Part II. All that it is proposed to do here is to outline the policy apparently pursued at various times up to 1914, and the frustrations encountered.

The first step taken was probably the first and only master-stroke. The great Ascotan deposit, which had been supplying the principal German refiners, was acquired in 1898, and the Ascotan agreement – which tied these German refiners to supplies from this deposit and to Borax Consolidated – was secured at the same time, as already related. This committed the company to working the Ascotan deposit, in the rarefied atmosphere of the high Chilean Andes, and to sending the material, via a mill and depot at Cebollar, down to Antofagasta by rail to be shipped to Europe. A stake in similar projects at Salinas in the high Peruvian Andes, with a depot at Arequipa and rail transport to Mollendo on the coast, and at Suriri, far away on the Chilean-Argentinian border, with a depot at Chilcaya and no railway line or any other obvious form of transport to anywhere convenient, was also brought into the merger. Interests at Chillicolpa, Cosapilla and Pintados were also acquired and further purchases were made subsequently, but apart from intermittent operations early on at Salinas, no attempt was ever made to mine them systematically. (See map IX, page 239.)

The Suriri-Chilcaya acquisition, however, soon brought with it a monstrous lawsuit which dragged on until 1911 and cost Borax Consolidated thousands in fees to lawyers of suspect loyalty and efficiency. Salinas was also disputed by a lawsuit of rather less intensity and duration, and a half-hearted attempt was even made to contest the company's title to Ascotan. In these circumstances the early optimism expressed by Smith and Zabriskie – that the essentials of the South

American position had been brought under control cheaply and quickly – soon vanished.[6] Instead Smith in particular started to realize that the laws of these South American republics relating to the staking of claims were so loose as to permit something akin to anarchy. As he wrote in 1901:[7]

> The ease, facility, and inexpensiveness of tying up these South American alkali flats for boric acid, without as much as prospecting or looking at them, or going out into the country where they exist, is somewhat paralyzing and tends to keep us on the anxious seat until we have examined, which is expensive and difficult. Then, it seems, they have a considerable time in which to designate their boundaries. In fact, those Indians are able to practically secure a floating grant without cost and compel us to prospect for them.

Smith had already grasped one of the realities of the situation – the entry of the comparatively rich Borax Consolidated group into South America had put it in a position comparable to that of a tourist surrounded by importunate souvenir-sellers. There were obviously a large number of these men, quite apart from people like Norman Walker, who play a prominent part in this history, and their wares had to be examined carefully for two main reasons. However scattered and remote most of it might be, South American borate of lime was on average significantly higher in boric-acid content than North American material. Also, many of these vendors of claims knew that they had other possible purchasers in Europe who were anxious to undercut Borax Consolidated's trade, and they could therefore play on the original fear that drew the company to South America.[8] As Smith believed[9] that it would not take much capital for a rival to install a plant in Europe and produce from South American raw material, his reaction to this problem was predictable. The company must at once expand its surveying staff so that as far as possible the offers made, and in addition every square mile of likely territory, could be examined for borate.[10]

This was a vast programme, and unfortunately, it could no longer be confined to Peru, Chile and a limited piece of Bolivia on the western side of the Andes. The Belgian company, Compagnie Internationale de Borax – who were rapidly bankrupting themselves in trying to fight Borax Consolidated's entry into the European market – moved into Argentina in these years, and others were also preparing to develop Argentina's deposits. Thus by the time (1905) that Borax Consolidated had won the battle against Compagnie Internationale in Europe, and taken over its Argentinian interests also, the group found itself with about half a dozen mines or deposits in Argentina in addition to an ever-lengthening list of deposits that it had purchased in Chile and Peru. Naturally enough, Smith began to wonder, as early as 1898,[11] how they could continue to find the money for this. It had been found hitherto entirely out of the First and Second Debenture issues of the 1899 merger, but this started to impose a limitation on South American expansion. As Baker wrote, in answer to a 'hot tip' from Coghill, a friendly borax-refiner, in 1903:[12]

> There may be some sixteen 'mines' in the Argentine Republic containing borate but so far as most of them are concerned they might as well be in the moon. If we are to purchase all such deposits at the price the owners expect, we should be issuing our Sixth debentures before long, but I think our Seconds will cover anything we shall be likely to want.

Cynicism had replaced earlier enthusiasm, and it was now realized that apart from shortage of money it was no longer necessary to buy up every deposit that came along. Even if some of the good ones had been in the hands of rivals, they were often so remote from a railway line as to be unlikely to earn a profit. On the other hand, a new type of money-drain appeared. As Baker put it, instead of paying out large sums on legal fees it was sometimes cheaper to buy up rival claims rather than fight them, even though this sometimes involved buying out as many as three sets of rival claims to each mineral deposit.[13] The souvenir-vendors had in fact moved on to a more sophisticated type of business when straight sales of mining claims were starting to flag; and this in turn prompted Borax Consolidated to start lobbying for a tightening up of state laws governing the registration of mineral claims.

However, hardly had this policy gone into action when the state governments in turn began to show signs of killing the foreign goose that might, with a little intelligent government co-operation, lay golden eggs.[14] Late in 1913 the Chilean Government expressed the intention of levying an export tax on borates. The Chilean Minister in London told Baker that his government thought that the company ought to contribute something substantial towards the country, as they were depleting it of its valuable raw materials. Baker countered by saying that if Borax Consolidated's South American business had visited upon it all those expenses with which it could legitimately be charged, the profit left would barely cover the proposed export duty, and that Borax Consolidated could produce more cheaply elsewhere (meaning Turkey).[15] Nothing daunted, the Chilean Government went ahead, and a month before the outbreak of the First World War, introduced a Bill imposing a 30s.-a-ton export duty on companies with less than 70 per cent Chilean shareholders.

By this time, however, Borax Consolidated had already partly covered itself by obtaining a decree from the neighbouring government of Peru that would exempt it from any export tax for eighteen years,[16] as by 1913 the plain fact had emerged that the commercial quality of Ascotan borate was deteriorating,[17] and that in future a minimum of 10 per cent sulphate in the material must be expected. It had therefore arrived at the fateful month of August 1914 with the intention already formed of switching gradually from the Ascotan complex in Chile to the Salinas complex in Peru, and with the uncomfortable realization that the young republics of South America (which had hitherto tried to make what profit they could out of foreigners on a 'private enterprise' basis) were now veering towards more jingoistic nationalism. It cannot be said, however, that the operations of Borax Consolidated at any other place in South America outside Ascotan were greatly upset by this tendency, as in truth they did not amount to much in 1914.

Turkey

In Turkey, in these first fifteen years of Borax Consolidated's existence, the position was comparable to that in Chile, but the influence of politics was very much more to the fore. The two French companies which were the predecessors of Borax Consolidated in this area were mining a few miles inland from the Aegean coast of Asia Minor during the reign of the shrewd but devious Sultan Abdul Hamid II. Two tendencies of his reign were a reaction away from the Young Turk movement back towards palace despotism, and the gradual replacement of French influence – which had predominated at the Sublime Porte through most of the nineteenth century – by deliberate German infiltration, following German unification and Bismarck's subsequent policy of *Drang nach*

Osten. The two French companies, and after the merger Borax Consolidated, had to do business with a sophisticated but corrupt civil service, controlled by the 'Palace Pachas', in order to preserve their mines at Sultan Tchair and Azizieh and the surrounding concessions which had been acquired to protect these main assets. To obtain what was required also included keeping on the right side of several expatriate or Levantine middlemen, such as Ernest Whittall or Edward Pears, who seemed to have the entrée to palace confidence by various unofficial side-doors, and also by using Aristide Tubini, one of the most skilled and trustworthy of them, as the company's agent at Constantinople.

However repellent this system might seem to modern efficiency experts, it appears to have worked extremely well for Borax Consolidated until 1908, when the Young Turk revolution deposed Abdul Hamid II and soon set up a predominantly military government, in which Enver Bey, the Minister of War, rapidly came to the top. Neither he nor his colleagues, however, were allowed sufficient time in which to throw off the evil effects of more than a century of decline from the great age of the Ottoman Empire, before the 'Great Powers' – with whom Abdul Hamid II had at least succeeded in playing successful diplomatic chess – moved in to seize outlying parts of the Empire. The military left the mines alone until 1914, but Borax Consolidated found itself carrying on there in a growing atmosphere of uncertainty. The board appears to have been blind to the increasing German menace in that area, but it was certainly troubled by the disruption caused by mine labour being drafted to the Army and by sudden interruptions in the export route to Europe, notably with the closure of the Dardanelles in 1912. Still more, the resurgence of Turkish nationalism that came with the Young Turks held out obvious portents for the more distant future for any foreigner trying to do business in Turkey. Baker (and probably his generation) were doubtless so used to the old Turkey that they may have found the idea of an emerging westernized Turkey almost impossible to grasp – as incomprehensible, in fact, as some of them found Turkey's entry into the War against the Allies, and her subsequent defeat of the Allied landing at Gallipoli.

Against this very marked political background Borax Consolidated's mines in Turkey may be said to have put up an encouraging performance in the years 1899–1914. This must be ascribed largely to the bold extrovert personality of their qualified and experienced mining engineer at the time, Charles Bunning, whose only obvious fault may have been that he had learnt a lot about mining and about managing non-British personnel, but nothing about treating far-distant directors in London with the deference that they seemed to regard as their due. However, in spite of Bunning's lapses into *lèse-majesté*, Baker obviously recognized his worth. So he survived, and his production and his costs were so good that he provided Borax Consolidated with a valuable card to play against South American governments, and against the company's numerous small rivals on that continent. Borax Consolidated could – and latterly did – threaten to push South American borate of lime out of the European market, with the Turkish product so conveniently available at the back door of Europe.

Fortunately, their bluff was never called, for bluff indeed it must have been as the Turkish supply, from their concessions around Susurluk village, was necessarily limited, and was liable to be closed down at short notice by political troubles, notably the outbreak of the First World War in 1914. The Turkish mines struggled back into relatively uninterrupted production after the Atatürk revolution of the early twenties, and were to prove a useful source of raw material, mainly for boric-acid production, for several decades thereafter. But that is a later story.

Europe

In addition to the Turkish mines, the two French companies brought with them into the merger of 1899 some useful factories, depots and sales interests which were to form the foundation of Borax Consolidated's strong network in Europe. As has already been mentioned, The Borax Company had a factory capable of producing about 3,000 tons per annum of borax and boric acid, exclusive of other products, at Maisons Lafitte near Paris, while Société Lyonnaise had a leasehold factory of just over half that capacity at Lyons and another smaller factory in Vienna.[18] Apart from the trading connections inherited from these two companies – which stretched into Germany, Spain, Austria and Russia, as well as France – the promoters of Borax Consolidated had also brought into the merger the tied consortium of refiners in Germany created by the 'Ascotan agreement'. This initial European structure rested on raw material from the South American and Turkish deposits, leaving the American assets as a separate, self-supporting unit under the direct management of Smith. Its main rivals in Europe were Lardarello, based in Italy, and Compagnie Internationale de Borax, which was founded shortly after Borax Consolidated, and centred on refineries in Belgium and Northern France, while drawing raw material from 'tied' mines in Argentina.

However, in spite of their attractive-looking assets, The Borax Company had not made a profit for several years, and there were suspicions that all was not well also at the Lyons and Vienna establishments of Société Lyonnaise (as later proved the case). Accordingly the very first meeting of the board of the new Borax Consolidated, Limited, held on 11 January 1899, set up an 'Advisory Committee' in France, consisting of Vial and Pradel (who were managing Société Lyonnaise), Jacquier, their banker, and a slippery couple called Monchicourt and Geisenheimer.[19] The latter two had formerly held contracts from The Borax Company to act as managing agents in France, since the board had consisted almost entirely of Englishmen. The Advisory Committee which they now joined lasted for only two and half years until 29 August 1901; and, although no evidence survives, it seems doubtful whether the individuals on it ever broke off the pursuit of their own personal interests long enough to meet as a committee, or whether the advice that they gave, individually or collectively, was ever of any use to Baker and the London board.

One of the few surviving documents of this period is a reply by Geisenheimer, dated 10 May 1901, to a letter (no longer traceable) from Baker which probably had asked the reasons for the high cost of running the Maisons Lafitte Factory. Instead of a list of commercial and technical reasons for the failure to make a profit over so many years a somewhat unusual reply came back, which began:

> When we entered The Borax Company and commenced to look after the factory we were, like you, struck by certain expenses which, at first sight, appeared to be made solely in a spirit of philanthropy or from motives of obligingness. We have, however, had to recognise since then that all these expenses, dating mostly from the time of M. Desmazures, were really most useful to the company and almost indispensable: that is to say that by their means some costly annoyances and claims were avoided.
>
> As you are aware, the Maisons Lafitte factory is situated in a locality inhabited during the summer by Parisians, and all the year around by English people interested

in the races. The locality has nothing of an industrial character about it, and the inhabitants as well as the authorities look on our Works with an evil eye.

Complaints about dust, noise, withering of vegetation, and poisoning of wells had formerly been of almost daily occurrence, and there had also been the occasional lawsuit. 'It is impossible to conciliate the rich and non-working: but it is possible to conciliate the authorities . . .' and 'we have always allowed some monetary aid either to the fire brigade, to the Mutual Aid Society, or to the choral societies'. Here Geisenheimer solemnly set out lists of the amounts paid, and added that in answer to a recent complaint from the Mayor and an order from the Prefect prohibiting 'shooting our residues on the bank of the Seine', 'happily the inspectors looking after the sanitation of the Seine-Oise factories were our intimate friends, so that, backed up by their authority, we were able to prove to the Prefect that he was mistaken'. Geisenheimer went on to emphasize another problem caused by the factory being sited in this particular locality:

> The racing community were in the habit of paying very dear for everything; the workmen are frequently tempted to leave the factory and work in the stables. In order to retain our best men we are obliged to give them certain advantages which they do not get elsewhere, such as stability of employment, medical care for themselves and their families, and a few amusements.

Among the 'amusements' mentioned were 'the fact of being affiliated to Musical Societies', which afford them distraction in the evenings, or of having a few square metres of garden for the cultivation of vegetables'.

However, it soon became obvious that the major part of the opposition to reorganization in France was going to come from the bankers supporting Société Lyonnaise rather than from the discredited management of The Borax Company.[20] Jacquier, Falcouz & Co. were the original bankers of Société Lyonnaise, and claimed to own more than 20 per cent of the capital. They had introduced the Banque Privée to stock-exchange placings and other special business of Société Lyonnaise, and Vial and Pradel were the two Banque Privée representatives who sat on the Board of the Société, so when Baker had made up his mind, by July 1902, that 'it appeared judicious to our Directors to sell their refining operations in France to a company constituted according to French law, and not merely as a formal matter but for certain very solid and practical reasons,' trouble flared up.[21] Clearly, Borax Consolidated realized that the delegation of powers to the board of a subsidiary company, operating under a different system of law, meant far more risk of losing control than the existing façade of an 'advisory committee', and that the directors of the proposed French subsidiary must be chosen with care. So, much to his annoyance, Jacquier was not appointed, but both Vial and Pradel were offered directorships. This meant trouble, and as early as June 1900 these two had written to Baker in as unprejudiced a manner as they could simulate to suggest that it would be difficult to cover the scattered French market from one factory, and that although a new factory at Le Havre would be a good idea, it should take the place of Maisons Lafitte and not of Lyons. Lyons, they asserted, was in an advantageous position to serve the markets of Spain and Southern France, and once abandoned would be taken over by a competitor.[22] They attributed deteriorating results at Lyons to interference in the management by Borax Consolidated, and the struggle to keep Lyons in operation went on for another five years or so before the factory was closed finally.

The final breach with Vial and Pradel came in 1902.[23] According to indirect references, Borax Français, the proposed subsidiary company, must have been incorporated officially in France at some time in mid-June 1902, and one of Baker's letters[24] indicates that Vial and Pradel accepted his invitation to join the board on 16 June. By 9 July, however, they had changed their minds[25] 'owing to various provisions in the statutes'. They stated that it was quite evident that 'Borax Français' was a misnomer, and must have English directors, as no Frenchman would accept the post under such conditions.[26] Baker was probably right to say that the real reason for their volte-face was that 'they quite thought that they would have not only control of our business but of our funds there [i.e., in France]'.

Monchicourt and Geisenheimer also parted company with Borax Consolidated in France, but very soon they were leading a sales campaign on behalf of its rival, Compagnie Internationale de Borax, to drive the English company back into the sea.

Those who remained on the Borax Français board were Bodington, Borax Consolidated's lawyer in France, who also became chairman and took over Monchicourt and Geisenheimer's former agency duties; Daniell, a Borax Consolidated executive who later became company secretary; and M. Devisscher junior, an engineer who had joined Borax Français from Compagnie Internationale.

To Devisscher belongs a tale. According to a 'comeback' letter from Vial and Pradel to Baker on 30 July 1903,[27] Devisscher junior's father, Jean Devisscher, must have been running a borate business in Belgium at the time when the Borax Consolidated merger was under negotiation. Vial and Pradel claimed that they could have brought Devisscher senior into the merger, but these negotiations had been left to Messrs Monchicourt and Geisenheimer, and had fallen through.[28] 'This resulted,' they continued, 'in the CIB being floated, and since that time prices have had to be reduced to an unprecedented figure, leading to the present bad state of things in France.'

Whether this account of the formation of Compagnie Internationale is accurate or not, by August 1900[29] it had acquired the former Tres Morros borate mine in the Argentine, and by 10 May 1901 it had built and was operating a factory at Loth in Belgium, and was completing a large factory on three hectares (about seven and a half acres) of ground at Coudekerque in Northern France.[30] It was already starting to push its product at different places in France, Holland and Spain.[31] By the end of 1902 Daniell reported to Baker a general tendency among Borax Français customers to desert in favour of Compagnie Internationale,[32] and listed five who had done so. The position, apparently, would have been worse if Borax Consolidated had not also reduced its prices: but by this time (as is obvious from the surviving correspondence) the whole situation had developed into an out-and-out price war with no holds barred.

The CIB was being financed mainly by Crédit Liègeois, a Belgian bank,[33] but why they should consider this expensive bid to race Borax Consolidated into the European market a good bet, and why young Devisscher should desert his father and Compagnie Internationale to become the rather inexperienced refinery manager of its rival, remains a mystery. For those to whom the bank's highly speculative behaviour also seemed a mystery Smith offered the following explanation, written in characteristic style:[34]

> The Belgian Banking way of doing business is an inspiration. They are not bankers in any sense. They take the public's money and speculate with it. If it turns out all right the bankers make the money. If it turns out to be a bad investment, the public lose.

In this case the public lost. Daniell had already hinted at the subtle way in which CIB might be neutralized when he told Baker in September 1902[35] that Borax Français were encouraging Devisscher junior to dribble his Compagnie Internationale shares on to the Stock Market in order to undermine the price at which CIB shares were shown in Crédit Liègeois's balance sheet. Crédit Liègeois, faced with heavy losses on their portfolio of securities, decided that they must call in their advance to Devisscher senior, which now stood at Fcs 330,000.[36] Devisscher had no option but to sell his Compagnie shares, and Borax Consolidated were ready to negotiate. Although the sales war with Borax Français continued for a time, a settlement was reached at the end of the year 1904, when Crédit Liègeois' share in CIB and the shares of T. Leblanc (partner in Devisscher's former business) were purchased for a total of Fcs 2,544,712.[37] There was also the undisclosed cost of the Devisscher shares already taken over and Baker admitted on 12 December 1904 that the price to be paid was much higher than he had anticipated when they had first approached the matter some weeks before, but that 'needs must when the Devil drives'.[38]

The indirect cost to Borax Consolidated of this price-cutting war – which had bedevilled the first five years of its existence – must have been extremely heavy.[39] However, there is no evidence that the new French subsidiary, Borax Français, was ever forced into an overall deficit, and unlike the barren struggles to obtain deposits of questionable value in South America, the real gains made from enduring and ultimately winning this long-drawn-out struggle were of lasting value to Borax Consolidated.

Daniell, now in London, visited the Belgian company's establishments in the first week of 1905. He reported[40] that the Coudekerque factory was well kept up, that the cost of maintenance and repairs was quite low as compared with the Maisons Lafitte factory at Paris, and that manufacturing costs per unit were roughly only Fcs 98.50 at Coudekerque, compared with Fcs 146.37 at Maisons Lafitte. He concluded that at Coudekerque there was sufficient output to handle the provincial and export sales of Borax Français, leaving the Paris deliveries to be produced at Maisons Lafitte. Nevertheless, it would probably be advantageous to produce the entire quantity if possible at Coudekerque, especially as Dunkerque (its port) was exceptionally well placed for shipments to all French ports, and to the Rhine and German ports. These were prophetic words, although at the time they meant nothing more than another nail in the coffin of the Lyons factory, for there was no question yet of closing Maisons Lafitte or Loth near Brussels, which was remodelled and rebuilt in 1906.

Borax Consolidated, through Borax Français and Compagnie Internationale, now held a dominating position on the ground in France, Belgium, Holland and Spain. Nevertheless, the fact that they were still not in a position to reap maximum profits is indicated in a brief sentence written by Daniell to Baker at the time of the Compagnie Internationale takeover:[41] 'I am inclined to think that you will not wish to raise prices in France until you have seen Lardarel.' The Lardarello company were in a powerful position in the European market, and during the years between 1905 and the outbreak of war in 1914 an attempt was made to come to some sort of accommodation with them.

As early as 1898 Smith had given Baker his opinion that there was a physical limit to the growth of boric-acid production in Tuscany, and that therefore Lardarello could not be dangerous competitors, although they could still disturb the market, even with their small output. By 1901, however, Smith was expressing surprise that the consumption of Tuscan boric acid in glass-manufacture in the USA – which had hitherto been very small – was increasing.[42] He doubted whether Borax Consolidated could produce a comparable product for glass-manufacturers owing to the lime content of their raw material, but he did not seem much concerned about Lardarello's activity in the USA.

The 'Leghorn people', however, were much closer to Baker's doorstep than to Smith's, and even Smith had to admit to Baker in July 1903 that figures received indicated larger production and profits from the Tuscan producers than he had expected.[43] All the same, it would be worth offering to bring them into the merger only if 'the properties could be negotiated on a rational basis'. Some months later Baker wrote that he would be having a meeting with Lardarello within a week.[44] There is no record of whether this meeting ever took place, but the matter became urgent in 1905 when Baker wrote[45] to Smith that the Tuscany people were thinking of putting up a factory in France for refining their goods, and that if this were done the price of borax in France would probably remain at its existing level instead of improving as they had hoped. Lardarello had expressed the wish to discuss matters with him, and Smith commented that undoubtedly the same opportunity would never occur again of securing these interests, which he assumed that Baker was pursuing.[46]

Lardarello were certainly not negotiating from weakness. Nine-tenths of their profits came from the sale of crude boric acid, and their annual distributed profits between 1895 and 1902 had varied between £155,000 and £380,400, and were £325,000 in 1902.[47] They were aiming at forming a company with two other Tuscan boric-acid producers, Fossi and Vernier, with whom they were obviously on very friendly terms.[48] If approached tactfully, they might have agreed to link this local association with Borax Consolidated. However, it seems that Baker scared off these small Tuscan producers by proposing a 'big brother' arrangement whereby Borax Consolidated would buy up the whole of their annual output at one fell swoop, and provide any necessary capital required to implement the plan. When, only a month later, Baker heard that Lardarello had appointed the old intriguers Monchicourt and Geisenheimer as sales agents for Europe, he exploded in a most abusive letter[49] and threw up the negotiations as a failure; Smith had already stated his company's view of this couple a few months earlier when he wrote that he had heard that they were 'again looking for somebody to rob'.[50]

However, early in 1907, at a time when Tuscan acid was invading the market in Austria and Russia as well as in Belgium and France, Baker swallowed his anger and reverted to the theme of a joint company with Lardarello, Fossi and, this time, Durval. But once more he proposed what amounted to a dominating position for Borax Consolidated, and again there is no further record of the proposal. All the same, half-hearted approaches between Borax Consolidated and the Tuscan producers continued right up to 1912, when a temporary and limited agreement was at last achieved under which Borax Consolidated would buy the surplus production of the Italian producers, who would make no further direct sales for export from Italy.[51] Finally, in July 1914, there is a cryptic sentence in a letter from Baker to Zabriskie:[52] 'Looking to our present relations with the Tuscan producers we do not think it would be wise to take any steps at present with regard to their material,' and this, naturally, was the last that was heard of the matter until the 1914–18 war was over.

Both the sales war with Compagnie Internationale and the lesser skirmishes with Lardarello were bound to have repercussions on the market in Germany and

Austria. Admittedly the Ascotan agreement of 1898 with the leading German refiners, and the acquisition of Société Lyonnaise's small factory in Vienna, had given Borax Consolidated a solid foundation in the most thriving industrial areas of Europe at the time. However, it was a foundation liable to be shaken by sudden cracks appearing in the cement of the complicated agreements which held it together.

To put the matter in perspective, it must be stated that the Ascotan agreement with seven powerful German refiners was not the only agreement limiting Borax Consolidated's freedom to be flexible in competition in Central and Eastern Europe. There was also a syndicate of Austrian refiners, who drew their raw material from Borax Consolidated, and there were several outlets for this material in Poland and Tsarist Russia.[53] These Polish and Russian refiners Borax Consolidated typically proposed drawing together into a syndicate to be supplied by them through one agent centred in St Petersburg, and somehow they succeeded quickly in gaining acceptance of their proposal by the Russians.[54] The Société Poudre, however, described as the main customer in Russia, was soon complaining that the agreed formula price for borax in Russia was always likely to be at least £6 more than the English price;[55] that this price that they were committed to pay to Borax Consolidated prevented them competing with the Polish refiners in the sale of their refined products; and that it was impossible to enforce a uniform price over a vast land like Russia. Possibly as a result of this difficulty, the Polish refiners were eventually drawn into the Russian agreement as 'allied refiners' by negotiations which were eventually concluded in 1904,[56] and which expressed among other things the intention of fixing new prices for the whole of this enormous area.

Meanwhile a letter of 1901 mentions similar discontent among the Austrian syndicate members, that they had to pay 15s. a ton more for raw material than refiners outside the country.[57] Their discontent with Borax Consolidated came to a head in 1907,[58] when Hell & Sthamer – who ran a factory at Aussig, in territory then within the Austrian Empire – complained that the scheme meant that Borax Consolidated could sell crude material to anyone in Austria and yet not be limited in quantity in sales of refined products. All that Baker could lamely reply was that the terms proposed were not in any way unfair to the Austrian refiners, and that he was simply desirous of placing the business in Austria on exactly the same basis as in other countries.

The real truth of the matter was that Borax Consolidated had had to build a new factory in Vienna, and probably wished to expand sales to procure a worthwhile return on capital outlay.[59] The lease of the former Société Lyonnaise factory at Brigittaplatz, Vienna, was due to expire in 1905. As early as 1902 M. Dupont (who had taken over as general agent) was pressing Baker to acquire a new site.[60] Baker said that such a costly decision could not be hurried, and Dupont exploded with rage at the frustrations and red tape to which he had been subjected by the London office, and resigned as manager.[61] Some of the sentiments expressed in his letter perhaps contain eternal truths about bureaucracy in any age:

> On the whole the attitude of your London office was such as to require the most minute details without giving sufficient liberty or power of action for the managing of affairs.
> . . . It is easily possible to forget, by mistake, first to obtain your sanction for something or another and, after, to get reproaches and, what is worse, perhaps to commit errors for fear of not doing the right thing.

The Mule Team, en route to Daggett railway station, leaving Borate, the colemanite mine in the Calico Mountains (c. 1895)

Below: Old Dinah, *a traction engine comfortable on the level but not on the steep gradients* (1894)

Francis, *one of the two Heisler tank engines used on the Borate and Daggett Railroad*

The Borate and Daggett Railroad, approaching the mine at Borate

John Ryan stops to quench his thirst alongside his buckboard. On the move again in Death Valley. Both pictures date from the 1890s

Daggett: the American Borax Co. plant making crude boric acid direct from low-grade ore. A corner of the many acres of solar evaporation ponds is shown in the foreground (c. 1906)

Below: Anglo-American flavour in the Memorandum and Articles of Association of Borax Consolidated Ltd, 1899

WE, the several persons whose names and addresses are subscribed, are desirous of being formed into a Company, in pursuance of this Memorandum of Association, and we respectively agree to take the number of shares in the capital of the Company set opposite our respective names.

NAMES, ADDRESSES AND DESCRIPTIONS OF SUBSCRIBERS.	Number of Shares taken by each Subscriber.
A. WILSON, 　　　2, Dartmouth Grove, Blackheath, 　　　　　　　　　　　　*Knight.*	One Preference
F. M. SMITH, 　　　101, Lansome Street, San Francisco, 　　　　　　　　　　　　*Capitalist.*	One Preference

His firm were willing to continue as the selling agents of Borax Consolidated, but with advertising particularly in mind, requested Baker not to be too pressing about submitting all details to him before incurring any expense.

Baker wrote Dupont a conciliatory letter, but, alas, the firm were to prove themselves unsuited for the role of martyrs for managerial freedom. The bi-borax sales venture which they had been advocating in 1902 had sold so little by mid-1903 that Baker had to stop them spending any more on advertising this product.[62] Land for a new factory had been purchased at Stadlau near Vienna in September 1902,[63] and Dupont suggested that his brother, F. M. Dupont, should be manager. In spite of three years' training in England and France, he soon proved a failure in Vienna. As Daniell wrote,[64] 'The new Works seem to have disorganised everybody, mentally and physically . . . Dupont has lately taken to consulting a lawyer on every step he takes or proposes to take which is quite unnecessary!' More seriously, Daniell had to write to Baker again, in June 1904, about the deplorable production results in Vienna. Finally, by January 1905, Daniell's suspicion was confirmed that Dupont's slackness over supervision also extended to the accounts, where defalcations, ingeniously hidden, had been laid at the door of the Czech accountant, Pollak.[65] The resignation of Dupont, the works manager (which was accepted with alacrity), was followed by that of Dupont, the sales agent.[66] The Stadlau works were taken over by a Dr Hecker, a chemist from a large German 'associated refiner' at Gernsheim. He soon proved such a success that in 1910 he succeeded the late Iltyd Redwood as chief technical executive in Britain and Europe. The end of this epoch came in August 1914, when as Baker wrote, somewhat quaintly:[67] 'Unfortunately Dr Hecker has been called away to Germany, being liable for service, and we should much prefer that he complete the plans we have made for the erection of a factory [for borax glass] here.' To Baker, in fact, as to so many business-men at the time, the war would seem for some months yet merely a tiresome and unexpected interruption in the normal run of business affairs.

Meanwhile, in these fifteen years between 1899 and 1914 Borax Consolidated's position in Germany, based on the Ascotan agreement, had been subjected to severe strains. This was inevitable in view of the narrow basis on which it rested. It may not have been foreseen in 1898 that so many other possible sources of supply would be discovered in South America quite so quickly. German refiners appeared who were not parties to the original agreement, and to deal with 'friendly' refiners – who might wish to draw their material from Borax Consolidated – the status of 'associated refiner' was created.[68] As Herr Bohm (manager of A. G. für Chem. Rheinau) told Rosenstern (the agent of Borax Consolidated) in 1900, he had never yet had, nor did he expect to have, the slightest difficulty in finding all the raw material that he required, although it might be a little more expensive than that obtainable under the agreement.[69] 'Further', wrote Rosenstern, 'we agree with Mr Bohm that your intention to make all German borax refiners a happy family has not proved a success'; and Bohm was not going to join unless a proper selling syndicate was formed instead of a mere agreement regulating the selling price.

The agreement also came under pressure from another factor, the impact of price-cutting by Compagnie Internationale de Borax between 1899 and 1905, and by Lardarello subsequently. This caused reductions in the selling price for the refined product which were inclined to push smaller producers into a loss position. It was difficult also to isolate these arrangements with German refiners from the effects of price disturbances in closely allied industries.[70] Thus when a Herr Winckler, with powerful financial backing, started to form a union of German enamellers,[71] and some years later was rumoured to have acquired an independent source of supply of borates at Aquas Calientes in Chile, nerves obviously became strained in Borax Consolidated.[72] However, although it would be natural to assume that the rigidities of the Ascotan agreement were ceasing to have much effect on German refiners in the few years before war broke out in 1914, and that cut-throat competition was taking its place, it is surprising to find Baker writing at the beginning of 1913[73] that the 'associated refiners' were bound to Borax Consolidated by a most complete agreement, tying them down in every way and not mentioning any breaches of this agreement. Baker does not seem to be the kind of man accustomed to self-deception, and the only possible conclusion is that the inborn German sense of discipline among conservative-minded industrialists, together with the German cartel habit, had insured Borax Consolidated's trade position in Central Europe against the growing chaos in international trade in borax. However, an even greater inborn German loyalty – to the Fatherland at war – was about to sweep away these arrangements, and with them most of the position that Borax Consolidated had laboured so hard to build up in Europe in the fifteen years since the merger of 1899.

REFERENCES

1. F.M.S., 16.6.99.
2. Minute Book, 11.1.99.
3. Prospectus issued on 12.1.99.
4. Gerstley, J.: *History of Borax Consolidated, Limited* (unpublished).
5. M.D., 9.7.07.
6. F.M.S. to R.C.B., 17.2.98.
7. F.M.S. to R.C.B., 26.10.01.
8. F.M.S. to R.C.B., 27.5.01.
9. F.M.S., 27.5.01.
10. F.M.S., 26.10.01.
11. Undated letter from F.M.S. to R.C.B., 1898.
12. M.D., 4.3.03.
13. M.D., 14.7.14.
14. M.D. to F.M.S., 5.12.13.
15. M.D. to Lord Chichester, 10.7.14.
16. M.D. to F.M.S., 5.12.13.
17. M.D. to Managers Association, undated.
18. Prospectus of Borax Consolidated, Limited.
19. R.C.B. – letter from Daniell – under 1898/99 but not dated. Also Minute Books, 11.1.99.
20. R.C.B. – two letters from Daniell dated 6.3.01 and 13.9.02.
21. M.D. to Vial and Pradel, 9.7.02.
22. R.C.B. from Vial and Pradel, 5.6.1900.
23. R.C.B. from Daniell, 15.7.02.
24. M.D., 16.6.02.
25. M.D., 9.7.02.
26. M.D. to Daniell, 9.7.02.
27. R.C.B., 30.7.03.
28. R.C.B., 21.11.1900.
29. R.C.B., Hell & Sthamer to Borax Consolidated, 20.8.1900.
30. R.C.B., letter from Geisenheimer, 10.5.01.

31. R.C.B., 14.10.02.
32. R.C.B. from Daniell, 8.9.02.
33. *Ibid.*, 5.8.02 etc.
34. F.M.S. to R.C.B., 11.10.02.
35. R.C.B. from Daniell, 8.9.02.
36. R.C.B. from Plouvier, 9 & 23.1.02 and 6.2.02.
37. Minute Book, 6.1.05.
38. M.D., 12.12.04.
39. R.C.B. from Daniell, 1.9.03.
40. *Ibid.*, 14.3.05.
41. *Ibid.*, 3.1.05.
42. F.M.S. to R.C.B., 23.12.01.
43. *Ibid.*, 27.7.03.
44. M.D. to Zabriskie, 25.3.04.
45. M.D. to F.M.S., 3.1.05.
46. F.M.S., 30.12.04.
47. Report of 29.4.03 attached to R.C.B. from Daniell, 1.5.03.
48. R.C.B. from Vernier, 4.7.05.
49. M.D. to Daniell, 29.3.05.
50. F.M.S. to R.C.B., 10.11.04.
51. M.D. to Daniell, 26.8.12.
52. M.D. to Zabriskie, 14.7.14.
53. R.C.B. from Hell and Sthamer, 11.7.07.
54. R.C.B. from Daniell, 22.8.1900.
55. R.C.B., 16.3.01.
56. R.C.B., 11.1.04.
57. R.C.B., 16.3.01.
58. R.C.B. from Hell and Sthamer, 11.7.07.
59. R.C.B. from Daniell, 11.6.04.
60. M.D. to Dupont, 17.3.02.
61. R.C.B. from Dupont, 3.5.02.
62. M.D. to Dupont, 12.6.03.
63. R.C.B., 13.9.02.
64. R.C.B. from Daniell, 1.9.03.
65. *Ibid.*, 10.2.05.
66. R.C.B., 1.2.05 and 14.4.05.
67. M.D., 11.8.14.
68. M.D. to Zabriskie, 5.2.11.
69. R.C.B., Rosenstern to Borax Consolidated, 25.6.1900.
70. *Ibid.*, 9.10.01.
71. R.C.B. from Billwaeder (Hamburg), 5.9.1900.
72. M.D. to Zabriskie, 11.12.08.
73. M.D., 3.1.13.

8

Borax in Britain after the merger

Following the merger of Redwood's and Pacific Coast Borax in 1896, Britain did not become a mere overseas colony of Smith's borax empire as might have been expected. It gained – because of its choice as the official and legal headquarters of the new Borax Consolidated, Limited in 1899 – an enhanced importance, particularly in financial matters. English company law in 1899 had not yet even reached the stage of the Companies Act of 1907, but it was stricter than American practice, and the original issues of shares and debentures and the subsequent increases in capital in the first decade of the century were all governed by English requirements. This necessarily restricted Smith's ambitions to raise further capital on the London market, through Borax Consolidated, in order to finance the more daring enterprises on which he embarked in other directions in the USA.

Smith was apparently quick to realize these restrictions, for he wrote to Baker in June 1900[1] proposing to merge his interests outside borax 'with the American company's interests, in a new company to be placed on the New York Market'. 'Of course,' he continued, 'this would in no way affect the English company or the personal control of it. The principal advantage to me would be in having collateral that I could use or realise upon. I would be controlling the New York company, the New York company controlling the English company.' Presumably anticipating polite but firm opposition from Baker, he concluded by adding that it was 'just a passing thought'. In the following twelve months two other 'passing' references to this proposal are traceable in his letters to Baker,[2] but it seems obvious that he was getting no encouragement to persist in it. He admitted eventually that 'the suggestion you [i.e., Baker] make is a new one and a good one to me, that it [the company] is English for the sake of getting money and it is controlled in America. This is a fact and I have largely overlooked it.'

Thus Borax Consolidated remained within the financial and ethical restrictions of English law, even if 'control' remained in theory with the majority shareholder, Smith, in America. It is perhaps merciful that, owing to a quirk in Smith's character, 'control' over all the new group's multifarious affairs did not pass in practice to him also. He was perfectly content to leave everything in borax outside the American borax world – which was in his blood – to Baker, provided that a show was made of mutual consultation.

Borax Consolidated started out in Britain in 1899 with the twenty-year-old works at Belvedere in Kent, which were rated as capable of producing 1,500 tons of borax a year 'exclusive of other important products'.[3] It also owned the three works of the former Mear & Green Ltd. Of these Kidsgrove in Staffordshire and Connah's Quay in Flintshire had rated capacities well above Belvedere, while the third at Tunstall (also in Staffordshire) was a small one.

No definite plan for rationalization of production emerged in the years that followed. Baker was thinking about constructing a single large refinery in England[4] right up to the start of the War, in 1914, and prior to the acquisition of the Compagnie Internationale de Borax Smith had suggested to Baker that their French factory at Coudekerque could give him the central refinery that he wanted.[5] Thus by luck rather than design there was productive capacity still available in Britain in 1914 when Picardy – and Coudekerque with it – became a battlefield. In fact, instead of any master plan to reduce the excessive number of small borax factories in Britain, there seems to have been an endless game going on in the years 1900–14 to knock out or neutralize the various small rivals who still survived. The deadpan seriousness with which these various events were recorded by all except one blithe spirit, James Gerstley, adds a touch of the ridiculous to the recital of them some seventy years later. This is exemplified by the ease with which any little man who led the company a long dance could be transformed from 'that ubiquitous thorn' (Coghill, in this instance) or a perfidious trickster into a 'good fellow' (a common expression in those days).

As regards Borax Consolidated's own British works, the first to close, in November 1903, was Connah's Quay, Flintshire,[6] where both plant and staff had been causing trouble for some years. Production was poor, and in April 1900 Iltyd Redwood, the irritable new superintendent of works, started an investigation which resulted in the chemist in charge, Hughes, being suspended, and a foreman dismissed on suspicion that the records were being 'cooked' and boric acid was being disposed of privately.[7] Hughes's defence was that the plant was poor, and Redwood apparently agreed in reporting it as 'badly designed, badly built and constantly troubled by tidewater' – while as to the men of Flint . . . ![8]

Connah's Quay, however, may well have appeared a model factory when compared with Tunstall, where in September 1900 Redwood wrote to Baker begging to remind him that on 6 March last the National Boiler Insurance Company had established that the Tunstall boilers were already about forty years old and required renewal within the year.[9] It was the next to close. The sale of the site proved difficult; but by August 1904 the board decided to get rid of it at any price,[10] and the land was eventually sold to the well-known potters Alfred Meakin Ltd, on 15 June 1906, for £1,000 only. There is no mention of what happened to the boilers. Industrial archaeologists did not then exist to cherish antiques of this type.

About this time also there was a brief flirtation with the owners of the small factory of the former Boron Products Company at Inchbrook, near Stroud in Gloucestershire, to which Borax Consolidated had been supplying raw material. As early as February 1901 a Mary Brownrigg had written to Baker:[11] 'We [i.e., she and her sisters] all think that the rent you offer is a very small one, considering the circumstances in which we are now all placed, owing to the marriage of two of us. We are prepared to accept an offer of £300 a year rent subject to certain conditions' – principally a three-year lease with option to purchase at the end of the first year.

Baker by reputation could rise to occasional heights of gallantry, but the minute book records that the ladies, after negotiations lasting until early 1904, were able to secure only a one-year lease from him at £200 a year.[12] He could not escape reproof from an outraged father, and in May 1906 he had to reply to the Rev. Charles Brownrigg of Magdalen College School, Oxford,[13] pointing out again with regret that Borax Consolidated had lost at the rate of £2,000 a year by trying to manufacture at Inchbrook instead of at their own recently enlarged works elsewhere. 'This regret is emphasised by the fact that the Inchbrook works belong to ladies, and you have my assurance that, but for this fact, . . . we should have relinquished the works sooner than is the case.' These incidents at least provided a delightfully polite interlude, 'Jane Austen' in style, in the otherwise tough and treacherous world of borax at the time!

There were several offers of factories to the new borax combine by little companies who could see no chance of independent survival in these altered circumstances. In February 1900 W. T. Glazebrook & Co. offered their borax works at Liverpool, an offer which was declined.[14] This company was in fact taken over by the Liverpool Borax Company, who later offered their business to Borax Consolidated in 1908.[15] This offer was also declined, and Liverpool Borax survived, to be used as an outpost for Borax Consolidated's main competitors abroad for many years to come.

The most substantial concern to suggest their acquisition by Borax Consolidated was H. Coghill & Son, who owned factories at Newcastle in Staffordshire and at Liverpool, and who were among the first to qualify for the status of 'associated refiner' in the early months of 1899. There were three brothers running the business, one of whom, Frank Coghill, seems to have taken Baker into the closest confidence and for the best part of five years to have deluged him with handwritten letters, full of gossip about the doings of other small refiners in Britain, and their sources of supply in South America. Coghill claimed in 1899 that their sales for that year would amount to over 3,000 tons, and that therefore their allocation of only 2,600 tons under the 'Ascotan agreement' was going to embarrass them.[16] Possibly for this reason, Baker raised no objection to their continuing their previous practice of trading in Italian boric acid in addition to their own products. However, like most others bound to the Ascotan agreement, they soon found themselves hit below the belt by the price war which broke out between Compagnie Internationale and Borax Consolidated, and they were too small to stand up to this type of competition.[17] In May 1902 Frank Coghill had to write to Baker to propose a fusion of interests in the form of a purchase of the Coghill business, adding that the brothers were prepared to sign on for five years as managers.[18]

Probably to their surprise, however, Baker replied within a few days that he was afraid that Borax Consolidated could not entertain any scheme for the purchase of borax properties just then.[19] They already had too many. So Coghill struggled on, and somehow Baker persuaded them to continue as reluctant 'associated refiners' for the rest of the pre-war period.[20]

Among other British refiners, who were far from docile, most of the trouble in the period from 1899 to 1905 came from a Glasgow concern called Joseph Townsend Ltd, who were from time to time in league with an amusing 'pirate' called Fleming.[21] Their letter-heading shows that the Townsends were a large family, with three brothers on the board, with representatives also of their sisters' interests, and were established as early as 1850. At the turn of the century they were making 'Preservide', Glauber salts and Epsom salts, in addition to borax and boric acid. A few days before the incorporation of Borax Consolidated, C. W. Townsend, the head of the firm, wrote to Baker offering his help with the new company 'after 20 years' experience in the trade'; and, during the first year or two Townsend provided Baker several times with useful information about the borax world.[22] Townsend's were affiliated members of the Ascotan agreement, but by December 1900 they were finding themselves too fettered on prices by the arrangement to be able to compete successfully with Italian boric acid. By September 1901 Frank Coghill, who had offered himself as an intermediary, warned Baker that Townsend were about to break with Borax Consolidated unless their pro-rata sales quota was raised to a level where they could show a little profit.[23] Baker saw fit not to accept this advice, but Townsend's carried out their threat, and were followed by another Scottish refiner, Sir John N. Cuthbertson, who terminated their agreement also in May 1902. This was at the height of the price-cutting war with Compagnie Internationale, and although Cuthbertson soon went to the wall, Townsend's proceeded to give Borax Consolidated some nasty shocks when they went off to join the band of much smaller and weaker refiners – notably the Liverpool Borax Company, H. C. Fairlie & Co. of Falkirk and Hughes Bros of Birkenhead – who were drawing on supplies of low-price raw material imported from South America through Liverpool and Glasgow.[24] Information supplied at the end of 1904 by Laird & Adamson, one of the leading firms of brokers in Liverpool, showed that Borax Consolidated's deposits at Ascotan and Salinas never accounted for much more than about two-thirds of the South American borate of lime, estimated by these brokers to have been imported through Liverpool in the previous five years. Their estimates showed also that the amount imported, which came from the Chilcaya deposit and other ones – sources which were not supplying Borax Consolidated – grew steadily during this period.

However, the garnering of this type of information had a comic and unreliable side. For example, Baker had two firms of brokers supplying him with confidential information from Liverpool on South American imports.[25] Characteristically, he became increasingly petulant when prompt and accurate information failed to arrive. One of these firms was therefore driven to remind him, early in 1904, that: 'We are Borax Consolidated, and enemies of outsiders, who give instructions to tell us nothing and to hide everything from us. . . . It is really astonishing how we get to know what we do.'[26] If Baker wanted precise verification, therefore, he must expect to wait some days for it. Most of the information came from bribing warehousemen, some of whom had been instructed deliberately to mislead Borax Consolidated's spies. Finally, the height of the ridiculous was reached when Pridham, a senior excutive, after a visit to agents in Glasgow, had to inform Baker: 'In fact some of our informants have simply "made up" shipments in order to obtain remuneration when they were short of money . . . !'[27]

It is symbolic of the times, therefore, to find Gerstley informing Baker in mid-July 1902 that the latest consignment of 'outside' material from South America had been taken up by their various small rivals at £5 10s. 0d. a ton, the lowest price ever recorded,[28] and a fortnight later that Joseph Townsend had 'slipped 320 cases of boric acid into Japan last week at £23/10/- a ton against our £26'. Coghill was starting to complain of Townsend's competition on the home front, and information was arriving from abroad that Townsend was capturing many foreign markets, where Borax Consolidated was handicapped by excessive carriage, insurance and freight costs.[29] No one seemed to guess at the time that – as happened in the successful struggle against Compagnie Internationale – the enemy had feet of clay which were bound to bring him down sooner or later, and when Borax Consolidated accepted the invitation in February 1906 to purchase Joseph Townsend,[30] it found that factory costs were high, and that the quality of products being sent out was abysmally low. Charles Townsend was, however, taken on as manager, and was still there in 1914, in spite of his poor response to the requests from Baker to improve his products and his profit margins, and his failure to carry out instructions to close down the unprofitable Glauber salts section of the factory. In fact by April 1914 things had got to such a pass that Baker was starting to consider closing the whole place down,[31] and only the outbreak of war (and the consequent shortage of imported boric acid) saved production there for a few more years.

However, as far as can be traced Townsend and his company never committed the supreme sin, in Borax Consolidated's eyes, of trying to invade Smith's protected market in the USA. The number one intriguer against Borax Consolidated, one Fleming (Borax Consolidated never conceded even as much as his initials or a 'Mr' in numerous references to him), comes on the scene first as a director of a small refinery at Warrington called W. A. Bishop & Co., but was even then playing very much the part of a lone wolf. In November 1899 Zabriskie reported that Fleming was sailing back to England to float a company to exploit the 'enormous' borate deposit that he had obtained in South America, 'capable of producing 20,000 tons a year'.[32] He planned to use motor-cars – which were then in their experimental stage – to transport the material from the marsh where it lay to the nearest port, and Zabriskie commented wryly that possibly Fleming had not read the latest articles concerning airships. A few weeks later it became clear that the property that Fleming had in mind was the Chilcaya (Suriri) deposit,[33] over part of which Borax Consolidated held a disputed claim, so they began to take notice. Two months later a chemist working for Coghill reported that alleged samples of Chilcaya material from W. A. Bishop & Co. were the best that he had seen for a long time, and contained 45·8 per cent boric acid.[34] Howard & Sons, one of the larger refiners, said that they had been 'invited' by Fleming to join in exploiting the deposit, and that Fleming was claiming that he held the title deeds and had been permitted by the government of Chile to lay a monorail to the deposit.

Two years later, however, in July 1902, a report appeared that a receiver had been appointed to wind up W. A. Bishop & Co.,[35] and they were only one of at least four concerns who claimed a title to parts of the Chilcaya deposit; their claim was a small one of only two hundred and fifty acres. Overstretching their resources to develop this miniature deposit, hundreds of difficult miles from the nearest port, probably toppled them rather than the startling news which broke three days later. For, on 17 July 1902, the *Westminster Gazette* announced: 'Ernest L. Fleming of Weversham, Cheshire, exporter of washing crystals, was charged at New York yesterday of falsely invoicing borax as soda. He was remanded on bail of £500. The duty on imported borax is ten times as much as the duty on soda.'

Baker had had private information that Fleming might attempt something of this kind,[36] and undoubtedly the New York office of Pacific Coast Borax had put the US Customs on the watch for Fleming's consignments of washing crystals. Fleming got to know that Pacific Coast Borax knew, and attempted to buy them off.[37] He let it be known that if the company would reimburse him the amount he had lost in Oregon (i.e., probably over a colemanite deposit) and his loss in South America, as well as ceasing to encourage the present proceedings against him, he would for ever retire from the borax business.

This incident, combined with Fleming's expressed obsession with breaking Borax Consolidated in both Europe and the USA, gave the Democrat press a wonderful opportunity to attack high American import duties and a supposed borax 'monopoly', as a number of newspaper cuttings of the time show. Fleming's defence that borax was a form of soda (i.e., a sodium salt) failed, and he lost his case.[38]

In October of the same year (1902) Fleming changed his line of attack and called on the Cuevitas Trading Company[39] (not realizing that it was actually a South American subsidiary of Borax Consolidated) in order to try to persuade them to supply a syndicate of refiners with raw material to enable them to defeat the 'Borax Trust'. He showed a picture of an elephant with the word borax written just over its tail, which seemed to give him great joy, and behaved in a manner that suggested that he was temporarily unhinged.

By April 1903 he was starting to collect capital to build a factory on the Manchester Ship Canal.[40] In December Redwood caught sight of him buying up plant when Bishop's borax works came under the hammer,[41] and by March 1904 Fleming's new 'Warrington Borax Company' was reported to be making borax with the intention of selling it in the American market.[42] Fleming was still hoping for a Democrat victory and for abolition of the American import duties. But for all his bravado he does not seem to have caused any further trouble, in either the British or the American market, and in February 1906 he wrote to offer Borax Consolidated a licence for an electrolytic process for borax-production.[43] This was not accepted, and by 1909 he had disappeared from the borax scene.[44] The only other refiners left in Britain who were not associated with Borax Consolidated were Liverpool Borax Company and H. C. Fairlie.

Borax Consolidated therefore reached 1914 in Britain with the former Redwood works at Belvedere, the former Mear & Green works at Kidsgrove, and the former Townsend works at Glasgow in operation, and with Coghill's two works tied to its apron-strings by a supply and sales agreement.

Electricity was installed at both Belvedere and Kidsgrove works in the comparatively early year of 1901,[45] but apart from this plant development does not appear to have been noticeably dynamic anywhere. On the contrary, in 1903 Redwood wrote to Gerstley[46] that the roof at Belvedere had become so bad – owing to the exceptionally heavy guns at near-by Woolwich Arsenal – that nearly the whole roof would have to be removed, which meant a matter of about three hundred panes of glass. As it was, every now and again a large piece of heavy glass would break away, and he added that a piece weighing eight or nine pounds coming down on a man's head might make short work of him!

A memorandum by Locke, who was in charge of sales in the United Kingdom in 1901, gives some indication of what was going on in the market-place.[47] The estimated output of Belvedere at that time was 1,750 tons a year, but owing to business lost by two of the company's agents in 1900, a quarter of this would have to 'hunt other buyers'. In addition to outside competition the company's own discount customers were also competing with it, by selling off part of their allotted consignments to competitors. Other refiners, he thought, were producing on the whole a better product than Belvedere's, and some were giving better credit terms – it seems that sales department comments change little from one generation to another. The company's foreign agents were 'useless', as they had lost interest in the business, presumably because the merger required so many transfers of allegiance. Borax Consolidated representatives never came near them, and usually other makers were able to supply the trade and allow buyers a little 'inside'. But there was business available on the Continent if the quality of Borax Consolidated goods could be improved, and agents kept up to the mark by an occasional visit from a traveller. These were perhaps the views of someone feeling the effects of Mr Baker's eagle eye in regard to travelling expenses and selling costs generally, but in all probability were near the truth.

However, a main reason why sales were disappointing during the first five years after the merger of 1899 arose from an unexpected quarter. The widespread use of borax and boric acid as a food-preservative came under attack, and it was much restricted. At some time just before the turn of the century protests against the use of these substances grew stronger, and emanated ostensibly from doctors and public health experts.[48] A mass of paper was produced, proving or disproving the alleged harmful effects of these preservatives on human health; but underneath all this there could have been another motive.

Baker pointed directly at this when he wrote that the Germans were trying to cover up their attacks on these preservatives by stating that their new law prohibiting the use of boric acid as a preservative in meat had been passed solely from hygienic considerations, whereas in Germany itself it was perfectly well known that the law had been passed as a sop to the Agrarian Party.[49] This was dominated by the powerful junker landlords of Prussia, who naturally would not wish their market undercut by cheaper imported meat. The German law had been promulgated after experiments by a Dr Rubner, which were strongly disputed by a Professor Liebreich, and this dispute stimulated practical action in Britain also.[50] In 1899 a Royal Commission was appointed; in 1901 it made the following specific recommendations:[51]

> 1. The use of any preservative in milk should be prohibited.
> 2. The only preservative lawful to be used in cream should be boric acid or mixtures of boric acid and borax in amounts not exceeding 0·25 per cent expressed as boric acid.
> 3. The only preservative permitted to be used in butter and margarine should be boric acid or mixtures of boric acid and borax to be used in proportions not exceeding 0·5 per cent expressed as boric acid.
> 4. No preservative of any kind should be permitted to be used in dietetic preparations intended for the use of invalids or infants.

Borax Consolidated were pleased to see a definitive report which made borax and boric acid the only legal preservatives in an important area, and that they had routed salicylic acid and other possible alternatives.[52] The preservation of meat from putrefaction – which Smith rather tactlessly referred to as 'the embalmed meat question'[53] – concerned, of course, the growing meat business in the USA and to a lesser extent in agrarian Europe more than it concerned Britain. It is not clear what the law was on this subject in the USA at the turn of the century, although just before this borax was being used extensively as a preservative in curing and packing meat.*

---

*On 4 January 1898 Smith wrote: 'We estimate that 35% of our borax sales is to meat packers, and the meat packers use 95% of the borax used in curing meat for the export trade'.

The scientists were divided on this matter. Professor Redwood, Sir Benjamin Ward Richardson and others had been convinced of the harmlessness of boric acid as a food-preservative, but Professor Armstrong and others now considered it to be injurious.[54] Dr Oscar Liebreich, professor of medicine at Berlin, was engaged to make a practical investigation, and he conducted research on dogs and a few of his own patients over three years from 1899 to 1902. A two-volume report described the results, which were that borax and boric acid – even in much larger doses than used in food – could not be considered as pharmacologically injurious substances. The dogs cheerfully and quickly excreted all the borax from their systems, and no injurious build-up occurred, and no ill effects were reported on the humans.

However, some twelve months after Professor Liebreich had announced his conclusions Dr Wiley of the USA announced that he was going to test borax entirely on human beings instead of on animals. Smith hailed the idea with delight and wrote: 'I have much faith in the final outcome. Dr Wiley is friendly to borax.'[55] This impression continued during and immediately after Dr Wiley's tests, and was supported by interviews that Dr Wiley gave to the Press and to Dr Liebreich.[56] In these he took the line that although he could say nothing until the results of the tests had been studied and collated, it seemed most unlikely, after the long period in which boric acid had been used as a food-preservative, that it could be generally injurious to health.[57] When his report appeared the conclusions were that the administration of boric acid at a level of about one-sixth of an ounce or five grains a day was adverse to health. The tests had been conducted with the help of twelve students who with the exception of water were required to use no food or drink, other than that provided by Dr Wiley, for a period of six months. However, Sir Henry Willcox, the British government expert on poisons, met one of the students, who told him that when Wiley's back was turned they sometimes broke out and went on a binge down-town.[58] This and their subsequent headaches and disorders would show up in the medical tests the next day. Whatever the significance of this story, Wiley's report made no immediate difference to the legal position about the use of boric acid in food in England.

In the USA, in contrast, the Food and Drug Act passed in January 1907 included prohibition of the use of borax and boric acid as a preservative in all food intended for inter-state trade, and sales by Pacific Coast Borax alone – that had exceeded 1,700 tons in a year within the United States – disappeared with a suddenness that for a time caused serious depression in the borax industry as a whole.

In Europe the use of boric acid as a preservative, particularly in margarine and butter, continued over a long period; however, in Britain a Ministry of Health committee reporting in 1926 on the use of preservatives and colouring-matter in food placed boron products among those which should not be used in food, and thus the findings of the 1901 Royal Commission were reversed.

During the Second World War the use of boric acid was reintroduced in Britain, and from April 1940 for the next fourteen years up to 0·25 per cent boric acid was used in margarine. During this whole period of food-rationing it is estimated that somewhere between 35 and 40 million people were each consuming about 0·6 grains per day of boric acid with no reported harmful effects. However, a number of highly publicized accidents in America – in which babies and young children suffered severely through gross misuse of boric acid, or from mistaking it for other products – were reported, and from 1954 the medical authorities in Britain decided it was advisable to discontinue its use in food. With the advent of improved methods of storage and the widespread use of industrial and domestic refrigeration, food-preservatives became increasingly unpopular and unnecessary.

REFERENCES

1. F.M.S , 14.6.1900.
2. F.M.S., 13.7.01 and 29.7.01.
3. Borax Consolidated, Limited Prospectus (12.1.99).
4. F.M.S., 27.5.01.
5. F.M.S. to R.C.B., 25.4.02.
6. Minute Book, 27.11.03.
7. R.C.B. from I. Redwood, 10.6.01.
8. R.C.B., 8.6.01.
9. R.C.B. from I. Redwood, 13.9.1900.
10. Minute Book.
11. R.C.B. from Mary Brownrigg, 5.2.01.
12. Minute Book, 8.1.04.
13. M.D., 23.5.06.
14. R.C.B., 27.2.1900.
15. M.D., 15.4.08.
16. R.C.B. from Coghill, 16.11.99.
17. R.C.B. from Coghill, 19.11.1900 and 4.12.1900.
18. R.C.B. from Coghill, 15.5.02.
19. M.D. to Coghill, 21.5.02.
20. M.D. to Coghill, 12.4.04.
21. R.C.B. from Pridham, 24.9.04.
22. R.C.B. from Townsend, 13.10.1900; 13 and 15.12.1900.
23. R.C.B. from Coghill, 21.9.01.
24. R.C.B. from Pridham, 24.9.04.
25. R.C.B., 28 and 30.1.04.
26. Prescott to Pridham, 4.3.04.
27. R.C.B. from Pridham, 24.9.04.
28. R.C.B. from Gerstley, 14 and 30.7.02.
29. R.C.B. from Pridham, 29.8.02.
30. Minute Book, 1.2.1906.
31. M.D. to Townsend, 28.4.14.
32. R.C.B. from Zabriskie, 27.11.99.
33. R.C.B. from Zabriskie, 18.12.99.
34. R.C.B. from Coghill, 26.9.1900.
35. R.C.B. from Gerstley, 14.7.02.
36. M.D. to Zabriskie, 9.5.02.
37. Zabriskie to F.M.S., 17.7.02.
38. R.C.B. from Gerstley, 8.8.02.
39. R.C.B. from Locke, 1.10.02.
40. R.C.B. from Gerstley, 23.4.03.
41. R.C.B. from I. Redwood, 2.12.03.
42. Prescott to Pridham, 4.3.04.
43. R.C.B., Fleming to Borax Consolidated, 23.2.06.
44. M.D. to Zabriskie, 10.2.09.
45. R.C.B. from I. Redwood, 10.4.01.
46. I. Redwood to Gerstley, 18.8.03.
47. R.C.B. from Locke, 31.1.01.
48. Gerstley, J.: *History of Borax Consolidated* (unpublished, c. 1945).

49. M.D. to Gerstley, 8.4.02.
50. R.C.B. from Gerstley, 8.8.02.
51. J. Gerstley, *op. cit.*
52. R.C.B. from Giles, 27.11.01.
53. F.M.S. to Baker, February 1899.
54. J. Gerstley, *op. cit.*
55. F.M.S., 17.10.02.
56. R.C.B., Reports of 13.7.03 and 7.9.03.
57. R.C.B. from Gerstley, 28.7.03.
58. J. Gerstley, *op. cit.*

9

China — new adventures in old haunts

Before we pass on from Britain to relate the rather tougher struggles going on in the world of borax in the USA in these years from 1899 to 1914 it is pleasant to turn briefly to an unexpected revival of interest in the far-off world of the borax industry's distant past – Tibet and its neighbour China.

This started when one of the directors, that 'good fellow' Halsey, wrote to Baker on 24 July 1899:

> You must know Savage Landor and his Tibetan adventures, at any rate by repute. He is just off again and has heard by mere accident that I am interested in borax. He says that there is a considerable barter trade in the article from South-West Tibet with India – that the source of supply can be obtained for a vegetable. Would you like to have a talk with him? . . .

There is no evidence that Savage Landor (grandson of the long-lived poet of that name) ever called on Baker, but someone, some time in 1899, left Baker an undated note saying that Landor had requested that details of the information required about China be sent to his residence at 24 Ashley Place, S.W.1. Landor mentioned among other items that the labour there would carry 60 lb weights a distance of 100 miles at a cost of 12s. 6d. (about 62p). He said he knew nothing about the business, but had an idea that he would like to make some money out of this if he could, and added that his most intimate friend was the man through whose hands all the tincal (crude borax) passed the previous year. Landor was well known as a writer of best-sellers containing colourful accounts of his personal adventures in Tibet: like others he strove and failed to reach Lhasa, the 'Forbidden City'. Probably Baker did not warm to the idea of any business involvement with Landor, for there are no further traces of the explorer in Borax Consolidated's records.

Maybe directors had learnt by now that transport is a main factor in assessing the profitability of any borate deposit, so that only the adventurous Mr Smith was thinking in terms of borax deposits in Tibet.[1] He suggested: 'Your people [i.e., the British], through Colonel Younghusband, are exploiting that country and there is supposed to be considerable borax there although quite inaccessible. It might be desirable to make a cursory examination and possibly get a borax concession.' The matter appears to have got no further.

As has been related, Sir Alexander Wilson, Borax Consolidated's chairman at that time, had held important commercial posts in India, and in 1902 he circulated an interesting cutting from a Calcutta newspaper which read:

> It appears from a report of the trade of the United Provinces with Tibet that, during the last official year, there was a large increase in the imports of borax, which amounted to over 27,000 maunds (about 1,200 tons). . . . The borax from Tibet, we believe, all goes to the Continent. The Tibetans, who bring it into British territory, make a secret of the place where it is actually mined or dug up . . . the borax is transported over the hills in small bags which are laid over the backs of sheep and goats, which are also sold in India after the borax has been disposed of.

This is interesting evidence of unchanged methods and of the sporadic demand in Europe for Tibetan tincal which continued into the twentieth century.

An active interest by Borax Consolidated in the commercial possibilities in China started at the end of 1902, when Redwood informed Baker that they had had an order for five tons of borax similar to a 'sample of Chinese borax' they had received,[2] but they had been unable to produce a similar article. Redwood described the Chinese borax as very hard and compact and decidedly brown in colour, and he doubted whether it was possible to imitate it with the borate of lime available to Belvedere.

It seems that this 'Chinese borax' was never made in China, and was actually produced in Europe for the Chinese market. In 1807 Chaptal (the French chemist and industrialist and member of the French Senate), wrote[3] (a translation from the French):

> There are two sorts of purified borax sold in commerce; one is termed Dutch borax and the other Chinese borax; they are both manufactured in the same factories; but the former is the product of the first crystallization, while the second (crystallization) furnishes the latter. The first consists of beautiful detached crystals about one inch in length and nearly of an equal diameter. They are white, semi-transparent, and of a greasy fracture. The form of the crystals is that of a six-sided prism, somewhat flattened, and terminated by trihedral, sometimes hexahedral, pyramids. The second is composed of laminated crystals, covered with a mealy or farinaceous powder, which seems to be of the nature of argil.

It therefore seems probable that the Chinese, long accustomed to the appearance of the borax they had received from Europe, found this difficult to obtain, when Tibetan tincal was ceasing to be used as the raw material for refining in Europe, and that new sources produced a borax with a somewhat different appearance.

Returning to the twentieth century, Smith felt that the proper centre for supplying the Far East, including the Chinese market, should be the underemployed works at Alameda on the Pacific Coast. He was headed off this idea, in so far as Japan was concerned, as the former French companies had established interests there; but Baker was quite prepared to hand this Chinese problem over to him, and suggested that he might be able to produce a suitable product from colemanite at Alameda.[4]

Smith responded with a characteristic burst of short-lived enthusiasm during which as usual an unconventional approach to the subject appeared which may shock connoisseurs of Chinese ceramics:[5] 'I do not see why there should not be a considerable trade. They are largely crockery manufacturers and there are a good many people over there and some of these Chinamen in San Francisco are very intelligent and we will make it to their interest to develop this trade!'

Baker sent over a sample of Chinese porcelain for Ryan's education,[6] but while Ryan was replying to Smith in August of the same year (1904), about Smith's tentative proposal to appoint a broker called Fair to handle the sale of Chinese-style borax to be manufactured at Alameda,[7] Pridham in London was writing on the same subject to Baker at Inverpolly, Scotland, where he was waiting for 12 August and the grouse. Pridham said that he had met recently a Mr Ed Little, manager of Brunner, Mond & Co.'s business in Shanghai. He described him as a very able business-man who had had twenty years' experience in China, and for twelve or thirteen years was missionary for the American Missionary Society. This capable blend of the spiritual and the secular had in Pridham's eyes all the qualities necessary to fit him for the purpose that Borax Consolidated had in mind.[8] For he added that 'this gentleman is wide awake and up to all the dodges of the Chinese Mandarins, Viceroys etc. and, speaking Chinese, was able to interview personally all the higher officials and knows exactly when and how to make a little present'. This, he stressed, was always necessary when dealing with the Chinese officials, as they received absolutely no pay from the Chinese government.

Along with Ed Little, Pridham met his brother Owen from the Eastern Trading Company, Borax Consolidated's agents in Shanghai, who explained that the Chinese could not be persuaded to buy English refined 'white' borax because they were convinced it lacked the strength of brown borax, and he added the remarkable statement that there might be some truth in this, as analysis showed that 'Chinese borax on rope yard contained 110 per cent of crystal borax'. Borax crystallized on rope seems to have originated in India as a suitable form for those using it as a flux in metalwork.

For all the strength of Pridham's recommendation, the subject of supplying Chinese-style borax from America drops out of the records entirely at this point. This was a period of great uncertainty in China just after the Boxer Rebellion, the collapse of the Empire, and the notorious grasping conflict among the Great Powers over the future of Chinese territory. Pridham reported that the duty on foreign borax imported into China was already the equivalent of 15 per cent, and likely to escalate rapidly to the equivalent of 37½ per cent.[9] Worse still, the currency of the country, based on the silver standard, was extremely unstable. The tael, which was equivalent to an ounce of silver, changed in value between three and five times a day, so that to fix prices in taels carried a great risk of loss when settlement day came. All this Ed Little felt might result in the Chinese developing their own source of borax made from tincal in the Szchuan and Kwantung provinces.

However, in London a reminiscent note by Mr James Gerstley says:

> A special form of crystal borax was introduced by us for the Chinese market. . . . Their imports consisted of an extremely hard and dirty-looking crystal, a form which the conservative Chinese insisted on maintaining. Thanks to researches at Kew [the Royal Botanical Gardens], we succeeded in obtaining the vegetable source necessary to impart this stain . . . and in consequence have enjoyed a substantial share of the Chinese trade.

This was probably about 1905. Older employees at Belvedere Works can remember a post-war revival in the late 1940s when the rods in the old crystallizing tanks (as in the *De Re Metallica* illustration four hundred years earlier) were

Lila C, the Pacific Coast Borax colemanite mine in Death Valley (1907–15)

Lang, the Sterling Borax Co. colemanite mine in Ventura County, California (1908–22)

One of the earliest advertisements for Smith Bros. Borax using W. T. Coleman & Co. as selling agents

The Borax Company, Paris (c. 1890), whose early advertising theme was 'borax in the bath and boudoir' – shades of Turkey!

From a brochure of 1896 entitled '200 Borax Best Recipes'

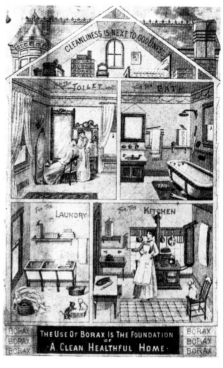

Early 20 Mule Team Borax advertisements

Pacific Coast Borax Refinery at Bayonne, New Jersey (c. 1900)

The merger of Anglo-American and French assets is reflected in the early bilingual share certificates

painted with caramel paste to colour the borax crystals the yellow brown needed for China. But this was of short duration, and thereafter China developed its own sources, which coupled with spasmodic imports of the normal borax of world trade have been able to satisfy their needs.

The areas to which Little referred – namely, the provinces of Szchuan and Kwantung – did not become sources of borax, and there is nothing to indicate that borax was produced in China before the Second World War. An area in Szchuan at Tzu-liu-ching on the Yangtse river and about 150 miles west of Chungking has been known as a salt and natural gas producing area for over a thousand years, and has been mentioned as a source of borax. Recent inquiries[10] confirm that although borax may be present chemically, this has not been a source of production. Kwantung in southern China (capital, Canton) was an area through which imported borax came by sea, but this was not a source of mineral. The brine lakes in central western China, Tsinghai Province (Chinghai) among the Tshaidam lakes, and the area farther north which contains Lake Koko Nor, are all an extension of the same volcanic regions which extend through Tibet. The reports that these areas contain borax may be correct, but much of them is unexplored, and owing to their extreme remoteness it is not surprising that they have not been used as a commercial source of borax.

The borate sources which have enabled China to become virtually self-sufficient in the substance since the Second World War are the deposits in north-east China formerly known as Manchuria (now Lianoning Province), which contain extensive deposits of a mineral called ascherite (a magnesium borate). This ore is relatively low-grade and contains about 12 per cent B_2O_3, and is shipped through the port of Yingkow to the borax-refineries of China.

REFERENCES

1. F.M.S. to R.C.B., 6.4.04.
2. R.C.B. from I. Redwood, 8.12.02.
3. Chaptal, M. J. A.: *Chimie Appliquée aux Arts* (Paris, 1807), Vol. IV, p.252.
4. F.M.S. from R.C.B., 4.5.04.
5. F.M.S. to R.C.B., 20.5.04.
6. F.M.S., 1.6.04.
7. F.M.S. from Ryan, 23.8.04.
8. R.C.B. from Pridham, 11.8.04.
9. *Ibid*.
10. Private communications and meetings with Prof. Y. F. Su, East China Chemical Institute of Technology, Shanghai, 1982–3.

10

Colemanite rivalries in the USA

The acquisition of Coleman's borax interests in 1888 and the change from 'cottonball' to colemanite as the raw material had for the moment left no serious rivals to Smith's borax empire in the USA. The home market was expanding, and the Bayonne and Alameda refineries were in a position to supply its demands, in the East and West respectively, with the help of sales offices in New York, Chicago and San Francisco. Moreover, Smith had some able assistants to rely on to keep the business running smoothly, while he devoted much of his time to his other interests, and to the expensive private pursuits that his wealth laid open to him.

Following the departure of Mather senior in 1896, Smith's first move had been to transfer Steve Mather to New York from the Chicago sales office, where Thomas Thorkildsen was left in charge. Steve Mather was well entrenched in Chicago with many interests outside his business, and he made it clear to Smith that after he had indoctrinated a successor for New York he wished to return to Chicago. As future events were to prove, Mather was a man of exceptional ability, and recognized as such by Smith, but they were not of similar outlook, and there is nothing that indicates that Smith wanted him close at hand on a permanent basis. For this Smith chose one of the small group that came from the Candelaria area of Nevada, who had served him so loyally in the West, and within a year Chris Zabriskie had moved to New York and been trained by Steve Mather, who then returned to Chicago. However, this was not before the direct exposure of Thorkildsen to Smith had provided a further change in the organization.[1] Thorkildsen was a tall, self-opinionated young man who had come from the San Francisco office and had some mining background. He despised Smith, considered him mean (a 'tight-wad'), and made a poor job of concealing his feelings, as was shown by his correspondence. A final scene, when Smith suggested that Thorkildsen was favouring certain customers, brought his resignation and a resolve to enter the borax business in competition, to which Smith is said to have replied characteristically, 'I'll wipe you out,' an objective he never achieved.

In the following year, 1898, Thorkildsen called on Steve Mather to tell him that a new colemanite deposit had been found at Frazier Mountain in Ventura County,

California, and suggested they went into business together. Mather was undoubtedly restless under Smith, but felt it was not the right time to move; he watched Thorkildsen build a refinery in the Union Stock Yards at Chicago assisted by ore supplies from Stauffer's Frazier Mountain Mine and he stayed with Smith for another five years. However, in 1903 overwork brought about illness and a nervous collapse,[2] and Mather entered a sanatorium in Wisconsin in June, where he was under treatment for six months and during which he received no salary. Early in the New Year he felt well enough to send Smith his resignation and to join Thorkildsen, a move which Smith had probably been expecting for some time. The Thorkildsen & Mather Company was in business, and they were both on the way to becoming wealthy men.

In 1900 the net profit of Pacific Coast Borax had passed the magic million-dollar mark, but by 1903 a decade of easy expansion was over, and there was a dramatic decrease in total sales of borax and boric acid in America.[3] This was true in particular for Smith's company, due to the campaign and adverse reports about the use of these substances in the preservation of meat, which culminated in their prohibition by law as food preservatives in January 1907. In addition, a contract which had run for seven years to supply Stauffer with colemanite ended in November 1901, following their entry into borate-mining as serious competitors.

Smith's answer to these adverse factors was to concentrate on the sale of borax for use in the home, and it was in this context that he made the fighting remark 'I'm going to place a fourth pound of package borax in every house in New York.'[4] He thus became increasingly enthusiastic about this subject, as well as about the introduction of borax soaps. The amounts spent on advertising escalated, and he forecast that their growing package trade was a line that could be held, whatever future discoveries of borate ores by others might do to their bulk trade in borax and boric acid.[5] He therefore proposed spending 150,000 dollars on extensive advertising of 'Twenty Mule Team' package borax during 1903, and supported enthusiastically the idea of taking a live twenty-mule team on a tour of the main cities.

When Smith wrote to Baker in 1903[6] he mentioned that Zabriskie and Ryan had been to St Louis to arrange for 'our animated trade mark' to drive over the World's Fair grounds daily:

> My idea is then to drive them through overland to New York with a good press agent in charge, salesmen, etc., driving them through the most thickly populated districts and making an attempt to drive them down Broadway, New York. If they stop us from doing this, we will at least get a good ad.

Over the next few years this team manœuvred down the crowded streets of many large cities, amid great public interest and with rewarding results to the company and its distributors, although exposure to a world and cities which these men from the desert had never expected to see proved in certain cases to be a somewhat intoxicating experience. 'Borax Bill' Parkinson, who drove and trained the mules for these performances, became such a prima donna, insisting on a team of 'swampers' to wait on him, that sadly he had to be recalled. It was an expensive exercise, and at the end of 1906 Smith wrote[7] 'As soon as it reached New York and had a fling down Broadway, I thought it best to take it off the road. The mules were sold and the wagons shipped to Ludlow, California.'

As regards the borax soap project, it was 1905 before the plant started, and things then went unhappily. The motive behind the new venture, said Smith, was that the soap would be popularizing borax as well as marketing it.[8] This shrugged off the fact that the market was already dominated by established soap-manufacturers. In order to sell borax, the borax content of the first soap produced was pushed as high as 10 per cent. This was a rosin soap for laundry use, named 'Boraxaid', but the overdose of borax caused it to form crystals on the surface, and it also contained a high proportion of cottonseed oil, so that many of the first batches turned rancid. This would be enough to make the average entrepreneur hesitate, as over ten thousand dollars had been spent on the new soap plant, and Smith had informed Baker in June 1904 that no less than thirty-one salesmen were already on the road 'mainly preparing for the soap business'.[9]

Smith, however, pushed on undaunted. The proportion of borax was halved, and a saleable product was produced.[10] A 'semi-toilet' soap, 'Queen of Borax', 'Boraxaid Soap Powder', 'Washing Machine Soap Powder' and 'Twenty Mule Team Soap Chips' all met with little success. A better answer was found with the introduction of 'Boraxo' (perfumed) and its variant 'Grime Off' (unperfumed); these started as bath powder alternatives to bath salts, but were later established as successful hand-cleansers, a role which 'Boraxo' maintains to this day. Since they contained about 75 per cent powdered borax, they also made their contribution to increasing borax sales.

Smith was not sure that they would make a great deal of money out of soap directly, but he explained that the circulars put out with the soap advertised the package borax, which should expand sales. However, the Board away in London had by now learnt that they must keep this particular segment of Smith's activities under close observation, and by March 1906 they had noted that annual advertising expenditure on borax soap and packaged borax had reached the large figure of $300,000, and that although volume had increased this had resulted in a loss of about $140,000 on the soap business and a smaller loss of about $17,000 on the packaged borax business.[11] They also noted with satisfaction Smith's statement that he proposed to reduce the advertising to less than $100,000 in the following year, but in spite of this cut sales in this sector were maintained, and profits soon began to be made.

Annual sales of packaged borax increased from 500 tons before 1902 to over 3,000 tons four years later, and although there was a decline in bulk borax sales, total refined borax sales of Pacific Coast Borax increased during this period from 10,500 to 12,500 tons annually (thanks to the use of borax in the home).[12] At the end of 1907 Smith reported that he had spent a quarter of a million dollars on the soap factory with little to show for it.[13] Nevertheless, packaged borax sales continued to grow as a profitable activity, and Smith's judgement about this aspect of the business was vindicated. Much has happened in advertising since the era of radio and television began, but for their day the methods initiated by Mather and carried on by Smith and his sales force, which succeeded in making the '20 Mule Team' trade mark known in American homes from coast to coast, showed great originality.

There was a further setback to borax sales in the United States in 1907, following what Smith referred to as 'The Financial Panic' (recession, in today's terms), but packaged borax held up well. In reviewing sales for 1910 Smith reported that out of a total of 16,000 tons of borax and boric acid, 25 per cent was now used as packaged borax; industrially, enamel accounted for another 25 per cent of the market; and medical uses of boric acid and borax had risen to almost the same level. Borax Consolidated had decided to start production of boric acid at Bayonne in 1900,[14] and this proved a well-timed decision in competing with boric acid imported from Italy. Rival borax-refiners such as Stauffer and Pfizer were manufacturing a whole range of chemicals and Smith arranged that a Mr Parker should be brought out from England to study diversification.[15] He recommended a long list of substances, such as nitric acid, soda ash, hydrochloric acid and at least twenty more, for production at the company's Alameda works. Smith started by wishing his company to be '*the* chemical concern on the West Coast', but his enthusiasm took only four weeks to evaporate. The trouble was not only the high capital cost at a time when sales were static, but it was also an inappropriate moment at which to inflict a chemical factory on the Alameda neighbourhood, where the vegetable-growers in particular were already complaining about acid fumes from the pyrites used in the Pyne copper-smelter on a site adjoining Alameda works, a smelter in which Smith was financially interested. So the idea of diversification into chemicals on the west coast died.[16]

However, worries for Pacific Coast Borax also lay in other directions in the form of financial and mining problems, preoccupation with which occupied much of the unflagging energy of Smith and his associates. In 1899, around the time of the London merger, Smith obviously thought that he had the colemanite supply position in the USA more or less taped and that in any case nothing more need be feared from the old marsh borax deposits.[17] He had wisely bought all the dwindling borax production from these old cottonball Nevada deposits in order to avoid the depressing effect this might have had on market prices. Ben Edwards – an early member of Smith's old Nevada team and still in Candelaria, where he ran the local store – organized this activity, which continued for a surprisingly long time. It was not until 1910 that the last seventy odd tons was shipped and Smith commented that at the current price of borax it was no longer an economic occupation even for Chinamen.[18] Indeed, the future appeared to lie entirely with colemanite, which in America had been discovered in three main areas of California – Death Valley; the Calico Mountains some 120 miles south of it; and, more recently, in Ventura County farther west, and near the Pacific Coast.

In 1900 Pacific Coast Borax's sole source of production of colemanite was still the Borate mine at Calico, but soon afterwards it was realized that its life would perhaps not exceed four years.[19] Although the largest bodies of ore were found to be at the lowest levels, the mine was already down 600 ft at one place, and this revealed a decline in the quality of the ore,[20] as it contained a higher proportion of silica and calcium sulphate, much of which could not be upgraded to economic values by roasting.

The short life of the mine was the subject of persistent rumours, and as sole source of supply to a company which was making great efforts to increase its market, the position was clearly becoming risky. By good fortune new underground discoveries continued to be made and production was increased to meet the existing market, which needed between 25,000 and 30,000 tons a year;[21] but by April 1904 Smith reported that there was only about one year's satisfactory supply left, though the road to the new mine in Death Valley, Lila C., would be ready within that time.[22]

However, the transport problems associated with man's conquest of Death Valley (which were already a legend) were not to be so easily overcome, and it was just as well for the salvation of Pacific Coast Borax that Billy Smitheran – Superintendent at Borate – found, mined and shipped another 55,000 tons of colemanite in the further twenty-four months that preceded the final closure of Borate in October 1907.[23]

At the time of the London merger Smith's experience of borate-mining stretched back nearly thirty years, and although he was now a well-established city dweller, with a growing reputation in finance and commerce, he retained his flair for the rough and dirty game of outwitting rivals in mining and mining claims, and he retained all the old instinct for the signs of a likely prospect. Immediately after his return from London in January 1899 he had set off on an extended tour of the Death Valley region and of the claims staked out there nearly fifteen years before by Coleman's men, and reported in great detail on what he had found.[24] He came away well satisfied that nothing of importance could be obtained by others outside his company's boundary lines; if anything were still to be found, he believed it would be confined to odd ledges of colemanite in the hills. To make certain of this he established a permanent company camp on a good supply of water eight miles from Amargosa, and left a trustworthy watchman there. He was much impressed with a hill of borate called Monte Blanco, which he described as a 'quarry' proposition; but he came out of the place which he later described as 'dubbed with that gruesome name "Death Valley"'[25] by an easterly route through the Funeral Mountains. This he did specifically to look at Coleman's Lila C. claim, which he had not seen before, and which he mentioned as being on the east side of Furnace Creek not far from a place known as Ash Meadows. The mine was 'an exceptionally good one', and he measured the 'croppings' in three different places, and found 24 ft in width of what appeared to be very superior ore.[26] He suggested also that in case this property should become exhausted at any time it would not be a serious problem to extend the road from this point around through another divide into the upper part of the Furnace Creek range, where the company had any number of good properties.

In these few sentences Smith had – perhaps unwittingly – laid down the mining policy that his company was to follow for the next quarter of a century. His reports and letters show, however, that he was already fully aware of the transport problems involved if the Lila C. area were to be used, and this may also have been the real reason for his confidence that others were unlikely to wish to disturb his claims to those remote areas. But transport problems apart, it was as well that Smith established and maintained a firm programme for future mining development so soon after the merger, as without that and his company's inherited hold on Death Valley colemanite deposits rivals could have made life even more difficult than it became in the course of the next few years. Indeed, rivalry and a static market from about 1905 onward had a noticeably depressing effect on borax prices in the USA.

By 1900 competition centred mainly round the two firms Charles Pfizer and Co. and the Stauffer Chemical Company. They had both come into borax at the chemical-manufacturing end – unlike Smith, who had come in at the mining end – and their requirement was raw material at the lowest possible cost. The only other borax-refiners of consequence in North America were one at New Brighton, Pennsylvania, associated with the American Borax Company's mine at Calico, and Thorkildsen's Chicago refinery dependent on Stauffer's mining activities,[27] but these formed a relatively small part of the market. In these early days of colemanite discovery in the USA, however, it could be all too easy for refiners to obtain their own 'captive' mines, and Smith was anxious to prevent such moves, at considerable cost if necessary. He wrote to Pfizer in April 1898 that he (Smith) was keeping them out of ownership of borate properties in the USA where they most desired them, and would be most likely to invest.[28] It should therefore be an easy matter to keep them out of borate properties in South America, especially at a period when there was such a high protective tariff. Pfizer in fact had leanings towards the Arequipa deposit in Peru, but Smith got young Pfizer on his hook instead with a long-term supply contract.[29]

The Stauffer Chemical Company was a tougher proposition, and Smith's efforts to pin them down to dependence on Pacific Coast Borax for all their ore requirements never succeeded. As a step towards influencing their policy he was able to obtain a one-thirtieth interest in their closely guarded share capital, which in April 1904 he bought in his own name from John Howard.[30] The only other shareholders were a wealthy Frenchman called de Guigne, who owned the majority $66^{2}/_{3}$ per cent interest (4,000 shares) and Stauffer and Wheeler, who each owned 900 shares against Smith's 200. Stauffer was acknowledged by all, including de Guigne, to be the driving force in the company. On Smith's own admission, Stauffer was not friendly to him;[31] Smith's shares were transferred to Pacific Coast Borax, but his hopes of influencing the company's policy by increasing this shareholding never materialized, and he sold the shares three years later. Stauffer & Co. had a number of chemical works 'on the Bay' (i.e., at San Francisco) and their letter heading shows that they were manufacturing quite a long list of chemicals apart from boric acid. These included sulphuric acid and other sulphur products, muriatic acid and nitric acid. They had started boric-acid production in 1895 in the Potero at San Francisco, and for the first seven years there was a tolling arrangement, by which Pacific Coast Borax supplied the ore and sold the boric acid and the two companies shared the profit.[32]

Meanwhile, Pacific Coast Borax's successful colemanite mine at Calico attracted considerable interest and activity by others. Smith wrote to Baker as early as 1898: 'There are so many low-grade propositions that if we undertook to get them all we would have to invest more money than it would warrant us in doing and, secondly, when it comes to tariff legislation, these outside companies are of great assistance in securing high tariff.'[33] Smith remained cautiously optimistic, and after visiting Daggett in March 1899 he wrote that he saw nothing in these outside properties that looked like serious competition, as the ore was too low-grade.[34] Smith, however, knew enough about mining to be cautious where he might have been more optimistic over sales and finance. He was 'favourably impressed' with the Gem borate property belonging to his former Nevada borate rivals, the Calm brothers, situated five miles south of Daggett. In fact they discovered little of value in that area, and soon moved elsewhere. The Calms – Max and Charles – were another group with a firm intention of 'downing the Borax combine', and had well-established markets in the east.[35]

The second largest operation in the Calico area (see map VI, p.61) which later became the American Borax Company, was based on a process started by a Dr Humphries and developed by a chemist Henry Blumenberg.[36]. In a break with tradition the process made crude boric acid from low-grade borate ore which was readily available close to Daggett, consisting of a silicious mud shale, in layers green, blue and purple in colour, the blue being the best. It contained between 2 and 10 per cent boric oxide (B_2O_3) but averaged less than 4 per cent. This was treated in 20,000-gallon digesters with sulphurous acid (obtained by burning sulphur) and the borate ore was thus converted to boric acid. This was leached out in solution and the liquors were concentrated by solar evaporation in ponds covering an area of about eight acres, from which crystals of a crude boric acid containing about 49 per cent boric acid were 'harvested'.

The American Borax Company, which was founded in 1900, was managed by Blumenberg and owned by his former employers, Dawes and Myler, who also owned the Standard Sanitary Company, which manufactured bathtubs and the enamel for them.

They also owned a borax refinery in New Brighton, Pennsylvania, to which the American Borax Co. shipped the crude boric acid. During their years of production – mainly 1904 to 1907 – they obtained their ore from the Columbia mine about seven miles north-west of Daggett, where it was brought by narrow-gauge railway to the processing plant. Solar evaporation confined operations to the months May to September and the most crude boric acid they produced in a year was 640 tons. They fought hard in trying to compete with the colemanite mining, but the cost advantages of colemanite were difficult to match. Another company, the Western Mineral Co., started similar operations to make crude boric acid at the Bartlett Mine just south of Calico, and there were two other patented claims just south of Borate, the Centennial mine and the Palm Borate Co., but none of these produced anything of significance.

Elsewhere in 1899 new discoveries of colemanite were made in the Frazier Mountain area of Ventura County, California.[37] Stauffer were the pioneers in this area, and formed the Frazier Borate Mining Company, which in 1902 had begun mining and shipping high-grade ore (about 34 per cent B_2O_3) by traction engine to the nearest railroad at Bakersfield about sixty-five miles away, and thence to their refinery in San Francisco and to Thorkildsen's in Chicago. Then in 1904 the Calm Bros (Columbus Borax Co.) started a second mine about two miles north-east of Stauffer, having abandoned the Calico area.

However, in spite of the deteriorating grade of ore at Borate, Smith remained unimpressed with this alternative Ventura area. On a visit there in 1904 Smith and Corkill examined the Weringer property which was situated between the Stauffer and the Calm properties, and they decided it was not worth even a tenth of the $45,000 demanded.[38] At the same time they took the opportunity to find out something about the other mines. The Calm mine (Columbus Borax Co.) was barred to visitors, but they heard that little ore of any consequence had been shipped. They succeeded in gaining access to the Stauffer mine (Frazier Borate Co.), and Smith shrewdly surmised that, although the mine was shipping about 4,000 tons of colemanite a year and evidently making a profit owing to the proximity of cheap timber, the flat location of its vein made it unlikely that it would be able to produce for many years longer.

Apart from American Borax Co. in Calico, the future seemed to lie with Pacific Coast Borax's Lila C. mine in Death Valley, which would soon be ready to open, and where seven years' supply of ore was already in sight. Smith could in fact afford to brace his nerves and hang on until transport to and from Death Valley was available, which is what he set out to do. Meanwhile, although the borax price had maintained a steady 7¼ cents per lb for nearly five years it started to decline in 1905 under the pressure of competition, particularly between Pacific Coast Borax and Stauffer, and it stood at 6 cents in 1907. Then, as soon as the new railroad to Lila C. was completed, in August 1907, the price declined dramatically to 4½ cents, and following the Democrats' cut in the tariff in 1909 it sank to the all-time low of 3 cents in 1910.[39] The borax mines were to be badly hit by these price-changes. In 1907 both Western Mining and America Borax closed down their operations at Daggett, and output had started to decline at Stauffer's Frazier Mountain mine, as Smith had forecast, and mining ceased the following year.[40]

For a brief few months Smith and the Lila C. mine seemed to be about to have it all their own way.

However, in August 1907 two old-time Western prospectors, Shepherd and Eppinger, were working near Lang in Tick Canyon, some forty miles from Los Angeles, trying to wrest a living from an old Spanish gold-mine. Shepherd found some glistening white crystals, of which a sample was sent to Blumenberg at Daggett. Blumenberg identified them as high-grade colemanite, and hopped on the first train to Lang. The bush telegraph brought 'Wash' Cahill[41] from Smith's company stationed at Ludlow fast on his heels, with John Ryan soon to follow, but Blumenberg got there first, and lode claims* were staked out.[42] In addition, Thorkildsen was also soon on the scene, and he used the earlier placer claims of two prospectors called Cook and Hopkins to challenge Blumenberg and establish his own claim. The prospectors each sold their interest on a basis of deferred payment, Shepherd and Eppinger through Blumenberg to the American Borax Company[43] (whose principal shareholders were Dawes and Myler), while Cook and Hopkins sold to Thorkildsen.[44] A battle between Blumenberg and Thorkildsen ensued; but after some months it seems that Thorkildsen's placer claim to the mineral rights was accepted, and an agreement to form the Sterling Borax Company emerged on 15 January 1908. This now included a 40 per cent interest for Stauffer, 40 per cent for Dawes, Myler and Blumenberg (American Borax Co.) and 20 per cent for Thorkildsen and Mather.[45]

Stauffer was back in borax-mining, and this made a strong combination of raw-material miners, refiners and end-users of borax in the enamel trade. Thorkildsen was elected president and general manager in charge of the mine, and as an ex-employee of Smith's with a chip on his shoulder, he set out to make life difficult for his old employer and for the Lila C. mine. This was quite an easy objective, as the Lang mine was only five miles from a branch of the Southern Pacific railroad, leading straight to Los Angeles and the coast, and the freight cost per ton was six dollars less than from Lila C. Smith, of course, had his spies out, and thought that Lang was yet another mine that might cut out suddenly, but instead it survived and prospered. By 1910 output had reached 12,500 tons a year of good-quality colemanite, which represented about 30 per cent of the American market.[46] Once the hideout of the bandit Tibercio Vasques, who was hanged in 1875, Lang again became active as a mining camp. It housed as many as a hundred miners during its peak, and continued to produce until 1922, after which it provided many a setting for Western movies.

Since they got together in 1904, Thorkildsen and Mather had been successfully developing their business, and it was said that Mather never wearied of poaching on Pacific Coast Borax's sales preserves. Through their mining interests in California, first with Stauffer at Frazier Mountain in Ventura and later at Lang near Los Angeles, they always managed to obtain sufficient ore for their refining needs without having to depend on their former company. In October 1908 Dawes and Myler bought out Blumenberg's small interest in the Lang mine and made a deal with Thorkildsen and Mather to sell them their 40 per cent interest in Sterling Borax Company, together with their refinery.[47] Mather and Thorkildsen thus held a 60 per cent interest in Sterling Borax Co. and they had two refineries, one at Chicago and the one at New Brighton, Pennsylvania – now called

---

*See footnote in Chapter 3 (p.44).

the Brighton Chemical Company – while the selling organization remained the Thorkildsen and Mather Company. The Lang mine was by now a serious threat to Smith's company, and Sterling Borax were showing a lively interest in selling to some of the larger European customers at that time supplied by Borax Consolidated, as well as supplying the German refinery which Stauffer had built in 1909 at Gernsheim in Worms.

In the context of the financial troubles which confronted Borax Consolidated, arising from the new railroad to the Lila C. mine, there was an urgent need to control the situation. Borax Consolidated, Limited were therefore eager to negotiate with the Sterling Borax Company, but it was not until the autumn of 1911 that they found (perhaps surprisingly) that the parties concerned were in a mood to sell. The further decline in the price of borax in 1910 had diminished the importance in Stauffer's eyes of being basically in borax, and Thorkildsen, who had been in close touch with Zabriskie and Ryan for a number of years and had played hard to get, probably now convinced himself and Mather that it was a good time to sell.

At the end of 1911 separate agreements were made with the two shareholders for Borax Consolidated to acquire their holdings in Sterling Borax for a total sum of $1·8m.[48] Stauffer received $125,000 in cash and a further half-million dollars in the form of notes repayable over five years and carrying interest of 5 per cent. There was also a contract to supply them with ore, including supplies to their refinery in Germany. The agreement with the Thorkildsen and Mather Co. provided for the payment of $1,175,000 over an extended period, and again there was a contract to supply their refineries with ore and Thorkildsen remained President of Sterling Borax. To the outside world the change in ownership passed unnoticed and the company continued to function as before. Even Roy Osborne, the Superintendent at the Lang mine, did not know about it until some years later.

The Calm brothers (Columbus Borax Co.) had high hopes of selling their colemanite mine at Ventura and small refinery in New Jersey to Pacific Coast Borax. In January 1907 Max Calm told Zabriskie that they could mine 1,000 tons a month, and had blocked out thirty years' supply of ore reserves.[49] They would be prepared to sell for $1·25m., and he asked that Smith – who was in the West Indies – should be told that their properties were not on the market, but the 'offer was only due to the pleasant relations between us these many years'. Actually the mine was producing at a rate of about 150 tons a month, and after a visit by Ryan and Corkill, Pacific Coast Borax were not inclined to make an offer. By 1909 the mine closed, and was later reported flooded.[50]

There were some other scares and threats to the supremacy of the Lila C. mine. In 1907 Blumenberg turned up in Death Valley, where he reported the discovery of a fifteen-foot ledge of colemanite in Furnace Creek Wash, and after trying unsuccessfully to sell the deposit to German bankers he started working it;[51] but in the face of falling prices he did not succeed.[52] Once Stauffer became interested in the Lang Mine production at Frazier Mountain had declined rapidly, and it closed in 1909. In December 1911 a German financier called de Fries (not the Borax Consolidated director) floated the National Borax Company, which acquired the Calms's Columbus mine and business (now called the US Borax Company) with the intention of reopening the mine and later combining it with a mine belonging to W. H. Russell, about half a mile south-west of the Calms's property and near the township of Stauffer.[53] The new enterprise planned to supply a refinery in Germany. However, the Russell mine closed at the end of

1912 owing to mining difficulties, the plans of the National Borax Company did not materialize, and through failure to meet payments the Columbus mine reverted to the US Borax Company.[54] In August 1913 Borax Consolidated approved the purchase of the US Borax Company from the Calm brothers for about $240,000, thus acquiring a closed colemanite mine, a small refinery which they closed later, and a small share of the market.[55] Owing to a legal dispute, completion was delayed until 1915, when it was financed by a dollar issue of 6 per cent gold bonds. The Russell mine was acquired by Stauffer at the end of 1913, but it was never reopened.[56]

Thus Borax Consolidated had won the battle of the colemanite mines, but, as both Sterling Borax and The Calms Company had been bought mainly with loan money,[37] it all became part of a heavy burden of debt in difficult circumstances to be outlined in the next chapter.

REFERENCES

1. Thorkildsen–Mather letters 1897 (US Borax files).
2. Shankland, Robert: *Steve Mather of the National Parks* (Alfred Knopf, 1970), p.37.
3. F.M.S. Ann. Reps. between 1900 and 1907.
4. F.M.S. to R.C.B., 9.11.04.
5. F.M.S. to R.C.B., 17.10.02.
6. F.M.S. to R.C.B., 28.10.03.
7. F.M.S. Ann. Rep., 1905/6.
8. F.M.S., 16.8.03.
9. F.M.S. to R.C.B., 2.6.04.
10. F.M.S. Ann. Reps., 1903–10.
11. Minute Book, and F.M.S. Ann. Rep., 1905/6.
12. F.M.S. Ann. Reps., 1901–5.
13. F.M.S. Ann. Rep., 1906–10.
14. Zabriskie to R.C.B., 27.2.1900.
15. F.M.S. to R.C.B., 25.5.01, 10.3.04.
16. F.M.S., 20.6.03.
17. F.M.S., 1.12.1900.
18. F.M.S. Ann. Rep., 1909/10.
19. F.M.S., 3.11.99.
20. F.M.S., 16.2.1900.
21. F.M.S., 25.5.01; F.M.S. Ann. Rep., 1901/2.
22. F.M.S., 27.4.04.
23. F.M.S. Ann. Rep., 1906/7; also Smitherman's Annual Mining Report for 1906/7.
24. F.M.S. to R.C.B., 7.4.99.
25. F.M.S., 28.11.03.
26. F.M.S. to R.C.B., 21.4.1900.
27. Yale, C. G.: 'The Production of Borax in 1904', in *US Geological Survey* p.9 (Washington, 1905).
28. F.M.S., 26.4.98.
29. R.C.B. to F.M.S., 26.4.98.
30. F.M.S. to R.C.B., 6.4.04.
31. F.M.S., 1.9.03.
32. Steiss, A. J. and O'Donnell, V. H.: *Stauffer Chemical Company – Work in Borax* 1956 (Revised).
33. F.M.S. to R.C.B., 24.3.98.
34. F.M.S. to R.C.B., 2.3.99.
35. F.M.S., 12.3.98.
36. (i) 'American Borax Company – Mines and Plant' by Fred Corkill, 3.1.06. Borax Hse archives.
    (ii) Yale, C. G.: 'The Production of Borax in 1903', in *US Geological Survey* pp.11–14 (Washington, 1904).
    (iii) Bailey, G. E.: 'The Saline Deposits of California', in *Cal. State Mining Bur. Bull.* 24 (1902), p.39 and pp.69–70.

37. (i) Yale, C. G., *op. cit.*, p.10.
    (ii) Struthers, J.: 'The Production of Borax in 1902', p.8 in *US Geological Survey* (Washington, 1903).
38. F.M.S., 27.4.04 and 12.5.04.
39. M.D., 6 and 13.11.07; F.M.S. Ann. Reps., 1906–10.
40. Yale, C. G.: *US Geological Survey: Mineral Resources 1908*, pp.604–5 (Washington, 1909).
41. Cahill, W. W.: 'Recollections' – Notes by Ruth Woodman (US Borax archives).
42. Ryan to F.M.S., 2.9.07.
43. Ryan to F.M.S., 26.3.08.
44. Ryan to F.M.S., 2.1.08.
45. Ryan to F.M.S., 10.2.08. Zabriskie to F.M.S., 23.9.08 (gives shareholdings).
46. F.M.S. Ann. Rep., 1909/10.
47. Ryan to F.M.S., 19.9.08.
48. Minute Book, 26.9.11 and 24.10.11.
49. M. Calm to Zabriskie 28.1.07, and Zabriskie to Ryan 19.2.07.
50. F.M.S. Ann. Reps., 1909/10 and 1910/11.
51. Ryan to F.M.S., 19.9.08.
52. M.D. to Zabriskie, 14.4.08.
53. M.D., 16.12.11 and 8.1.12.
54. M.D., 21.1.13, also Yale and Gale, 'The Production of Borax in 1913, in *US Geological Survey*, Part II, p.522 (Washington, 1914).
55. M.D., 13.8.13 and Minute Book, 6.1.14 and 27.4.15.
56. Yale and Gale, 'The Production of Borax in 1914', p.287, in *US Geological Survey* (Washington, 1915).
57. Minute Book, 26.9.11, 24.10.11 and 18.2.13.

11

The Tonopah and Tidewater Railroad saga

The financial predicament in which Borax Consolidated found itself during the last few years before the outbreak of the 1914–18 war in Europe was due partly to its determination to purchase the mines owned by its principal colemanite rivals in the USA, but mainly to the type of railroad which had been built to connect its new Lila C. mine to a trunk railroad, and thence to the refineries.

This mine, in the course of development work which had been going on at the site since 1899, was found to consist of two main colemanite veins.[1] The upper vein was about 2,000 feet long and from 6 to 18 feet wide, with a very steep dip to a maximum depth of about 300 feet, and with the lower vein lying about 25 feet below it. During 1903 and 1904 John Ryan and Fred Corkill were working there, and by the time it was decided to start operations, some three thousand feet of shafts, drifts and tunnels had already been driven and the mill was almost complete. The ore was pronounced to be of very high grade, far higher than that of the old Calico mine, the mine was capable of producing 30,000 tons a year, and the payrolls show that a total of nearly 100 men were employed. However, eight years had elapsed between the acceptance of Smith's view that Lila C. should be prepared to succeed the Calico mine and its coming into operation. Also, it was a full three years between an assessment that the Calico mine had only about a year's supply of colemanite left and June 1907, when Lila C. eventually started up. The delay was caused by several changes in plan about the way in which the ore would be transported from the mine, and the final decision when it came involved a project of immense difficulty with its own inevitable delays.

Ever since the end of the Civil War in 1865 a boom in railway-building had been taking place, which by the year 1900 resulted in a mood of careless optimism about railway construction. The two trunk railroads nearest to Death Valley, the Atchison, Topeka and Santa Fe through Daggett and the Southern Pacific through Mojave, were both well over a hundred miles from the Lila C. mine. Smith, since the time when his company had built the Borate and Daggett Railroad to supersede mule teams on the short twenty-mile stretch between Daggett and the Calico mine, no longer thought in terms of mules. Reporting on his Death Valley trip of 1899, he put forward the idea of building a railroad into

Death Valley from Mojave[2] – that is to say, along the former Twenty Mule Team route. In April 1900, however, when he paid a second visit to Lila C., he had changed his views in favour of a route through Ash Meadows and, keeping well east of Death Valley, down about 135 miles to a little mining town called Manvel.[3] This place was up a branch line from Goffs on the Santa Fe railroad, which was later extended to Ivanpah. Smith described this route as almost one continuous valley, which with the one exception of the State Line Divide had no engineering difficulties, and this was to dominate his plans for the next five years. Smith reached the sound conclusion that no railroad need ever enter Death Valley itself, which he thought was too low a depression for railroad purposes, and that all the ore bodies that would ever need to be developed were situated at a much higher elevation. Noting that a route for a new railroad from Salt Lake to the south passed close to the route he had outlined and had already been surveyed, his plans for the company were merely to build a private branch line, connecting Lila C. with such a trunk railroad. This was not just to save expense but also to deter others from building public railroads into the Ash Meadows country, so that only Smith's company would be able to benefit from the freight arising from further developments there.

At the end of 1902 prospects in the area had started to become clearer. Senator Clark, the copper king from Montana, had been given a concession to build a trunk railroad from Salt Lake to Los Angeles, and Smith wrote: 'It is quite apparent now that the Clark road will go to the south of the State Line Pass and, in that case, we will have to go carefully into the proposition as to the wisdom of building a road ourselves.'[4]

Nevertheless, Smith, with all the experience of the failure of the traction engine *Old Dinah* to transport ore from Calico mine to Daggett station, first reacted to the problems posed by the remoteness of Lila C. by planning construction of a road for traction engines.[5] Work started in the first half of 1903 on grading a road from Ivanpah up the easy Pahrump Valley, and then westward over the State Line Pass and down to what later became Death Valley Junction. Ivanpah had become the northern terminus of the California Eastern Railroad, which was linked to the Santa Fe main line to Los Angeles at Goffs, about fifty miles south. But by September 1903 Ryan, the superintendent, was having difficulties, and was not finding it easy to make a practicable grade over the pass. However, Smith went ahead and made a contract with a firm called Gibbs Engineering Company for a traction engine of advanced design which involved a gasoline engine generating the electric power which drove the wheels of the machine (a forerunner of the diesel-electric system).[6] The board in London approved these plans at the end of October 1903,[7] but only a fortnight later the first signs of doubt started to appear in Smith's letters:[8] 'If we should find that our traction proposition is not feasible we have the alternative of putting down rails ourselves or making some advantageous arrangement with either Clark or the Santa Fe people'. Then, late in April 1904, the traction-engine idea started to fade.[9] The engine had not been built, and in fact the design proved to be impractical. Smith was already thinking that a railway line might have to be laid down instead, and by early June the traction engine idea was dead. Smith wrote that to install a broad-gauge railway system from Clark's railroad (the San Pedro, Los Angeles and Salt Lake, then under construction) to the Lila C. mine would cost half a million dollars, or $600,000 if the grading already done was included.[10] The London board accepted the railroad idea immediately,[11] and Borax Consolidated rushed cheerfully into railroad-building without, perhaps, pausing long enough to consider the politics of the area in which they intended to become involved.

Smith's letters to Baker show how far his ideas had now expanded beyond the simple proposition that all that was required was the cheapest and quickest means for transporting borate from the Lila C. mine to the coast.[12] He wrote that Tonopah and Goldfield were going to be great mining camps, and he believed it to be of vital importance to secure the State Line Pass and a railroad reaching up towards that country. He said he had already taken steps towards incorporating a company to build this road, which was to be known as the 'Tonopah and Tidewater Railroad'. 'How does that sound?' he concluded. 'Broad enough, is it not? It will put some of those railroad people to guessing, at any rate.'

At this time the state of Nevada, north from Death Valley, was coming into prominence as yet another get-rich-quick prospect for gold, silver, copper and other types of attractive metal. Some of the big mining men around Salt Lake City and the mid-West were showing interest, and this acted as an additional incentive for plans to build railroads going south and west through mining areas of Nevada and South California to the Pacific coast. In 1901 two of these were nearing open strife, as headlines in *The Examiner* at the time show: 'Senator Clark of Utah defies Harriman, the promoter of the Union Pacific and Southern Pacific combination – Possible that blood may be shed when rival railroad workmen meet on the Nevada Sagebush plains.' After two years of war Senator Clark reached a compromise with Union Pacific at the eastern end of the Salt Lake route, and in January 1905 his first major railroad – the San Pedro, Los Angeles and Salt Lake – was in operation.

In 1904 Smith had been surveying alternative routes to link his railroad with a main line to Los Angeles. One was the traction-engine route from Lila C. to Ivanpah, and another was to go directly south to the Santa Fe at Ludlow. However, another possibility was to use the new Clark railroad, and a route north to Lila C. from Soda Spring (fifteen miles west of Crucero) was examined. When this railroad came into operation in 1905 Smith began serious discussions and at a meeting in April Clark suggested to Smith that the shortest and best route for him would be to join the Clark line at Las Vegas; Smith agreed, and left the meeting thinking he had a deal.[13]

Ryan confirmed that this would mean a much easier route to Lila C. – north-west across the desert from Las Vegas, around the Charleston mountains, and down through the Amargosa desert to the mine. But nothing, it seems, was put down on paper and signed, and the explanation of what followed may lie in the fact that Smith had been talking too much about the wider prospects for such a railway, and that Clark had been listening very attentively.

In May Ryan and Cahill, assisted by a consultant engineer named Clarence Rasor, arrived in Las Vegas and quickly completed the first ten miles of rail track out into the desert. However, when they started to connect up with the Clark line they received a sudden and unexpected shock.[14] Clark's General Counsel appeared and peremptorily ordered them off, refusing to give an explanation. Clark was in Europe, but when he disembarked in New York a month later Ryan was there waiting. The position became clear; Clark did not recognize any deal with Smith, and had decided himself to build a new railroad from Las Vegas to the mining area and to keep the profits to himself. Accordingly, in October 1905 he started to build the Las Vegas and Tonopah Railroad out to Beatty.

Undoubtedly Smith had been foolish to start work with nothing in writing when he was dealing with a robber baron of the magnitude of Clark, but it was an era when the rules of poker were perhaps closest to any expected code of conduct. Smith had lost valuable time, and rightly felt he had been double-crossed. As he himself said, he never did like paper-work or detail and he was suddenly jerked back to realities, for he had no legal ground for protesting, and two and half years had now gone by with nothing on the ground to show for his efforts, or means of coping with the now urgent problem of getting the borate away from Lila C. He went south and west of Las Vegas to Ludlow, the nearest station to Lila C. on the Santa Fe railroad, and made a quick deal to link up with them there by building a railway going due north to the borate grounds of Death Valley, and to the other less certain riches beyond. By August the team of Ryan, Cahill and Rasor were based at Ludlow, working on the new project. Smith managed to collect $30,000 from Clark for the grading work they had done, but he was now determined to compete with Clark in a game that involved more than borax.[15]

None could seriously question the wisdom of linking Lila C. and the other promising borax prospects just beyond it to the nearest trunk railroad. Nevertheless, the situation in 1905 still called for a cautious approach, which meant – in words that Smith himself used at the end of 1904[16] – returning to the original plan of laying a light railway for transporting the company's goods only, and probably earning a little from some return freight. But instead, as the succeeding sentences in this same letter and the content and tone of the prospectus introducing the mortgage debenture bonds of the proposed railroad company show, the whole venture was slanted towards the Bullfrog area and the gold and silver prospects of the area surrounding it. Lila C. and the surrounding borate area were treated merely as secondary activities to be linked in by a branch line en route to Bullfrog.

Life at Ludlow at this time was no picnic, but was no doubt improved by the presence of Ma Preston, well known to early borax pioneers at Daggett (see p.64). Cahill recalls:[17]

> Ludlow at that time was but little more than a railroad tank town, the Santa Fe's only business being with a mining company whose mines were located at what was called Camp Rochester, seven miles to the south. Ma Preston, known to many as 'the Queen of the Desert', operated a lodging house and saloon. Mr. Ryan, Mr. Rasor and I, together with our surveying party of about five men, all rented rooms at Mrs. Preston's lodging house and stayed there for several weeks, until facilities of our own could be constructed. Being midsummer, just about the hottest time of the year, we got hold of a bunch of cots, and as we all had our own bedding, slept on the sidewalk and on the street in front of the lodging house. Of course we paid Ma Preston for the rooms, the same as if we occupied them. In the meantime, if she had an opportunity to rent one of the rooms to an overnight guest, the opportunity was never overlooked. There was only one bathtub in the hotel, and it was ordinarily in such a condition that we followed Mrs. Preston's example and bathed by turning the hose on each other in front of the house.

The railroad was to be a broad-gauge one with its own full-scale locomotives and rolling stock, just like the other 'big boys' who were building in the area. This reflected Smith's whole grandiose way of thinking, since he ceased at the turn of the century to be solely and simply a borax producer. Like other tycoons of the period, he now owned his own private railroad car, named the 'Hauoli' (Hawaiian for delight). He much enjoyed travelling in style, and when he moved across the continent he was accompanied by his family, his secretary Evelyn Ellis, and a whole retinue of French maids and Chinese servants.[18] However, the unfortunate aspect of the new transport project was that, unlike his other outside activities in

"RAILWAYS IN THE AREA OF DEATH VALLEY AND SEARLES LAKE"

Map VII   LUDLOW TO LILA C, 122 MI.

minerals, real property and transport schemes, Smith and the needs of Lila C. mine involved Borax Consolidated in this particular enterprise.

Smith's new railroad company itself planned and carried out the construction of the railroad, and those involved endured much hardship, and day after day carried out their gruelling hard work in exceptional extremes of heat and dust over a period of more than two years.[19] The line eventually stretched north for just under 170 miles from Ludlow, with a spur line running another seven miles up to the Lila C. borax mine from Death Valley Junction, which was 122 miles from Ludlow. The main problem, inevitably, was water-supply. Most of the places selected for stations had no spring or any other vestige of water, which had to be brought by train. At one station, 'Val Jean', the Irish supervisor, Wash Cahill, wrote: 'We drilled a well there through solid cement, gravel, and stone, making sometimes less than a foot a day. After a month or two we'd gotten down 400 feet, and found it drier at the bottom than on the top, so we gave it up.' He added that the Amargosa river appeared on the map to be flowing alongside the rail route for some distance. 'But everyone familiar with the Amargosa Desert knows that the river has a habit of disappearing underground, and where it does remain on the surface, the water is either undrinkable or the river bed is bone-dry. Old timers assert that the only three fish ever seen in the Amargosa River all carried canteens.'

Another big problem was labour turnover. Originally it had been intended to recruit only white labour, but this idea soon melted away in summer temperatures around 135 degrees in the shade. As Wash Cahill said: 'It got to the point where we had three complete gangs. Gang No. 1, which had just quit and was on its way back home. Gang No. 2, actually at work. And Gang No. 3, on its way out to the desert to take the place of Gang No. 2 which we knew would be quitting any minute.' Attempts were made to get over the problem by hiring foreign labour, but these met with little success. Cahill also described how, having once hired Japanese from San Francisco, he went out from Ludlow one day to see how they were getting on. 'Out of the hundred we'd hired', he wrote,

> only eighteen were working. And when I say 'working', I don't mean that all eighteen were on the job. Nine of them held shovels in their hands. The other nine stood in front of them with buckets of water. Have you ever seen a Chinese laundryman spray clothes before he irons them? Well, that was the scheme. The nine non-workers would fill their mouths with water and blow it in the faces of their nine yellow brethren who were going through the motions of shovelling. Nine human geysers, regular old faithfuls.

Less than a week after they arrived a desert sandstorm blew down their camp, so as they just sat and did nothing the opportunity was taken to pack them off to San Francisco. A Western Union Cable (January 1906) sent from San Francisco to the Tonopah and Tidewater Railroad Company at Ludlow reads: 'White men scarce. Shall I send ten Mexicans?' The Mexicans came, and they stuck it longer, so that the job was finished over the next eighteen months. However, although in July 1906 Smith, acting in desperation, called for outside bids (at $30,000 a mile!) for crossing the Amargosa Canyon, the heat was insufferable, no bids were forthcoming, and work had to be postponed till winter.

The man in charge of construction of the railroad, John Ryan, was once described by Smith as a bulldog, and his tenacity was never more in evidence than on this project. He was born in County Antrim, Ireland, in 1849, and migrated to America at the age of twenty-two. Within a year he had joined Smith in

Columbus, Nevada, and was soon his right-hand man. While he was building the railroad Smith gave him the cash he needed for the payroll and everything else, and Ryan paid everyone by personal cheque through his own account. In the beginning this caused raised eyebrows with some suppliers, but soon everyone accepted Ryan's credit.

Ryan's request to the Board of County Supervisors of Inyo County before the railroad entered their territory in September 1906 gives a picture of one of his problems, and of his determination to keep this project moving:[21]

> We beg to petition to you with us to try to regulate the liquor traffic, so that it will be almost impossible for tents or saloons to either commence or do business along the line of our railroad in your County during construction. The Supervisors of this County (San Bernadino) have favored us to the extent of not granting a licence to a single saloon or liquor merchant on or near our line, since we commenced construction over a year ago, they deeming that it was to the County's interest as well as ours.
>
> A saloon on this desert, especially the class of saloons that follow railroad construction camps, are nothing more or less than a rendezvous for low grade characters, women as well as men, whose only occupation is to fleece the working man by thieving or other criminal means.
>
> Hoping that your Honorable body will give this letter and its object due consideration.

The need to economize on the railroad was much in the minds of Smith and Zabriskie, and advice was sought and willingly given by President Ripley of the Santa Fe Railroad. Zabriskie wrote to Ryan in September 1905[22] that 'Ripley will do everything he can in pushing our work', and he agreed to supply them with light rails which were being replaced on Santa Fe tracks. It was decided to modify locomotives from coal-burning to oil at a cost of $350 each, and Mr Ripley said this should be done by the Baldwin Locomotive people, who were the only ones who understood how to do it, and he would make the Santa Fe shops at Topeka available. The only touch of human comfort in the programme to be spared from the economy axe was the superintendent's private rail car[23] (later to be christened 'The Boron'), with kitchen and observation deck, which was being constructed by Woods and Huddart in the east; Woods was a 'dear friend' of Zabriskie, but had to be competitive.[24]

Smith had originally proposed that the railway company should be incorporated in America, the stock to be subscribed for by representatives, who would transfer it to the English company, so that Borax Consolidated would be the owner but it would be an American railroad.[25] Borax Consolidated could then guarantee the bonds which would be issued, and the line would be built entirely on the proceeds of the bonds without touching the share capital. It is possible to imagine quite a number of reasons why Smith was so keen to tap the English rather than the American public for funds. The American rail boom was virtually over, there was competition to face from two other railroads in the Bullfrog area and Smith may at that time have been already too far extended on money borrowed for his other projects to risk further blandishments on American stock exchanges and money markets.

However, English financiers also were obviously not enthusiastic over the prospects for yet another American railroad issue. On 1 June 1905 Alfred Emberson, the principal stockbroker used by Borax Consolidated, wrote a most tactful letter to Baker. In this he reminded him that the number of Borax Consolidated's debenture and shareholders was now relatively large, and that as they had been introduced by Baker and Emberson, it 'behoved' the two of them to try and gauge what would be the stockholders' view of the railroad proposition. At that moment they had a very high opinion of the managing directors of the company and the conservative policy that had been adopted by the management hitherto, but a departure in the way of a railway enterprise would naturally be the subject of close scrutiny. Then, after a further page of tact, he pointed out that there had been in the past considerable dissatisfaction felt by English investors about American railways owing to the tendency to underestimate the capital required, and that intense competition from other American lines was a usual cause of utterly inadequate profit forecasts, aggravated by unscrupulous non-cooperation when freight cars were expected to be fed into a rival line and moved on. He concluded his letter by sending an 'unpleasantly worded article' from the Press about Tonopah, which mentioned its unestablished reputation as a gold centre and suggested that it would be better if the name of the railway were changed to 'The Amargosa Railway' and the project confined to the section from the trunk railroad to Lila C., leaving the Goldfields section to be built later as a separate line.

Although Baker was already worried by similar reservations about the project, the only point on which he tried to fight was one suggesting that the section to be guaranteed by Borax Consolidated should be confined to the link between the trunk line (Santa Fe) and the borate mine at Lila C., and that it should not include the goldfields section farther on as they were not in the gold business. This probably made some impression on Smith and he did compromise by agreeing that Bullfrog should be the northern terminus, which was about one-third of the way to Tonopah from Death Valley Junction and the Lila C. branch line.[26] However, almost immediately the entrepreneur John Brock floated the Bullfrog Goldfield Railway to connect Tonopah with Bullfrog and this eighty-three miles of track thus completed the link from north to south which Smith had dreamed about when he thought up the name 'Tonopah and Tidewater'.

The prospectus advertising the First Mortgage 4¼ per cent Gold Bonds of the Tonopah and Tidewater Railroad was duly published in London, where the money was raised in November 1905, and it mentions an estimate of 'under £3,000 per mile' for the cost of construction of 170 miles of line. This, at £510,000, was something less than Smith's off-the-cuff estimate of £600,000, but by May 1907 it had become clear that this estimate had been exceeded substantially. Second Mortgage (Sterling) 5 per cent Bonds were therefore approved, to the amount of £250,000, in the second half of 1907, of which £175,000-worth were issued. The prospectus gave colourful reasons for the over-expenditure, including the consequences of the earthquake and destruction of San Francisco, the subsequent strikes and enormous increase in the cost of labour and material, and also unexpected difficulties which arose in making the long cutting through the Amargosa Canyon, delaying the completion by several months. The difficult part of the work, it claimed, had now been finished. This second issue of bonds again appears to have been taken up satisfactorily by the market, and the line to Lila C. via Death Valley Junction duly opened in August 1907. Actually Lila C. had started production in June, and to meet the urgent need for colemanite the Twenty Mule Teams had to be brought into action again for two months to cover some thirty miles to the station named Zabriskie on the partly completed line.[27]

The Smith-Baker axis at this time reflected the strains that accompanied much of the extensive involvement of British capital in the development of that huge area west of the Rockies before the First World War, when Wall Street was busy accommodating industry in areas less risky and easier to watch. Thus London's attitude was continually oscillating between encouragement and apprehension, the hallmarks of the wise venture capitalist. No doubt Baker saw only too well that Smith was caught up in the heady atmosphere of the Nevada boom, and for Smith to hold back would have been too much to expect when tycoons like Clark, Schwab and Brock were all setting the pace by building railways and buying mining stock themselves. In March 1906, when the Press had expressed doubts about completion of 'Smith's railroad' (as it was frequently called), he followed the example of other tycoons and demonstrated his faith in the railroad's future by giving them the headlines that Smith had purchased 100,000 shares in the mine 'Yankee Girl' which in fact (wisely!) was not a very substantial outlay.[28]

October 1907 saw a financial panic that spread from Wall Street across the United States. It hit Nevada within twenty-four hours, the banks closed their doors, and on Thursday, 24 October, Governor Sparks declared a three-day bank holiday. The following Wednesday the Tonopah and Tidewater Railroad sneaked into Gold Centre, and Smith firmly announced that this was to be the terminus, having made a deal with Brock's Bullfrog Goldfield Railroad (which had now arrived in Gold Centre from the north) to use its tracks north to Beatty and west to Bullfrog and Rhyolite. The mining camps at Greenwater and Rhyolite were soon hit, and 1908 was the beginning of the end of the Nevada desert mining boom.[29] Mining dragged on for a few years before the area becme a network of ghost towns, and although Clark's Las Vegas and Tonopah had arrived in Gold Centre a whole year before Smith's line, the Senator's enterprise was worst hit, and was destined to be the first to close. Indeed, for the next two decades borax-mining was to be the main activity which kept the region alive, although borax alone, owing to the depressed prices prevailing from 1908 to 1914, could not support a railroad on the scale of the Tonopah and Tidewater. Unfortunately for the owners, no miracles occurred to reverse the inevitable result.

Long before completion of the railway, the financial problem of raising capital for other purposes had started to perplex the board of Borax Consolidated. On 22 January 1907 a letter was sent to shareholders explaining that at the time the company was formed ordinary share capital of £600,000 had not been issued to the public but had been divided between the various parties interested in the companies which were consolidated. As the current value of the mines and other assets now far exceeded the valuation of 1899, it was proposed to bring the ordinary capital 'in closer accord with the value of the undertaking'. To do this each £10 Ordinary Share (then valued at £28) would be exchanged for 15 Deferred Ordinary Shares of £1 each and one Preferred 6 per cent Ordinary Share of £5, and 60,000 of these £5 Preferred Ordinary Shares would be kept in the hands of the company, to be issued at any time thereafter as might be deemed expedient.[30]

The directors' revaluation of assets was to be used to increase the issued capital of the company and in the absence of any Companies Act allowing this, an Act of Parliament was needed. In due course, with shareholders' approval, the Borax Consolidated Act (1907) was given Royal Assent on 4 July, but it was not until August 1908 that a Stock Exchange quotation was obtained for these new shares, in the face of what the company's brokers, Emberson and Hughes, described as 'almost insurmountable difficulties'.[31] General depression in Britain, the collapse of the Nevada gold-rush, and rumours about the ill-fated Tonopah and Tidewater Railroad were affecting the market, and in February Hughes advised Baker strongly against the issue of 30,000 of the £5 Preferred Ordinary Shares, and called attention to the company's adequate cash balances to supply immediate needs. However, the shares were issued in early April 1909,[32] just after Baker had left for the United States on a visit to Smith that lasted until mid-May. No record explains why Baker decided to ignore his financial advisers, but a loan due to Smith of $350,000 for a further railroad commitment was under discussion,[33] and this, together with emerging plans to acquire certain competing borax-mines in California and further cash needs in South America, could explain his action.

The Bullfrog and Goldfield Railway Company were soon feeling the impact of the goldfield that never was, and early in 1908 they were 'talking mergers' with the Tonopah and Tidewater Railroad Company as the Bullfrog area contained most of the important passenger traffic, which needed both these lines to connect with Ludlow as well as with the Santa Fe line, which in turn connected them with Los Angeles. Los Angeles, although at this time a relatively small sheep and cattle town, was the closest place of any size in the area, and there was also still the hope of mineral freight to be handled over both railways. The Bullfrog Goldfield was already losing more heavily than the Tonopah and Tidewater were yet showing signs of doing, and both sides appear to have had doubts. However, the main points of a merger which Borax Consolidated accepted in general principle in May 1908 were that Smith should put in $350,000 from his personal account by two equal instalments to pay off the Bullfrog Goldfield over-expenditure and that a holding company should be formed under the name of the 'Tonopah and Tidewater Railway Company' (as opposed to the original 'Railroad Company').[34] The obligations of the Tonopah and Tidewater Railroad, now a subsidiary, were extended to the management of the Bullfrog Goldfield subsidiary which, however – a significant point – remained open to foreclosure by its bond holders, and therefore still open to seizure by the railroad giants outside.

The merger pleased everyone locally, although the idea that the two companies would offset each other's losses with mutual financial benefits was wonderfully optimistic. From Smith downward optimism still prevailed, and London had little choice but to follow its initial stake. Borax Consolidated undertook, of course, to repay the $350,000 to Smith from its own resources. The use of the unissued £75,000 Sterling Bonds of the Railroad was discussed, but was considered too small an amount for a special issue of capital in London especially to finance a railroad to the speculative Nevada goldfields which was a project unrelated to the borax business. In the event the solution of this problem along with other financial worries had to await a major financial reorganization of Borax Consolidated's capital and a new issue of debentures some years later.

Meanwhile, from 1908, certain entries started to appear in the Borax Consolidated minute book with monotonous regularity at six-monthly intervals, being the unpaid interest of the First and Second Mortgage Bonds of the railroad company and also the premium due on the sinking fund policy, making a total of about £40,000 each year. Only on one occasion, in 1913, did the railroad company ever themselves pay the interest due, and there is no record that they ever refunded these 'loans' to Borax Consolidated. Baker pointed out that although the interest was guaranteed by Borax Consolidated, it was quite irregular for that company to go on meeting the Railroad's obligations as though the two concerns were one company; however, these remonstrances went unheeded.[35]

Although the Tonopah and Tidewater Railroad may have created financial gloom in London, life on the railway itself was not without fun. Opinions differed about the efficiency of Wash Cahill's management, but everyone agreed on one thing – he enjoyed a good time, and music and dancing were high on the list. His black cook, Jean Lucas, was chosen for something more important than cooking; he played the piano, and was not restricted by any nonsense about working hours. A description of a Ludlow party around 1910 reads thus:

> The dancing was not confined to employees only; everyone joined in the fun. In fact the T. and T. benefited considerably from the increased passenger revenues as it brought people in from far and wide to share in the amusement. Complications were bound to arise, and did; but solutions to the problems were usually found. When one dancing party was scheduled to start at 9 p.m. and it was discovered that the train was not due to arrive until almost midnight, Cahill was petitioned to try to improve upon the situation. Much to everyone's surprise there was not only an improvement – the train arrived at nine on the dot and the costume dance was a tremendous success. In fact, people were enjoying themselves so greatly it seemed a shame to break up the party, and Cahill was approached again – could not the northbound departure be delayed a bit so that all could have breakfast together?

By 1911 the actual loss during the first three years by the Tonopah and Tidewater Railroad Company had averaged about £19,000 a year, while the loss on the two railroads (i.e., Tonopah and Tidewater and Bullfrog Goldfield together) was about £31,000 a year.[37] By early 1913 the railroad's total debt to Borax Consolidated itself had increased to a level which was starting to worry the directors as to how to explain this item in the published balance sheet.[38] The railroad had obviously become a burden to be shed as soon as a worth-while offer appeared. Indeed, Baker was veering towards the view that if proper terms could be secured it might be wise to let Santa Fe have the railroad, provided a suitable long-term contract could be secured for the carriage of borate.[39] During 1913 the rival Las Vegas and Tonopah gave up its route from Beatty to Goldfield, and by 1918 they had abandoned the rest of it. By the end of 1913 Southern Pacific were also possible candidates to purchase the Tonopah and Tidewater railroad but by this time the situation had again changed.[40] The ore in the Lila C. mine was running out, which meant that a new rail extension had to be built from the railroad at Death Valley Junction through the Funeral Mountains to the proposed Biddy McCarthy mine, and completion of this lifeline became a matter of first priority.

Back in London, with mounting railroad losses, acquisition of further properties in California and a settlement of the Chilcaya dispute in South America, there was already a pressing need for cash, and Baker had decided in 1912 to follow the earlier advice of Frederic J. Benson & Co., a City finance house, and to create a new consolidated Debenture Issue, the proceeds of which would be used to pay off all Debenture Issues outstanding and to provide any additional funds the Company required then and in the near future.[41]

Emberson & Hughes, the stockbrokers, had no objection in principle to the proposed issue, but other circumstances rapidly whittled down Baker's original intention to use the new issue to pay off the debentures of both Borax Consolidated and the Tonopah and Tidewater Railway Company, for which £2·2 million would have been required. Delay in completing the acquisition of the Sterling Borax Company prevented a public issue, and the offer had to be restricted to existing shareholders and limited to £1,150,000 only,[42] which was totally insufficient to cover all the company's real requirements.

Had the issue been larger than this, however, Borax Consolidated would soon have found itself in an even more difficult position than actually developed. In spite of warnings from the City that the investment climate at that time of international tension and labour unrest was not at all favourable, Baker remained reasonably optimistic; but by mid-May 1912 the customary early rush for the new Debentures had not materialized.[43] Only two-thirds had been taken up by the company's shareholders, and it was intended to dribble the balance on to the stock market. Nine months later, in 1913, one-third had still to be placed[44] and late in May Baker wrote to Smith that, although the war situation (i.e., the First Balkan War, the Agadir crisis, etc.) had cleared up in some measure, the financial position in Britain had not altered to any extent.[45] He added that nothing was doing well with less than a 5 per cent yield, and that Smith would realize therefore that the chances for a 4½ per cent stock at that time were not very good.

At this most inappropriate time the problem of Lila C. mine became pressing. After Baker's sudden visit to the USA in May 1913 the board heard that the ore in the mine would last only until July 1914, after which they would have to start mining the Biddy McCarthy claim above Death Valley. This would mean another seventeen miles of railroad through difficult country from Horton (named after the Tonopah and Tidewater roadmaster Ben Horton) on the Lila C. line to the new mine. As Baker sadly summarized it in a letter a few days later:[46] 'I note that Mr Ryan says he hopes quick action will be taken as there is not much time to lose. The main action that can be taken on this side is to find the money which, at the present moment, is not an easy thing to do.' One way could have been to find the £62,000 estimated to be required for the new branch of the railway by issuing part or all of the remaining balance of £75,000 of Tonopah and Tidewater 5 per cent Bonds. However, the California Railroad Commission refused consent to the issue and to the construction of the new line because the liabilities of the Tonopah and Tidewater now exceeded its assets by more than $400,000.[47] An alternative proposal was then made to the Commission to form a separate railroad company, which would then issue bonds to build the extension.[48] Surprisingly enough, they agreed, in mid-January 1914, and the new company was hastily dubbed 'The Death Valley Railroad Company'.[49] The Commission, however, complicated matters by laying down that the money was to be put up in the form of $60,000 in stock and $230,000 in bonds. This meant introducing a bond issue of only about £47,000 on the London market, and at that time of international tension this was considered too small an issue to be viable. In the circumstances, therefore, the cash required for building the Death Valley Railroad had to be voted by the board out of Borax Consolidated's reserves,[50] and the issue of bonds was not made until a year later, after war had broken out in Europe.

The legal effect of this was that the extension could be built only as a private railroad, and could not receive full powers to carry traffic until bonds were issued.[51] Needless to say, little heed was paid to these subtleties on site at Death Valley Junction. Clarence Rasor did a survey in the autumn of 1913 and Ryan started construction on 1 March 1914 with a crew of 325 men and 150 mules. After only nine months, on 1 December 1914, the narrow-gauge track was opened and in full operation.[52] *The Technical World* pronounced it 'the most remarkable railroad in the world, in that it had a grade of no more than 1½ per cent in its entire twenty mile length [from Death Valley Junction to New Ryan] through rugged and tortuous mountain country.' Actually the grade in the last mile was 3½ per cent, which made it necessary to cut the train every time it came in with extra freight, and the sharpest curves were 24 per cent. During the following 10 years this railroad was a vital and most profitable link from the mines in the Funeral Mountains to Death Valley Junction, where the ore was calcined in a roasting plant – also built in 1914 – before being railed to the refineries.

Railway enthusiasts will be pleased that *Francis*, the 3 ft gauge Hesler tank locomotive, which worked on the Borate and Daggett Railroad with its twin *Marion* until it was closed in 1907, came out of storage fit and well and was used during the construction of the Death Valley Railroad until the middle of 1914, when the first of two Consolidation locomotives (2-8-0) arrived from Baldwin's. *Francis* was then sold to the Nevada Short-Line, and finished with the Terry Lumber Company at Round Mountain in California. What happened to *Marion*, the other locomotive, is unknown. Starting at Death Valley Junction, a three-rail track for the first 3·2 miles to Horton enabled the narrow-gauge trains to operate over the broad-gauge Tonopah and Tidewater Railroad tracks on the Lila C. branch line as far as Horton, where the Death Valley track then headed northwest to the mines at 'New' Ryan. The Death Valley railroad completed its journey at an average 15 m.p.h. and offered a daily – except Monday – service. The revised estimate of $350,000 for its cost of construction exceeded the original estimate by $60,000.[53] However, among the general financial problems outlined above, no one seems to have minded much. Emberson, from his sickbed at Vernet-les-Bains, and Hughes – destined shortly for the Royal Naval Reserve – were still trying to arrange the sale of debentures as late as the second half of 1914;[54] although by then it was a hopeless task, as the institutions had little money available for such doubtful luxuries.[55] However, within Borax Consolidated the endless problems caused by shortage of capital had taken second place to those posed by news of almost national importance which broke in the American press on 7 May 1913.

REFERENCES

1. (a) 'Mineral Resources – 1906', in *US Geological Survey – Dept. of the Interior* (Washington, 1907).
   (b) 'The Production of Borax 1911', in the *Dept of Interior US Geological Survey* (Washington, 1912). 'The Lila C. Borax Mine at Ryan, California' by Hoyt S. Gale, p.7.
2. F.M.S., 7.4.99.
3. F.M.S., 21.4.1900.
4. F.M.S., 5.7.02 and 16.12.02.
5. F.M.S., 14.9.03.
6. Apparently this was a novel idea – 'I am sending you herewith a blueprint of our traction outfit which will give you some idea of what it will look like' (F.M.S. to Baker, 16.11.03), also F.M.S., 20.1.04.
7. Minute Book, 30.10.03.
8. F.M.S., 16.11.03.
9. F.M.S., 27.4.04.
10. F.M.S., 2.6.04.
11. Minute Book, 24.6.04.
12. F.M.S., 29.4.04 and 4.7.04.
13. Glasscock, C. B.: *Here's Death Valley* (Grosset and Dunlop, New York, 1940), pp.175–7. Myrick, D. F.: *Railroads of Nevada and Eastern California*, (Berkeley, California, 1963), Vol. II, p.546.
14, 15. Cahill, W.: Manuscript notes sent to Ruth Woodman in February 1949 – U.S. Borax Archives, Los Angeles. Also Glasscock, *op. cit.*, p.176.
16. F.M.S. to R.C.B., 19.12.04.

17. Cahill, W., *op. cit.*
18. Shankland, R.: *Steve Mather of the National Parks*, Knopf (New York, 1951), p.29.
19. Myrick, D. F., *op. cit.* pp.548–56 and Borax House Archives.
20. Cahill, *op. cit.*
21. Ryan to the Hon. Board of County Supervisors of Inyo County, California, 6.9.06.
22. Zabriskie to Ryan, 9.9.05.
23. Zabriskie to Ryan, 3.10.05.
24. Zabriskie to Ryan, 24.1.06.
25. F.M.S., 29.4.04.
26. M.D. to Zabriskie, 5.4.05; also 1st Mortgage Prospectus of Bullfrog and Goldfield Railway.
27. *U.S. Borax 100 Years* (Los Angeles, 1972).
28. Myrick, D. F., *op. cit.*, Vol. II, p.467.
29. Myrick, D. F., *op. cit.*, Vol. II, pp. 489–94; Woodman, Ruth: *The Story of Pacific Coast Borax Co.*, Division of Borax Consolidated, Ltd. (Los Angeles, 1951), p.36.
30. Report of AGM of 28.2.08 and letter from Stock Exchange, 11.4.05; also Minute Book and R.C.B. from Emberson of 13.8.08, 21 and 25.9.08, 21.1.09, 12.2.09, and 12.7.09.
31. R.C.B. from Emberson, 13.8.08.
32. R.C.B. from Hughes, 12.2.09.
33. M.D. to Zabriskie, 26.3.08.
34. Minute Book, 13.5.08.
35. M.D., 20 and 22.10.08.
36. Myrick, David F., *op. cit.*, p.585.
37. M.D., 8.9.11.
38. M.D., 14.1.13.
39. M.D., 29.7.13.
40. M.D., 18.11.13.
41. M.D., 1.2.12 and R.C.B. from Benson 11.6.09.
42. Minute Book.
43. M.D., 13.5.12.
44. M.D. to Chairman, 11.12.13.
45. M.D. to F.M.S., 21.5.13.
46. M.D., 29.7.13.
47. M.D. to Zabriskie, 10.12.13.
48. M.D. to Zabriskie, 19.12.13.
49. M.D., 17.1.14.
50. Minute Book 20.1.14, 3.3.14 and 21.4.14.
51. M.D., 16.12.13.
52. Woodman, Ruth: *The Story of Pacific Coast Borax Co.* (1951), p.39.
53. M.D., 17.1.14 and 1.7.14.
54. R.C.B. from Emberson, 31.1.12 and 3.9.14.
55. R.C.B. from Hughes (Emberson's partner), 30.12.14 etc.

12
The Fall of F. M. Smith

In spite of all the financial problems that worried its small board, Borax Consolidated had been producing a steady level of gross trading profits, varying between £200,000 and £300,000 every year, since 1900, and in 1913, mainly as a result of the price increase of 1912, these profits jumped to £358,909, a level which they maintained in 1914 and 1915 before increasing again in 1916. Although this end-result never appeared unduly exciting in the pre-war years, it is obvious that those shareholders who took an interest in the company's progress were content with the conservative policy of this solidly respectable company, if the tone of speeches at Annual General Meetings is ever anything to judge by.

It must therefore have struck many of these shareholders as incredible when, early in May 1913, the following announcement appeared in the San Francisco and Oakland papers on behalf of a committee of five presidents of well-known banking and trust companies.[1]

> We wish to advise you that Mr F. M. Smith, by a trust agreement dated May 5 1913, conveyed to the Mercantile Trust Company of San Francisco all of his properties, consisting of stocks and bonds of eastern and California corporations and particularly the stock of Pacific Coast Borax Company, the Realty Syndicate, and the corporations holding the properties formerly owned by the Oakland Traction Company and the 'Key Route System'.

It would be outside the scope of this book to go into detailed discussion of the extent of Smith's interests at this time and of the reasons why he crashed, particularly as a quantity of contradictory information can be traced about the alleged facts. Briefly, he had invested extensively in real property and suburban transport systems in and around San Francisco, as well as in other smaller adventures elsewhere, and, as the official announcement quoted above went on to say: 'Mr Smith has for some years past financed his very large holdings upon a temporary basis only by the issuance of secured notes upon short-term maturities'. The *San Francisco Chronicle* put it more strongly: 'It was Smith's "endless chain" idea of high finance – issuing notes to pay notes – that is responsible for all his troubles. It was a sort of wheel of fortune, which the Borax King, like other sanguine miners, hoped might some day stop at a lucky number.'

As regards the actual extent of the crash, a letter dated 3 June 1913 can probably be trusted for accuracy, as it was from a Mr Brunner of the Paris branch of the Banque Franco-Américaine of New York to a Mr Stedman, of Bloomfield Street in the City of London, and states: 'Lately, Smith, the "Borax King", has got in financial troubles and his whole estate is in the hands of five trustees, subject to debts amounting to about $12,500,000. The value of the assets is probably not less than $30,000,000, but far from being liquid.' The value of the dollar was at that time something just under five to the pound sterling.

On a lighter note, and at a time when the newspapers of San Francisco and Oakland were reporting daily on the collapse of Smith's empire, an American consular invoice arrived for Smith originating from Federico Lesser in Antofagasta and covering the shipment of 'one crate with two live vicunas – value one hundred pesos – to Pacific Coast Borax Co. – San Francisco de California'.

That Smith, on the verge of financial disaster, still had the urge to win, especially when impeded by bureaucracy, is shown by his cable to his congressman friend, the Hon. Joseph R. Knowland in Washington D.C., as follows: 'Two vicuna arrived San Francisco for me via Kosmos line steamer August twelfth from Antofagasta, Chile. Kosmos line advises me United States officials refuse their landing on account of absence Chilian health certificate. They suggest government's permission to land for health examination here. Animals are a variety of deer, [!] interesting specimens, delicate and valuable. Can you help me. Must be settled today.' Sadly, a note on the file says 'Government would not allow vicuna to land, hence they were killed by the Kosmos S.S.C. instead of shipping back to Chile.' Things were clearly not going at all well for Smith at that time.

The immediate concern of the board and shareholders of Borax Consolidated in May 1913 was of course to know how deeply the personal financial troubles of its founder, majority shareholder and joint managing director had involved Borax Consolidated itself, apart from Smith's known interest in it, and to stop that known interest – which could still confer control of the company – from being taken over en bloc by an unwelcome outsider.

The first step was an immediate visit to the USA by Baker as soon as the news was made official. This can be regarded as the culmination of a long exercise that Baker had been conducting on behalf of the board ever since he first became acquainted with Smith's volatile character in 1896. It is in fact a tribute to Baker's tact and 'eternal vigilance' that when May 1913 came Smith was found to have been prevented so effectively, and over all those years, from causing financial damage to Borax Consolidated itself with his endless flow of bright but risky ideas. In fact, the only real – but comparatively light – damage caused was the financial effect of two of Smith's schemes, the form in which the Tonopah and Tidewater railroad materialized and the excessive advertising of package borax and soap. If he had been allowed his own way there might have been a lot more damage.

There was of course no objection to the frequency with which Smith encouraged Baker to 'make a market' for Borax Consolidated shares and debentures on the London Stock Exchange[2] so that he might get rid of some of his enormous holding at the highest obtainable market price. As he had written as early as 1898, in the time of the predecessor company: 'I have been considerably disappointed of late that there has been no market for these preferred shares as I have abundant use for funds personally.'[3]

Nevertheless, when it came to borrowing money from the American side of the business without first consulting the board, and in anticipation of dividend distributions not yet declared, he started to cross into illegal territory. 'I have felt at liberty', he wrote at the end of 1899,[4] 'to discount or anticipate dividends here to the extent of profits earned,' and proceeded to give a page of justification, including the admission that the funds had in part been loaned to companies in which he was interested. No direct protest by the board or the auditors is traceable; but probably there was trouble, as there is plenty of evidence that they soon realized that they had to protest, and protest several times, over unauthorized 'investment' of the company's reserve funds. Perhaps Smith foresaw trouble when he appended a note to the draft accounts for the first six months of 1899:[5]

> Among the earnings you will note that the interest account is quite large. It is through force of circumstances that I was able to make so good a showing, using money largely myself, in connection with the Realty Syndicate business, and being able to place it at a favourable rate 'east' . . . and outside of myself and people with whom I am directly associated, I did not place any money, except with banking concerns.

He then asked for a covering resolution from the board to meet this and similar transactions in future. From the tone of the long board minute[6] in which Baker is recorded as pointing out that 'the course is not regular', it is obvious that the board did not like the suggestion and resolved to take legal advice.

Early in 1904 a similar situation occurred.[7] The American accounts showed a debt of £84,916 due from Smith to the company. The auditors wanted an explanation, and Smith was asked to send over monthly statements of investments and advances made 'on behalf of the company' in future.[8] A statement was duly received,[9] and showed various loans secured on the shares of the Realty Syndicate, the ferry company and so forth. Smith again asked for *ex post facto* confirmation by the board of the loans thus made, and the Board sought counsel's opinion about this procedure.[10] What counsel's opinion was is not traceable, but presumably it was adverse, as de Friese wrote to Baker[11] suggesting that he – de Friese – put the matter to Smith as a personal friend, and that he could explain the entire matter to him in such a way as to remove as far as possible any annoyance he might feel. Smith took over two months to reply,[12] but when he eventually did so it was in an outwardly friendly manner. However, beneath the surface banalities the purport of his reply was that investment in the USA was more profitable than investment in Britain, and that the American stockholders had cheerfully accepted his personal guarantee of all loans that he had made, without asking that the directors' approval should first be obtained.

There the matter appears to have rested. But although Smith may have paid a little more attention thereafter to the strange susceptibilities of British directors and their auditors[13] – as when he reported details of 250,000 dollars invested in 1907[14] – he is found arguing against Baker on two occasions for a higher dividend to be declared,[15] and generally contemptuous of another strange British practice, that of providing depreciation against profits of the year.[16] Apart from these disagreements about what constituted normal routine, Smith made several proposals in these years to alter the general constitution and purposes of the company, which appear dangerous now in the light of the knowledge of what actually happened to his fortunes in 1913.

It has already been related how, on 4 June 1900, Smith wrote a personal letter to Baker[17] in which he put forward proposals to merge his 'interests' with those of the American company's into a company to be placed on the New York market, and how Baker managed to dissuade Smith from this 'passing thought'. Indeed, the idea of merging Smith's Realty Syndicate and other scattered interests with the sound borax business, and in effect shifting the borax headquarters from London to New York – so that Smith could use the borax part as 'collateral' for raising funds for further speculation in other directions – must have given the cautious Baker some uneasy moments.

Smith kept on chewing over this idea of an American-based company for nearly two more years. By July 1901 he had modified it to mean 'simply a company organized that will deal in such blocks of stocks as to enable it to control the borax companies or borax interests rather'.[18] By the end of the year, however, something revived his enthusiasm for the original idea,[19] and he wondered whether it would not be wise to move on lines similar to the Kodak Company – i.e., to move the headquarters to New York and organize an American company to take the place of the English company. The European interests could be handled from New York through agencies and representation, and in fact the whole business could be handled from America. Less than two months later, however, Smith suddenly dropped the whole idea.[20] He wrote that he saw too many obstacles in the way of attaining this result. It would take time and cost money, directly and indirectly, to educate the American investors in the merits of his borax group.

Mention of Smith's other interests leads to discussion of the second area in which Smith might have entwined Borax Consolidated in his own fortunes and eventual disaster. He did not offer participation in 'the Philippine (Mindanao) Project' to acquire 'sugar lands' and other interests in those remote American territories,[21] and offered only to Baker, on a personal basis, shares in the Realty Syndicate and the Oakland Traction Company.[22] Smith's interest in the Pyne copper smelter at Alameda, which produced copper and sulphuric acid from pyrites,[23] and his ideas about diversification into other chemicals,[24] have already been mentioned in Chapter 10 (page 117). The Pyne smelter was finally closed in 1904 by a combination of financial losses and environmental opposition from the local vegetable growers, but before this happened Smith had been trying to sell it to Borax Consolidated, together with the idea of further involvement in copper.[25] Fortunately the Board had not found it difficult to say no to their principal shareholder.

This then was not a favourable background for his next proposition to the board,[26] which was that as they had about them and could secure 'conservative mining men', why not 'reach out for anything that might be going instead of restricting ourselves to smelting propositions here on the Bay?' But the board had no great difficulty in fending off this proposition also, and they fell back on the now time-honoured dodge of pleading that the objects stated in the Memorandum of Association of the company would not cover investment of the funds of the company in enterprises other than borate-mining or propositions closely akin thereto.[27] Probably much to their surprise, they had to use the same argument only just over a year later when Smith, nothing daunted, came up with a proposal to acquire an interest in another copper proposition, this time in Panama.[28] This appears to have been his last attempt to draw Borax Consolidated into copper-production. Furthermore, in spite of all the money that Smith had quite obviously been investing in or lending to his other companies over more than thirteen years, neither the Balance Sheets, the Annual Reports, nor any other

Borax Consolidated document, shows any sign of writing off of specific investments after Smith's crash.

A possible explanation lies in the fact that although Smith placed his affairs in the hands of a committee of San Francisco bankers early in May 1913, the letter containing his resignation from all his appointments within the Borax group is dated 17 July 1914,[22] which was over fourteen months later, and regular correspondence with him about the conduct of the company's day-to-day affairs is traceable during this period. The figures already quoted show that he did not go officially bankrupt, as some have assumed, in May 1913, but that his assets, as ascertained at that time, still exceeded his liabilities.

Certainly, Baker put this interval to very good use. The danger, of course, was that the block of shares held directly or indirectly by Smith in Borax Consolidated would pass to a rival. Indeed, in this connection it is interesting to read that De Guigne, the rich man in the Stauffer set-up, called on Baker on 24 October 1913 and was anxious to know if the F. M. Smith holdings in borax were being disposed of.[30] Baker headed him off by referring him to 'Anderson or some other banker' – i.e., to one of the members of Smith's committee. Surprisingly enough, however, no letter from Baker to Frank Anderson can be traced before 21 April 1914, and this letter discloses that Smith had been trying to raise money to save his affairs by a note issue on the security of his Borax Consolidated shares. The negotiations had fallen through, and he had then come to London to try out the same manœuvre, in spite of the unfavourable economic conditions of the time. Obviously he failed, and by mid-May 1914 Frank Anderson, president of the Bank of California and chairman of the committee, was taking steps to realize the proceeds of Smith's Borax Consolidated shares.[31] Luckily for Borax Consolidated, the City firm of George Clare & Co. made an offer at the end of May to take over the deferred shares 'to redistribute to the public.'[32] Both Anderson and Baker were quick to encourage this offer and to say that they would be delighted to co-operate by encouraging existing shareholders and debenture holders to take them up. By 17 July 1914 the shares were practically all placed with the public,[33] including Anderson, chairman of Pfizer & Co., and other industrial customers of the company, so that all seemed set for a happy ending to the drama.

Unfortunately, the ending turned sour. Smith – now awake to reality – saw, after years of chasing less substantial phantoms, the value of his first real love, borax, and was willing to let all the rest go to save his borax shares.[34] He started legal proceedings to this end, and although he did not succeed, he nourished for the remaining seventeen years of his life a grudge against his committee and his old colleagues for removing his beloved borax interests from him.

By a strange and of course entirely unpremeditated coincidence, all the procedure for severing Smith and his interests from Borax Consolidated was completed just a few days before the First World War broke out on 4 August 1914. The USA was not involved actively, of course, until 1917; but the commercial balance of the world altered so much in its favour and away from Europe that it would be an interesting speculation to meditate on what advantage Smith might have taken of the war situation if he had not gone virtually bankrupt just before it arose.

It is not difficult to be hard on Smith, but it must always be remembered that without him the Borax Consolidated group would not have come into existence, and the worldwide and still growing borax industry would have been considerably the poorer. Moreover – and regardless of whether this was more due to Baker's vigilance than to Smith's own actions – he did not involve the Borax group in his fall, and the outside enterprises on which he broke himself – housing, transport and so on – were all in the category of endeavours useful to humanity rather than candy-floss, or wine, women and song. He may have lived initially by washing dishes and selling firewood, but he was not without education. His father and mother were strict Methodists, and he was brought up with the discipline of farm life at Richmond, Wisconsin. He followed his brothers to Milton Academy, and later, because he needed more discipline, he was moved to Allens Grove Academy. Like many of his type, Smith boasted that he learnt nothing at either place 'except a little knowledge of card games', but when he headed west from Wisconsin at the age of twenty-one he at least started out in style with a train ticket to Denver, a food-supply and a Bible.

Smith's yachting endeavours reached their peak in 1906, when his sloop *Effort* won the King Edward VII Cup off Newport, Rhode Island. His messages of congratulations covered a broad canvas, including Kind Edward VII, Charles Schwab, the steel millionaire, and, from California, Henry Blumenberg. Smith's first wife died in that year of 1906, and his second wife was Evelyn Ellis, his faithful secretary. Both ladies are described by those who met them as kind and lovable persons, and Evelyn was undoubtedly a woman of character and business ability, who went through the traumatic experiences which befell her husband with a loyalty and cheerful common sense that must have been a great support to him in the days that followed. Obviously, in spite of his drive and his ruthlessness with commercial opponents, he was a kind and humane character outside corporation rivalries. Clearly also he possessed those gifts not common among business magnates – a tremendous sense of humour and the capacity to take himself and other human beings not too seriously.

It is therefore appropriate to end this chapter with a brief picture of the Smith family at leisure, taken from a letter of his written on 3 July 1903, from his summer abode on Shelter Island near New York, during his heyday with the Pacific Coast Borax Company:

> We shall all be glad to have Mr & Mrs de Friese visit us. We enjoyed their visit very much before, and I think we will again. I was a little surprised to hear that Browne is coming over. We will be glad to have them out here with us, although I am afraid our shirt waists at dinner and tennis with our bathing suits on, may upset them a little, and give them the impression that we are not only natives of the wild and woolly West but crude and unpolished in our tastes. However I will make amends for all this when I get them aboard of the new steam yacht Hauoli, by showing them a little of the formalities of life, while spinning over the waters of the Sound; for where we are very informal at Presdeleau and break loose sometimes like wild Indians, we atone for it in a way by slight excess of formality on the yacht. I expect the new yacht will be in commission in a week or two, and I believe she will be a very creditable craft.
>
> Tomorrow is the Glorious Fourth with us. We have quite a house party on hand, and are expecting Zabriskie and another gentleman up tonight. Have already five other gentlemen in the house besides myself. The weather has become seasonable, and the gentlemen members of the family have taken kindly once more to the skin bath. I thought of you this morning before breakfast as our representative party lined up at the end of the lock, with their pink tights on, and took the regulation plunge.
>
> With kind regards, I am, Sincerely yours, F. M. Smith.

(Above) The refinery of Borax Consolidated Ltd at Belvedere, Kent (in operation since 1896) and (below) that of La Productora de Borax y Articulos, near Barcelona, Spain (in operation since 1922)

The cover and a page from the Tonopah and Tidewater Railroad timetable

FIRST DISTRICT

9 MIXED LEAVE DAILY	DISTANCE FROM LUDLOW	Time Table No. 11 July 1, 1910 STATIONS		DISTANCE FROM GOLDFIELD	10 MIXED ARRIVE DAILY
9.20 PM	.0	LUDLOW	N	242.69	2.30 AM
f 9.46	12.68	12.68 BROADWELL		230.11	f 1.58
f 10.07	21.49	8.81 MESQUITE		221.20	f 1.36
s 10.19	25.68	4.19 S. P. L. A. & S. L. XING CRUCERO		217.01	s 1.26
s 10.30	29.38	3.70 RASOR	D	213.31	s 1.15
f 10.40	33.38	4.00 SODA		209.31	f 1.05
f 11.00	41.82	8.44 BAKER		200.87	f 12.46
s 11.22	49.50	7.68 SILVER LAKE	D	193.19	s 12.26
f 11.43	59.00	9.50 RIGGS		183.69	f 12.04 AM
f 11.54[10]	64.35	5.35 VALJEAN		178.34	f 11.54[9]
f 12.14 AM	73.60	9.25 DUMONT		169.09	f 11.35
f 12.25	78.33	4.73 SPERRY		164.36	f 11.25
f 12.35	82.49	4.16 MORRISON		160.20	f 11.16
s 12.46	87.14	4.65 TECOPA	D	155.55	s 11.06
f 12.57	91.23	4.09 ZABRISKIE		151.46	f 10.56
f 1.09	96.44	5.21 SHOSHONE		146.25	f 10.46
f 1.31	106.00	9.56 JAY		136.69	f 10.25
f 1.38	109.11	3.11 EVELYN		133.58	f 10.18
2.02 AM ARRIVE DAILY	121.41	12.30 DEATH VALLEY JCT.	N	121.28	9.54 PM LEAVE DAILY
25.83		Average Miles per Hour			26.41

No. 1 locomotive on the Tonopah and Tidewater Railroad. Scrapped 1941

Above: Bullfrog Goldfield Railroad. The morning passenger train about to head south

A Bullfrog Goldfield Railroad share certificate. F. M. Jenifer was the General Manager of the Tonopah and Tidewater and Bullfrog Goldfield Railroads

Oakland Examiner

VOL. XCVIII. WEDNESDAY SAN FRANCISCO, MAY 7, 1913.—TWENTY-FOUR PAGES. WEDNESDAY No. 127.

F. M. SMITH'S FINANCIAL INTERESTS UNDER CONTROL OF TRUSTEE

Experts Appointed to Manage the $200,000,000 United Properties and Private Affai

BORAX KING AND FINANCIAL ASSOCIATES
F. M. Smith, the Capitalist.

TWO BOARDS SELECTED TO REORGANIZE BORAX KING'S VAST HOLDIN

Personal Interests in Realty Syndicate, C
land Traction, Pacific Coast Borax
Key Route Properties Will Be Mana
by a Committee of Five Financ

WILL CONGEAL UNWIELDY FORC

Second Committee of Five Will Take C
Management of United Properties and
Two Will Work Together to Create
mony in Great Association of Enterpri

FRANK M. SMITH, known far and wide as
Borax King," master of millions and leading s
in some of the largest financial undertak
recently effected in California, has opened a new ch
in the history which makes his career stand out ar
those of the most picturesque and successful of ch
made Western millionaires.
Mr. Smith after prolonged consultation with hi

The dramatic news of Smith's financial collapse

Smith decides to start all over again . . .

TO PROSPECTORS
BORAX MINE WANTED

You find it. I will find the buyer. Send me samples.
I will test them and report without cost to you. I will
assay your gold, silver or copper ores also without
charge IF YOU ARE A PROSPECTOR. Send me a
sample of any unfamiliar rock you are unable to classify,
if you think from your experience it might have value.

F. M. SMITH,
SYNDICATE BDG.,   OAKLAND, CAL.

REFERENCES

1. San Francisco and Oakland Newspapers, 7 May 1913.
2. F.M.S., 5.2.1900, 21.3.1900, 15.1.02.
3. F.M.S., 7.7.1898.
4. F.M.S., 28.11.1899.
5. F.M.S., 29.11.1899.
6. Minute Book, 23.2.1900.
7. Minute Book, 29.1.04.
8. F.M.S., 24.2.04 and 1.3.04.
9. Minute Book, 18.3.04.
10. Minute Book, 27.5.04.
11. R.C.B. from de Friese, 18.5.04.
12. F.M.S., 1.8.04.
13. F.M.S., 24.2.04.
14. Minute Book, 25.1.07.
15. F.M.S. to R.C.B., 6.1.01 and 9.1.05.
16. F.M.S., 6.1.01.
17. F.M.S., 14.6.1900.
18. F.M.S., 29.7.01.
19. F.M.S., 7.12.01.
20. F.M.S. to R.C.B., 25.1.02.
21. F.M.S., 7.2.01.
22. F.M.S., 1.8.01.
23. F.M.S., 31.3.02.
24. F.M.S., 9.1.03 and 8.5.01.
25. F.M.S., 19.5.03.
26. F.M.S. to R.C.B., 5.1.04.
27. Minute Book, 5.2.04.
28. Minute Book, 26.5.05.
29. Minute Book, 28.7.14.
30. M.D., 24.10.13.
31. M.D. to Deloitte Plender, Griffiths, 14.5.14.
32. M.D. to George Clare & Co., 26.5.14.
33. M.D., 17.7.14.
34. M.D. to Sir Frank Crisp, 5.6.14.

13

The First World War

In generalizing about the effects on the borax industry of the outbreak of the European war in 1914 it must be borne in mind that the industry, at that time consisted of a colossus, Borax Consolidated, surrounded by a few relatively small rival activities in North and South America and Britain, Lardarello in Italy, and a number of substantial refiners on the Continent, most of which were within the territory of the enemy. From the correspondence of the time it is arguable that for some months after the fatal day, 4 August 1914, the directors of Borax Consolidated in London were more worried about Smith than about the Germans. If this were so they were no exception to the general run of business-men at the time, who remained preoccupied with the attitudes and worries of peacetime. It was different at the beginning of the Second World War, when the devastating experience of the first great conflict was still comparatively fresh in the minds of many. In 1914 there was no such comparison. The last great war to reach the other side of the English Channel had ended almost a century before with the removal of Napoleon to St Helena, and, since then there had been only the remote Crimean War, the Boer War, and remote colonial ventures to ruffle the surface of Victorian peace and prosperity. The big fear still was business rivalry, and to Borax Consolidated F. M. Smith, who had played such a dominant role for a quarter of a century, was now at large as a potential rival – openly so from 1 October when he resigned officially from the presidency of the Pacific Coast Borax Company of Nevada.

Smith soon made it clear that he intended to become a very active rival. The committee of five San Francisco bankers which had taken over control of his affairs in 1913 had already disposed of the majority of his 160,000 deferred ordinary shares in Borax Consolidated to the public by 1914,[1] but he kept repeating that this action was taken against his express wishes, and that he intended to get back into the borax business as soon as he could find a mine. Baker did his best to be sympathetic with Smith in the misfortunes that the latter had brought on himself, but at last had no option but to write to Smith in no uncertain terms just before the first Christmas of the war:[2]

I have seen the Mining and Scientific press of November 21st last with an advertisement from you headed 'To Prospectors – Borax Mine Wanted.' As your co-adjutor and old friend during all these years I am distressed that you have adopted this course and sincerely hope that it has been done without mature consideration and that, on reflection, you will recede from it.

A very large number of people have invested their money in the business of Borax Consolidated, relying on the honour and integrity of those associated with and managing the Company. The fact that your Deferred shares have been sold by those managing your affairs does not alter the situation – the sale has been made for your benefit and those who have bought the shares are especially entitled to your consideration and protection and yet, having received their money for the benefit of your estate, you apparently are endeavouring to find the means of operating against their interests. I do not overlook the fact that the shares were sold against your wish but this does not alter the fact that your estate received their money. . . . I say nothing as to the consideration which is also due to myself and those who loyally worked with you for the benefit of the Company during the past fifteen years. My opinion is that you are in honour bound not to do anything that could injure those who have invested in the Company or who have been associated with you in its management. I hope that you will see things in this light on reconsideration. I feel sure that any other course could only bring regret and loss of the esteem of those whose regard is worth having.

With my best wishes . . .

But although Smith had played tennis at Shelter Island while wearing his bathing suit, he had certainly not played cricket like gentleman Baker and was determined not to 'see things in this light'.[3] His reply to Baker is not traceable, but Baker had to write to him again at length:[4] 'I have received your letter of January 15th and am very sorry that you do not see eye to eye with me in this matter of your going into the Borax business. . . .' He went on to develop his previous arguments, but there is no evidence that Smith even replied this time, and certainly no indication that he ever gave a moment's thought to turning away from a collision course with his old company. For some months more, however, Baker continued to hope that old friendships would triumph in the long run, and, at the end of December he wrote to Zabriskie:[5] 'I see that you are not hearing anything from Mr. Smith nowadays. He is of very queer disposition. However, when I come out I hope you will be able to arrange to go West with me when we can look him up and possibly remove any nasty feelings he may entertain towards us.' Smith's feelings, however, tended to get nastier. Also, now that Baker had become managing director on both sides of the Atlantic, various quaint facts started to emerge about the previous regime in the USA. Extracts from letters read: 'F.M.S. still owes the company $11,939 on the books. . . . He should not, in any case, as Managing Director, have had a debit balance with the Company';[6] 'Mr. Dumont seems to have entered the Company's service when he was over 70 years of age and appears to have continued in this service for 9 or 10 months during which time – as might be expected from a man of his age – he had two or three minor accidents and then was laid off by Mr. Smith with a pension of a dollar a day';[7] and 'it appears that he [Mr Smith] wishes to put in an expense account since the [San Francisco] fire in 1906 for $500 a year which will be $4,000. I shall write a reply to this that . . . it is a most unusual proceeding for a manager or director to put in a claim of this kind after so many years.'[8]

Very soon Smith started trying to entice away Pacific Coast Borax staff; John Ryan and Tom Cramer, their refinery superintendent at Alameda, both turned down offers. Cramer said Smith took his refusal in good part, and with a laugh he told him he did the best assays on borate in the West, and that he would continue to send him samples. However, Fred Corkill senior, who had now retired from Pacific Coast Borax and was in receipt of a pension, went to work for Smith.[9] Baker wrote that Corkill was conversant with every borate deposit that Borax Consolidated owned, and that the dangers of his prospecting for Smith were therefore obvious. Corkill found nothing during his first two prospecting trips on Smith's behalf, but 1916 saw Smith undaunted and, by his persistent attendance at Washington, being admitted to one of the new Federal leases at Searles Lake.[10] This was the first successful step in Smith's rehabilitation of himself in the borax world, and he proceeded to form the West End Chemical Company to run this Searles Lake enterprise. The evidence about how he managed to finance this bold move is scanty, but he seems to have carried it off by forming the new company as a subsidiary of the West End Consolidated Mining Company which he had organized back in the boom days of Tonopah and which continued to mine silver. The stock of that mining company was one of the few things which had survived Smith's financial débâcle, probably because it was held in the name of other members of the family.

Meanwhile, although to the borax industry 1914 might mean primarily the resignation of F. M. Smith, to the rest of the world it meant something even more devastating, and it was not long before the all-embracing effects of twentieth-century warfare on industry began to appear.

The uses of borax and boric acid in wartime, and the effects of wartime interruptions of the supply of the raw material for their production, were summarized by Baker in a letter to Sir Joseph Mackay of the Transport Department of the Government in March 1917, when shipowners were refusing to provide space for shipments of borate from the USA and South America.[11] He began by protesting that this constituted a serious danger for many industries in Britain which depended on Borax Consolidated for borax and boric acid. The enamelled iron, optical and fine glass trades, leather-makers and the pottery industry were particular cases. The Red Cross and hospitals needed boric acid for use in the manufacture of lint and dressings, while borax and boric acid were also principal ingredients in food-preservatives supplied to contractors to the British authorities for foodstuffs, such as butter, margarine, bacon, etc. No other sources of supply were open to them, as the other refiners of borate of lime in the country depended on Borax Consolidated for their raw material. A fortnight later Baker replied to a query from Sir Percy Bates[12] (who was director of commerical services in the Ministry of Shipping), setting out the amounts of borax supplied each year to various classes of consumers at this period. Surprisingly, Medical heads the list at 2,200 tons. Potteries were taking 1,600 tons, including 'amounts to other refiners'; Metal Trades 1,500 tons; Explosives 500 tons; Fine Glass makers 500 tons, but 'rapidly increasing'; and exports to Russia, Japan, China, India, South Africa, Australia and neutral countries were taking 10,000 tons.

War with Germany and Austria automatically put an end to relations with the German refiners, who had played such a big part in borax commerce hitherto, and removed the small factory at Vienna from British control. In the first few days of the War the German right hook through Belgium towards the Channel ports put the factory at Loth into enemy territory and nearly reached Coudekerque,[14] which had to be closed, and which could not be reopened until the battle of the Marne in September 1914 pushed the front back farther north.

Even then, owing to general disruption in these rear areas of the war front and wartime shortage of labour, Coudekerque could operate only intermittently for the rest of the War.

In November 1914 the Turks went to war on the side of the Germans and the supply of calcium borate from Borax Consolidated's Sultan Tchair mines was cut off; this calcium borate had been of great importance to the European industry in these earlier years, principally as the raw material for making boric acid. In particular, it was used in considerable quantities at Coudekerque and at Townsend's old works in Glasgow, and in smaller quantities it was exported to the American works at Bayonne. Now that supplies were cut off the solution was adopted of switching to the colemanite produced by the United States mines.[14] However, after early 1915 the situation worsened in several directions simultaneously. While the demand for boric acid increased enormously – including, for example, an unusual demand from the Red Cross in Russia[15] – the transport for moving the raw material, colemanite, diminished. Apart from there being insufficient American colemanite to satisfy regular customers in the USA as well as the extra customers in Britain and Europe,[16] there was also a large wartime demand for sulphuric acid for munitions and other wartime needs, and supplies in Britain were insufficient to keep going the remainder of industry, including boric-acid manufacture, on a pre-war scale.[17]

An attempt was made to overcome the raw-material shortage by using South American borate of lime;[18] but this started with the initial problem that the Chilean government had put a heavy export duty on this material,[19] which in spite of the representations of Lord Chichester (chairman of Borax Consolidated, now working at the War Office) via the Foreign Office, it failed to remove.[20] Above all, the shortage of shipping, caused by wartime priorities and the U-boat menace, sent freight rates up to prohibitive levels,[21] both from the USA and from South American ports. By February 1916 the rate that mattered most – from the Gulf Ports of the USA to Europe – had risen dramatically, and Baker wrote that the market would not bear this extra charge at the current price of borax.[22]

Naturally, policy shifted away from shipping raw material towards saving freight by turning that raw material into the finished article, boric acid, before shipping it.[23] As a result, a tremendous strain was now placed on the productive capacity of Bayonne works in New Jersey,[24] a factory always prone to labour troubles spreading from the Standard Oil works next door, with their much higher rates of pay. The Allies had to be kept going as well as British industry, although from the European point of view soaring freight rates and the high cost of production at Bayonne made it an expensive business to import boric acid from the States.[25] The shipping position got steadily worse, and ran down rapidly after the entry of the Americans into the War in April 1917, owing to diversion of transport to troops and supplies. A final blow came on 27 March 1918, when the Ministry of Supply announced that all shipments from the USA, whether of crude or refined borax products, would cease with effect from the end of the month. No help could be expected from the Italian boric-acid producers in Tuscany as relations with them had verged on the frigid for several years, and early in April 1915 the Italian government had banned the export of boric acid.[26] During the last twelve months of the War, therefore, this crisis situation for the borax industry in Britain boiled down to a continuous and unsuccessful argument with the Ministry of Shipping,[27] aimed at increasing the shipping space allowed for the carriage of borate of lime from Chile to Britain.

If shipping was the main worry of British borax concerns during the First World War, labour problems certainly ran them a close second. A generosity uncommon in those times was displayed by the promise of Borax Consolidated to take back at the end of the War all those who went off to serve in the Forces, and to pay them the difference between Forces pay and their former company pay if they were married, or half of this in the case of bachelors.[28] Baker's right-hand man and close friend James Gerstley joined up, although he was well on in his forties. After a dull start in the rear lines he volunteered for the Intelligence service, as he spoke fluent French and German. Instead he found himself posted to No. 34 Chinese Labour Corps in France, and finished as Captain Gerstley.[29]

Of the fate of most of them regrettably little is on record, but their colleagues who stayed on the factory floor, or who replaced those who had gone, gave their managements a difficult time from 1915 onward. This was in fact the beginning of a phenomenon new to managers at all levels, which was to become a permanent problem after the War. The unions of the time were still in their infancy, but all the same, proved more mature and adroit in putting on the pressure than did the employers. In Britain the men at the small Midlands works of Borax Consolidated at Kidsgrove put in a succession of claims[30] for increased basic wages and a war bonus to meet the rising cost of living, but after several concessions Baker wrote to the manager at Kidsgrove:[31] 'I am very sorry indeed to see that the workmen can be so unpatriotic as to refuse to fall in with the urgent request of the Government and also of Sir Douglas Haig, commanding at the front, that holidays should be postponed. . . .' In fact, like other employers at the time, he soon learnt that appeals to patriotism were unavailing in the face of living conditions in working-class streets in the middle of the First World War, with bereavement, deprivation and shortages added to the heady atmosphere that the prelude to the Russian Revolution was bringing to the discontented. By September 1917, after still further pay claims and the resultant closure of the cooperage section, male labour was drifting away from Kidsgrove works, and output was being reduced. Female labour had already made good some of the shortage, but it presented obvious risks. 'I understand from you on the telephone yesterday,' wrote Baker,[32] 'that you found him [a foreman of long service] kissing one of the women, which is quite a different thing from the position as he puts it. I understand that when he was Sunday-school teaching there was a similar complaint against him. . . .' He lost his job in both cases.

There were no women at the Belvedere works of Borax Consolidated on the Thames, as they refused to walk so far from their homes, and the quality of the male labour was very poor. This did not matter so much, as Belvedere was not producing the much-needed boric acid in those days.[33] It had its fill of labour problems, like Kidsgrove, and, by April 1918 there were, including the foreman, only ten men and one boy still in the place;[34] all that was hoped for was to keep the preservative and phosphate departments in production.

Over the ocean at Bayonne Works there were similar troubles, but in forms about as different as is American football from the British game.[35] There were riots at the Standard Oil works at Bayonne in July 1915, and the National Guard was called out. As a result there were fears that violence might spread to the adjacent borax works, which, however, got away with a demand for a 15 per cent wage-increase followed by a strike. By mid-August this had been settled on the basis of a 10 per cent wage increase. The Standard Oil men then started campaigning for an 8-hour day for the same wages as the existing 14-hour day and Baker wrote to the American management:[36]

> I am interested in noticing your remark that strikes are occurring all over the country, especially where it is an industry handling anything for the Allies. This looks like an engineered affair and there is no doubt that the Germans or Austrians are at the back of this – probably the Austrians as so many of them are employed in those areas where the strikes are taking place.

In October 1916 there was further violence at Bayonne extending to the borax works[37] where 'some of our people were injured' and a demand for a 20 per cent increase by the borax-workers followed by yet another settlement, this time for 12½ per cent. A year later there was trouble at Death Valley Junction,[38] where the ore from the mines was treated. Corkill reported that all the union men, including those who were making trouble, had been discharged, but that before leaving they surreptitiously threw bolts and cement into the gearwork of the new mill addition, hoping that it might not be seen and that the machinery might be wrecked when it started up. However, they had been watched, and no mischief had been done.

Baker commented generally at the time that every advance to these people seemed like blood to a tiger and that they would probably regret after the War that they had been so grasping.[39] He was proved right in the short run, even though eventually, some years after the War, the management had to concede some of the demands – notably for the closed shop, which had been regarded as preposterous in these early days.

In concluding this account of the effects of the 1914–18 World War, it seems relevant to mention an aspect of it which shows up again as strongly in borax records of these years as it did in a recent history of the zinc industry.[40] In view of the intricate and world-wide commercial links between Britain and Germany over the preceding half-century, it is possible that some business-men took the view that the war was a tiresome and unnecessary interruption of the normal flow of commerce, brought about by the ambitions and incompetence of politicians and generals. This is an aspect of the history of the First World War which is almost entirely missing from the Second, with its overwhelming sense of national purpose in fighting a great evil rampant in Europe and, later, in Asia. It may also be an explanation of why business appears to have returned to normal so rapidly after the conclusion of the 1914–18 war. It would not have been surprising if this outlook had influenced Borax Consolidated's attitude towards trading with Germany during the War. The company had been exporting formerly a large amount of both raw and refined material to countries which had become engulfed in the conflict[41] – notably to Germany and to Austria, where there was a large-scale enamelling industry. Baker's view was that the export business to these countries would probably fall into other hands for the time being, but that this would amount to only 'a temporary diminution of our trade in some directions'.

There were two partial loopholes available to circumvent this situation if the Anglo-American group, Borax Consolidated, had chosen to make use of them. The bulk of the group's export trade to Europe was conducted from Britain. So, when Baker in Britain wrote to Zabriskie in the USA on 15 September 1914, about a new government proclamation prohibiting the supply of goods directly or indirectly for the use or benefit of an enemy country, he was probably hinting at an indirect loophole when he added: 'While we may supply goods to such countries as Holland and Belgium we are not permitted to supply the enemy.'

The second loophole might have taken the form of direct export from the USA, which was not at war with Germany until April 1917, and where there was and is a large community of people of German origin. But the records contain absolutely no trace of the US side of the business – i.e., the Pacific Coast Borax Company – bringing any sort of pressure to bear on their British colleagues to bypass their country's embargo on trade with the enemy by use of either of these loopholes. On the contrary, unstinted co-operation and encouragement was given to the British side to abide by the British rules and beat the enemy, in spite of the difficulties and loss of trade encountered throughout this long-drawn-out war.

War profiteering, for which the 1914–18 War became notorious, was in any case beyond the reach of the borax industry at that time. Borax was not yet in great demand, and although demand for boric acid exceeded supply, no one could have hoped to make much of a fortune in that line, with its limited markets and profit margins. It is true that there were one or two light-hearted attempts to turn the situation to commercial advantage, as when Baker wrote to Colonel Reid, the deputy chairman, on 4 September 1914:[42]

> Dear Reid,
>
> Several references have been made in the newspapers as to the value of borated vaseline for foot sores for the troops but sufficient attention has not been called to the value of boric acid powder as a preventive of foot soreness. We are getting up some Red Cross envelopes which we shall supply at cost to the chemists so that they can fill them with boric acid powder and show them. The public will buy these and send them to the soldiers. It is, however, very advisable that someone should call attention in the papers to the relief that boric acid gives to the feet in marching. I had thought that Borax Consolidated might do this but it would be looked on as an advertisement. Therefore it is better it should come from another source. Would you mind sending the letters either in your name or, if you object to this, sign them 'An Old Campaigner' and enclose your card with the letters to the newspapers editors. I have had the letters prepared. . . . Our motives are philanthropic and commercial, mostly the former, for the few tons of boric acid we might sell would not be very important, but we know that if the soldiers will take the trouble to use the stuff it makes a wonderful difference to comfort in marching and keeps the feet sound which is half the battle. Yours sincerely, . . .

The war effort was also mixed with good commercial benefit when Baker wrote to Reid at the Southern Anti-Aircraft Command at Lewisham on 2 November 1916:

> I was at Belvedere yesterday and Wilson [i.e., the works manager] tells me that he is not getting the soldier labour that he was obtaining when you were in charge [i.e., of the local A.A. battery]. They are not now letting him have any men from the guns. If you know the present officer-in-charge, could you suggest to him that he might help us with a little more labour than he is doing at present?

And so the War passed on, with its terrible accompaniment of death and bereavement, until, in mid-1918, the news that Germany was about to collapse came almost unexpectedly and Borax Consolidated started to buzz with suggestions – as to the best way of building up stocks again, and getting back into business in Europe as soon as the War ended.

REFERENCES

1. Share Register.
2. M.D. to F.M.S., 23.12.14. Also Smith's advertisement in *Mining & Scientific Press* of 21.12.14.
3. See final page of Chapter 10.
4. M.D. to F.M.S., 1.2.15.
5. M.D. to Zabriskie, 20.12.15.
6. M.D. to Zabriskie, 8.12.14.
7. M.D. to Locke, 14.1.15.
8. M.D. to Locke, 4.12.14.
9. M.D. to Locke, 4 and 8.12.14 and M.D. to Ryan, 13.8.15.
10. M.D. to Zabriskie, 31.5.16.
11. M.D. to Sir Joseph Mackay, 28.3.17.
12. M.D. to Sir Percy Bates, 13.4.17.
13. M.D. to P.C.B., 15.9.14.
14. M.D. to Thorkildsen, 17.12.14.
15. M.D. to P.C.B., 9.4.15.
16. M.D. to P.C.B., 26.4.15.
17. M.D. to Lord Chichester, 3.5.15.
18. M.D. to Thorkildsen, 30.1.15.
19. M.D. to Zabriskie, 9.9.14.
20. Lord Chichester to M.D., 2.5.15.
21. M.D. to P.C.B., 8.1.15 *et al*.
22. M.D. to P.C.B., 21.2.16.
23. M.D. to P.C.B., 29.12.15.
24. M.D. to P.C.B., 22.5.16.
25. M.D. to Ryan, 10.9.15.
26. M.D., 4.5.15.
27. M.D. to Ministry of Shipping, 24.7.17 *et al*.; also Ministry of Shipping to R.C.B., 26.3.18 *et al*.
28. M.D. to Works Manager, Kidsgrove, 8.8.14.
29. M.D. to Zabriskie, 16.3.17.
30. M.D. to Swindells, 10.5.16 and Gerstley to Swindells, 30.11.16 *et al*.
31. M.D. to Swindells, 4.8.16.
32. M.D. to Swindells, 18.9.18.
33. M.D. to Chief Industrial Commissioner, 6.4.18.
34. M.D. to Wilson, Belvedere Works Manager, 25.4.18.
35. Gerstley to Zabriskie, 24.7.15; M.D. to P.C.B., 12.8.15; Zabriskie to M.D., 28.7.15.
36. M.D. to P.C.B., 17.8.15, 13.9.15 and 27.9.15.
37. M.D. to P.C.B., 13.10.16. Also M.D. to Zabriskie, 17.10.16 and 26.10.16.
38. M.D. to P.C.B., 1.10.17.
39. M.D. to P.C.B., 11.5.16.
40. Cocks, E. T. and Walters, B.: *History of the Zinc Smelting Industry*.
41. M.D. to *Financial Times*, 31.8.14.
42. M.D. to Reid, 4.9.14.

14

The Challenge from Lake Borax

Searles Lake, California was named after the brothers Searles, who were the first there in 1864 and returned in 1873 to patent 2,000 acres of it as a borax claim. They harvested the surface deposits of cottonball (ulexite), converted it to borax in boiling-pans and crystallizing vats on the lake side, and transported the borax by mule team to the railroad at Mojave. During their latter years of occupation of the area they had begun to realize that the 'solid' lake was rich in other minerals. They started to explore, and drilled to a depth of 600 feet.[1] However, when the Democrats reduced the tariff on imported borax in 1894 the Searles brothers were forced to close, and in 1895 their San Bernardino Borax Mining Company fell into the waiting arms of the acquisitive F. M. Smith. The drill cores and the samples of other minerals beside borax failed to inspire any curiosity among Smith's employees, and the operations remained closed after the takeover. As already quoted, it was Smith's opinion that 'We do not need to disturb ourselves about the surface deposits in this country. They have all been exploited, beyond a doubt some of them several times over'.[2]

If that were all, the history of the borax industry following the First World War would be a worthy and rather dull affair; but as events turned out, and after all kinds of difficulties, the brine of Searles Lake as a reservoir of chemicals was to provide a competitive challenge which the established borax-miners had never expected.

The creation of the high mountain range of the Sierra Nevada, which separates the interior from the Pacific Ocean, was followed by two glacial periods which in turn formed a series of valleys and basins below the eastern slopes, in the midst of which lies Searles Lake. Geologically this area is a recent formation, and only about four thousand years ago it was part of a chain of lakes and connecting rivers running from Owens Valley in the north through Indian Wells Valley and China Lake to Searles Lake and Panamint Valley, and then north-east to Death Valley. Searles Lake then covered about 400 square miles and overflowed at a level 600 feet above its present surface. Today the residue of the lake consists of a mineral salt body covering about 30 square miles to an average depth of about 150 feet, and for most of the year the surface is firm and can be driven across.

The salt lake is permeated with a brine which seeps continuously into its muddy base, and is one of those isolated pockets which form a very small part of the earth's surface as a whole, where the precipitation in the surrounding area no longer finds its way to the sea. The surface water dissolves its share of minerals on its way there, and then collects in a basin with no way out. In this warm and arid climate evaporation exceeds precipitation, and the dissolved minerals crystallize out continually in solid form. The 'salt body' of the lake consists almost entirely of about six main mineral salts, which between them contain only two metallic elements – sodium and potassium – and four main acid groups – chloride, sulphate, carbonate (and bicarbonate) and borate. There are only traces of other chemical elements, and substantially this 3,000 million tons of crystals consists of salts of sodium and potassium.

US Government drilling teams had started investigating some of the old marsh areas for minerals at some time in the first decade of this century, and among other spots, they probed Columbus Marsh, Death Valley and Searles Lake.[3] These investigations were carried out under the leadership of Hoyt S. Gale of the United States Geological Survey. Although there is no record of how the first venture began, or whose idea it was, it is possible that Gale's published reports gave C. E. Dolbear[4] – 'a chemist of impracticable ideas', as he was once described – the idea of starting the California Trona Company with the primary object of producing soda-ash from the 'trona' in Searles Lake.[5] 'Trona' is a natural crystalline double salt of sodium carbonate and sodium bicarbonate found in playa deposits in other, but relatively few, parts of the world, and is today a substantial source of soda-ash at Green River, Wyoming, and at Lake Magadi in Kenya.

Dolbear's company was incorporated under the laws of California on 19 February 1908,[6] with an initial capital of $1 million, in order to produce soda-ash from the brine from the big reefs on the eastern side of the lake.[6] It leased for the enterprise the land comprising the derelict Terrace Borax Mine and Lyons Association Placer Mine, the patented property of the San Bernardino Borax Mining Company – a subsidiary of Smith's Pacific Coast Borax company since 1895 – and also bought their buildings on this land. It was realized that processing of the brine would inevitably produce a mixture of products,[7] and limits were placed on the rights of each party to them. An arrangement was made whereby Pacific Coast Borax undertook not to sell soda-ash for ten years except to California Trona, and a similar but a vaguer obligation was placed on California Trona not to sell any borax arising from their operations except to Pacific Coast Borax. Not many years passed before this became an obstacle to the developing ambitions of the parties to this lease.

California Trona had borrowed considerable sums to start this undertaking, but it did not prosper,[8] and in 1909 the Foreign Mines Development Company, a British-domiciled subsidiary of Consolidated Goldfields, foreclosed the mortgage that they had held on California Trona since 1 August 1908. They then sent an overbearing individual called Guy Wilkinson to reorganize the operation. Wilkinson appears not only to have failed in this task but to have spent some £30,000 of Goldfields' money while doing so.[9] He also made himself thoroughly unpopular with Pacific Coast Borax, particularly by behaving as though he owned the lake and ordering off it their guest, the Kali (German Potash) Syndicate representative.[10] Accordingly, against legal advice, Pacific Coast Borax did their utmost to avoid renewing California Trona's lease in 1913. Angered at this Wilkinson pressed them to sell their patented land to his company, and threatened that he could produce borax as a by-product of the lake brine far cheaper than they could mine it, and that he could therefore wreck Pacific Coast Borax's business unless they co-operated with Trona. Baker replied with characteristic contempt:[11] 'I think the low class character of Mr Wilkinson's threat stamps the whole operation and, personally, I feel no anxiety whatever as to the result of Mr Wilkinson's borax operations.' Both Zabriskie and the chemist Dr Hecker agreed with these views, and, indeed, they were justifiable in these early years at Searles Lake.

Nevertheless Consolidated Goldfields pressed on. In 1913 Wilkinson disappeared from the scene and the company was reorganized financially as the American Trona Corporation with Baron Alfred de Ropp – a Russian from Petrograd, and as difficult to deal with as Wilkinson – as president and leading light. Before quitting the 2,240 acre San Bernardino site they proceeded to claim most of the rest of the Lake,[12] (which was variously calculated at 20,000–40,000 acres) on the strength of applications already made by California Trona. Soon they were putting up a new plant[13] and building the Trona Railway consisting of thirty miles of track from the Lake to connect up with the Southern Pacific Railroad.

This reviving enthusiasm, however, received a temporary setback when the Director of the United States Geological Survey used an Act of 1910 to withdraw, with effect from 21 February 1913, 'certain potash lands', including Searles Lake, from 'entry' as patented (owned) lands under the existing mining laws, and proposed that a new Bill should be put to Congress, making Searles Lake subject only to leases granted by the Federal Government.

The Geological Survey probably had two main reasons for taking this step. Early in 1913 the Press had been reporting 'the spectacular race of armed automobile parties across the Mojave Desert for the possession of valuable soda deposits at Searles Lake'[14] and armed conflict leading, it was hinted, to fatal results in some cases. Secondly, the presence of potash (recoverable as potassium chloride) at Searles Lake which Hoyt Gale's reports had publicized had called the government's attention to a dangerous feature of the pre-1914 economy of the United States – the fact that one of the world's largest agricultural producers depended on imports from Germany and France for over four-fifths of the potash that it needed. Its annual needs, according to a letter written in 1919 by the President of the Nevada Chemical Company to the Commissioner of the General Land Office,[15] amounted to 240,000 tons of K_2O (as it was now customary to express the 'potash', whatever compound was involved) in the year before the 1914–18 War, while American production, even four years later in 1918, had amounted to only 60,000 tons of K_2O, about 70 per cent of which came from Nebraska.[16]

It is not difficult, therefore, to understand the hostility that potential American producers of potash exhibited towards American Trona's claim to most of Searles Lake, from 1913 until the matter was settled by a series of decisions and compromises between 17 September 1917 and April 1918. The outbreak of the European war had reduced very rapidly the possibility of receiving further potash supplies from that continent, and the price of potash in the States rose accordingly,[17] thus offering glowing prospects of a quick profit in this line to new and untried American producers. At the same time it was claimed that the Foreign Mines Development Company, which controlled American Trona, was an alien corporation, being domiciled in Britain,[18] and some went further and claimed that it was backed by the Rothschilds, who were backing the German Potash (Kali) Syndicate, the main importers of potash into the States before 1914.

These facts are mentioned here because of what was to happen later, and because the Kali Syndicate's interest in Searles Lake is traceable right through from 1909 until 1941, when the whole story came out into the open. Clearly Germany was not going to give up its enormous sales outlet in the USA without a struggle.

For the time being, however, this hostility to foreign influence failed to hold up the onward march of American Trona. In September 1917 a Court at the Inyo County capital of Independence ruled that Trona's claims at Searles Lake were valid as against those of a number of rival American locators, and a General Land Office ruling by the Assistant Secretary of the Interior, Vogelsang, on 28 March 1918[19] laid down that, although Trona's cpaital might be owned by aliens, they were not an alien corporation, since the company was organized in good faith under the laws of the United States. Accordingly, they were granted a patent for 2,560 acres of Searles Lake with, in addition, 'a small area on the existing Trona reef'.[20] As part of the bargain they 'relinquished in favour of the United States Government every claim on the greater area which they held under provisional title',[21] in accordance with the policy of the new Potash Leasing Act, which had become law on 2 October 1917.

This Act parcelled the remaining area of Searles Lake (i.e., about 28,000 acres outside the San Bernardino and Trona Patents) into eleven leases of 2,560 acres each,[22] as eleven hopefuls had applied, and it tried to ensure their effectiveness as potash-producers by imposing some strict provisions as to the minimum capital outlay and 'prompt commencement of active operations'.

Few of the applicants were able to meet these conditions and avoid forfeiture of their leases. Some obviously went into their venture without previous knowledge of minerals and merely as a speculation, while the main difficulty – which pervaded all Searles Lake enterprises until about 1925 – was failure to find or devise a suitable process for obtaining reasonably pure potash (or indeed, any readily saleable mineral) from the brine. Even when there were hopes of producing acceptable unadulterated potash, the end of the War in November 1918 brought with it the certainty of the return of cheaper German and French potash sooner or later. In these circumstances Victor Barndt, a friend of Pacific Coast Borax, and his Nevada Chemical Company, and Colonel Merrill of Standard Oil and his Merlie Chemical Company, were among those who gave up, but two unexpected outsiders stayed the course for some years yet. Firstly, F. M. Smith, by persistent lobbying at Washington, had managed to obtain a lease and incorporated the West End Chemical Company to conduct the operation as a subsidiary of his West End Mining Company. He also obtained a lease for this last company, as well as one for his friend Chapin, and one for a financial backer, Carlston. Secondly, Burnham, a former instructor in chemistry at the University of California, also obtained a lease and started the Burnham Chemical Company. He developed a solar evaporation process, and although the vagaries of the weather created problems,[23] he managed to hang on until 1928.

These leases from the Government and American Trona's patented acres formed, therefore, most of the Searles Lake picture when the War ended in 1918 and the picture was completed by a corporation called Borosolvay, which was to become one of the most unsuccessful ventures of Borax Consolidated in its long history.

Borax Consolidated had not attempted to emulate the efforts of California Trona and its successor before 1914, and in 1915 merely decided to post a 'watchman', Jack Ernst, at Searles Lake to report periodically on what American Trona seemed to be doing. However, in that same year they received overtures from the Solvay Process Company of Syracuse, the biggest producer of soda-ash in the United States. Solvay had investigated Searles Lake brines as early as 1910, and had devised the outlines of a process, but the cost of the operation was far too high to compete with the potash imported from Germany and France. They already had a small potash project at Saldoro, Utah, but from 1912 onward active encouragement by the U.S. Government of the indigenous production of potash turned Solvay's attention once more to Searles Lake, where an additional attraction was the possibility of simultaneous and abundant recovery of their main product, soda-ash. Accordingly they approached Borax Consolidated and offered the Solvay technology in exchange for the coveted position on the Lake. Borax Consolidated first wanted to be convinced that Solvay's process – which had been patented in the name of two Solvay chemists, Dr L. C. Jones and Dr F. L. Grover[24] – would not result in soda alone being produced, and untreated potash and borax being thrown back into the lake. They sent their representative, Dr George Hecker (who had somehow managed to return from military service in Germany to the USA while hostilities still continued) and he, after witnessing trials with Dr Grover, reported that the process was a failure as a potash-producer, and that a more effective one was being devised jointly.

This removed Solvay's main bargaining card and the negotiations continued in an atmosphere of ultimatum from both sides which did not augur well for the future. Finally a written contract with Solvay was thrashed out point by point and signed on 21 October 1916,[25] which was the de facto beginning of a tripartite venture between Pacific Coast Borax, Solvay Process Company and San Bernardino Borax Mining, who were the nominal owners of the land. In November 1917 with the arrival of the Potash Leasing Act the venture became a bipartite corporation called Borosolvay, with Pacific Coast Borax and Solvay Process Company each taking 50 per cent of the $2,500,000 equity.

Meanwhile, from the day that the original agreement was signed, Solvay had pushed ahead with the construction work with such speed, and with so little regard to keeping to the agreed estimates, that Baker and his Borax Consolidated colleagues (who habitually behaved as though they might have been the natural ancestors of the later generation of budgetary control experts) were reduced to despair. Production started in April 1917, and the result of just under four years' production,[26] until the plant closed on 1 February 1921, was only 18,827 tons of potash,[27] an average of just over 400 tons a month compared with the 1,000 tons for which the plant had been designed. No borax was produced because Pacific Coast Borax discouraged their chemists from working out a process, which if successful would have encouraged other producers on the Lake to do likewise. This would have precipitated the problem which the advent of Searles Lake was now clearly going to cause for the borate-mining industry – the unavoidable production of cheap by-product borax at the rate of about one ton of borax for every two tons of potash.[28]

After closure early in 1921 neither Borax Consolidated nor Solvay ever restarted operations at Searles Lake. Their joint company had found the cost of producing potash from it very high, and Baker had started warning the company before the end of the War[29] that he could not see American potash standing up to competition from Europe when hostilities came to an end. Trona's reaction to this threat was to urge the United States administration[30] to consider putting a tariff on imported potash immediately after the War in order, ostensibly, to protect its infant potash industry. But Congress, with an eye to the agricultural vote, threw out the proposal, a decision which brought immense relief to the management of Borax Consolidated.[31] Their viewpoint, as stressed several times in Baker's letters,[32] was that they had gone into the potash business at Searles Lake purely to protect their borax business by outsmarting Trona at their own game, and that they were not interested in potash as such. In other words, a tariff on potash imports, thus enabling Trona to charge a high price for American potash, would enable Trona also to cut the price of their by-product borax to levels which would have ruined Borax Consolidated's conventional mines, works and business.[33]

At the end of the 1918 war the European Kali Syndicate did not immediately attack the American potash market, and they accepted a modest share at profitable prices, thus encouraging United States producers to continue. However, by 1920 the Kali Syndicate were ready to make their move, and they contracted to supply 250,000 tons a year of potash to the USA,[34] representing a substantial increase in market share, and the inevitable fall in prices followed.

1920 and 1921 were therefore hard years for the infant chemical enterprises on Searles Lake and on Owens Lake,[35] where six small companies also hoped to produce borax as a by-product of soda-production. This was the moment at which Solvay suggested to Borax Consolidated that the game was up for Borosolvay for the time being, and that the new plant should be closed, since the warehouses were full of potash and sales had faded out. Borax Consolidated, with its dominating interest in borax, readily agreed.[36]

A further, but involved, reason for Borosolvay closing down is suggested in Zabriskie's letter to J. D. Pennock, general manager of Solvay Process Company, on 8 August 1919. He had heard that a Mr Keene (a chemical engineer) was coming out from Consolidated Goldfields' London office to make a thorough investigation of American Trona's affairs, and that on his report would depend whether or not Trona would continue to operate. Zabriskie's argument[37] was that if Borosolvay shut down immediately 'this might have the result of this man Keene making an unfavourable report and of the Goldfield people deciding to abandon the enterprise. If this were done your company and ours might be placed in a position where we could gather in their whole proposition. What do you think of this?' This letter to Pennock shows that Zabriskie had moved from his contempt for Trona's efforts of a few years previously to a measure of anxiety that they might after all make a success of it and endanger Borax Consolidated's future.

American Trona had been producing potash for some time, but the costs were only sustainable because of a tenfold increase in price which occurred during the First World War. They were also up against a further problem that was to trouble them for some years to come, and one which had troubled Borosolvay throughout its brief existence. Borax in minute quantities is good for agricultural crops, but in larger quantities it is very harmful.[38] It was therefore necessary to improve the quality of Searles Lake potash, and to reduce the proportion of borax to a level tolerable to agriculture, which was 1 per cent or less. Trona eventually got the proportion down to less than 0·5 per cent, but Borosolvay never managed much less than 3 per cent, so that it is not surprising that users, in spite of the cessation of German potash supplies during the War, were reluctant to buy Searles Lake potash. Quite a number of crop failures, notably in corn, tobacco and cotton plantations, were reported,[39] and so much was this the case that at one stage the Southern States were campaigning to have the use of this potash banned altogether, and it seems that North Carolina actually did ban it.

However, in 1919 John Teeple, a consulting chemist with a Ph.D at Cornell University, joined American Trona as works manager. He realized the shortcomings of the process in operation, and commenced a programme of investigation and analysis of the conditions needed to produce at a viable cost a range of industrial chemicals from the reservoir which the Searles Lake brine offered. Teeple – ably assisted by R. W. Mumford, in charge of research, and others – applied scientific research at a level far above anything done previously. He also had the ability to engineer on a production scale, and to direct technical effort to sound commercial objectives. Those interested in a more detailed account should refer to Teeple's book published in 1929[40] as an American Chemical Society monograph, or the excellent summary in R. F. Multauf's recent account of the history of common salt.[41]

In spite of Teeple's work, his confidence in profitable production was not shared by the owners, and in May 1921 American Trona closed down for ten months.[42] Consolidated Goldfields felt at that time that they were straying from their traditional metal-mining field, and the price of potash or borax was likely to remain at an unprofitable level, so their Searles Lake business was offered for sale at eight million dollars.[43] Kali Syndicate, among the many to whom it was offered, turned it down for political reasons. As Dr Hecker later wrote to Baker from Berlin,[44] they did not want to take the risk of Dr Teeple, with his known anti-German views, going to the Government of the USA, only some three years after the War, and complaining that the country's potash fertilizer supplies were becoming the monopoly of a German-owned syndicate. It seems Borax Consolidated were not prepared to pay more than $3 million, and no sale took place.

In March 1922 the American Trona plant was restarted, and neither Borax Consolidated nor Kali Syndicate appear to have worried overmuch about this for a year or two. Meanwhile the potential of the work of Dr Teeple and his associates must have had a persuasive effect on Consolidated Goldfields Ltd, as instead of selling they came round by 1924 to a decision to build a new plant,[45] with about two and half times the capacity of the existing one, which was then producing about 100 tons/day of potash and 50 tons/day of borax.

Borax Consolidated's sole activity at Searles Lake at that time was represented by H. P. (Nix) Knight, a remarkably observant Pacific Coast Borax chemist, who was posted there to watch developments, and he reported[46] that in 1925 Trona produced about 31,500 tons of potash and 15,500 tons of borax. However, by 1927 Trona's new plant had been completed, production steadily increased, and a significant competitive force in both the borax and potash industries had been established. Borax Consolidated's '20 Mule Team' brand was being matched by Trona's 'Three Elephants' brand.[47] The new plant incorporated many new features which overcame the problems experienced in the old one, and by the time the Depression arrived in 1929 Searles Lake was producing over 20 per cent of the potash used in the United States and 40 to 50 per cent of the world demand for borax.

An analysis of Searles Lake brine shows the solids content to be 16·36 per cent of sodium chloride, 4·86 per cent of sodium carbonate and bicarbonate (soda), 6·85 per cent of sodium sulphate (salt cake), 4·74 per cent potassium chloride

Smith's cabin, where he lived in the Nevada desert (c. 1872)

Left: F. M. Smith towards the end of his career with Pacific Coast Borax introducing his youngest child, Frank junior, to one of the twenty mules

Below: Arbor Villa, Smith's mansion in San Francisco (completed 1895)

View of the mining camp at Ryan, California, with Death Valley Railroad locomotive (1918)
Death Valley Railroad, loading ore at 'Played Out Mine' near Ryan (1925)

The Death Valley Brass Band (c. 1910)

A poker game in the company store at Borate (c. 1900). 'Wash' Cahill standing right, Lew Rasor (Clarence's brother) seated left

Horace Albright and Steve Mather with 'National Park Service No. 1'

The tribute to Steve Mather placed in many National Parks and Monuments

(potash) and 1·5 per cent sodium tetraborate (borax), with very minor amounts of salts containing lithium, arsenic, iron, bromine, iodine, sulphide and phosphate. The chemicals which have been produced commercially are potash, borax, soda and salt cake with small quantities of lithium salts.

Without doubt the work of Dr Teeple and his team, once it was backed financially by Consolidated Goldfields, changed the politics of the borax industry dramatically. Those involved in the traditional mining of borax ores experienced a new threat, the potential of which was extremely difficult to assess and was to worry them for many years to come.

In 1920 on another part of Searles Lake and on a smaller scale, Smith's West End Chemical Company attempted to put their lease to good effect by constructing a plant designed to produce borax and potash by solar evaporation, but this was a failure and the plant closed the following year. However, later a new approach to Searles Lake brine was made by a chemistry graduate of the University of California named Henry Hellmers, who had now joined Smith. Hellmers ignored the potash and designed a plant in 1925 to produce borax and soda-ash[48] by first carbonating the alkaline brine, so that sodium bicarbonate is precipitated and separated by filtration; from this, sodium carbonate (soda-ash) is recovered by heating and driving off carbon dioxide, which is recycled. At the same time the acidity of the residual brine converts the borax to the highly soluble sodium pentaborate (($Na_2B_{10}O_{16}$)) and when fresh brine is added which makes the solution alkaline this is changed back to sodium tetraborate. The brine mixture is then cooled, and borax crystallizes out. This process was a commercial success, and with Stauffer acting as sales agent operations were gradually expanded, and West End Chemical Company eventually became the most successful of Smith's enterprises during his second borax career.

REFERENCES

1. Teeple, John E.: 'Industrial Development of Searles Lake Brines'. *Am. Chem. Soc. Monograph* 49 (Chem. Catalogue Co. Inc. N.Y., 1929).
2. F.M.S., 18.12.1900.
3. P.C.B.: Corkill to Ryan, 19.3.13; Locke to F.M.S., 17.4.13; Locke to Gerstley, 25.10.13.
4. P.C.B.: Hecker to R.C.B., 25.4.13.
5. P.C.B.: Zabriskie to Lesser, 6.9.18.
6. US Department of the Interior Memo of 28.3.18.
7. P.C.B.: Locke to R.C.B., 20.1.14.
8. Gerstley J.: 'History of Borax Consolidated, Ltd' (unpublished).
9. P.C.B.: Zabriskie to F.M.S., 20.8.13.
10. P.C.B.: Locke to R.C.B., 25.1.14.
11. R.C.B. to P.C.B., 31.5.13.
12. P.C.B.: Locke to F.M.S., 4.6.14.
13. US Department of the Interior Memo, 28.3.18.
14. Extract from unheaded newpaper cutting, 21.3.13.
15. P.C.B.: Victor Barndt to Hon. Clay Tallman, 10.12.19.; also Zabriskie to Kelly, 13.12.20.
16. *Chemical Trade Journal and Chemical Engineer*, 21.2.20.
17. Teeple, *op. cit.*, p.25.
18. P.C.B.: Dolbear's opinion quoted in Locke to F.M.S., 29.4.13.
19. P.C.B.: Horace Clark Attorney to Frank Wehe, company legal adviser, 7.2.24.
20. P.C.B.: Lesser to R.C.B., 10.4.18.
21. P.C.B.: Victor Barndt to Hon. Clay Tallman, 13.8.20.

22. P.C.B.: Barndt to Lesser, 14.11.18; also Department of Interior decision by Hon. Clay Tallman, 16.7.18.
23. U.S. Borax Archives, Los Angeles.
24. M.D. to Z., 27.6.16.
25. P.C.B.: *Facts concerning the Borosolvay Plant* by H. P. Knight, Borosolvay, 28.8.20. Attached to Dudley to Z., 28.8.20.
26. *Ibid.*
27. M.D. to Pacific Coast Borax, 11.3.18.
28. P.C.B.: Zabriskie to Knight, 5.6.20.
29. M.D. to Zabriskie, 7.8.18.
30. Cable from P.C.B., 26.7.21.
31. P.C.B.: Cable from Zabriskie, 15.9.22.
32. M.D. to Hecker, 7.9.20 *et al.*
33. M.D. to Zabriskie, 20.8.21.
34. M.D. to Zabriskie, 29.3.20, and article in *Chemist and Druggist* of 4.9.20.
35. P.C.B.: Cramer to Zabriskie, 23.9.21; Rasor to Zabriskie 25.9.21 *et al.*
36. M.D. to Hecker, 10.2.21.
37. P.C.B.: Zabriskie to R.C.B., 9.6.21, also uses this argument.
38. P.C.B.: Harris to Zabriskie, 29.6.20, about experiments with borax at an agricultural experimental station; A. de Ropp, Jr.: *Ind. Eng. Chem.* 10 839 (p.918).
39. M.D. to Zabriskie, 5.8.19, and to P.C.B., 2.9.19.
40. Teeple, *op. cit.*
41. Multauf, R. F.: *Neptune's Gift* (Johns Hopkins University Press, 1978).
42. P.C.B.: Zabriskie to R.C.B., 13.6.21.
43. M.D. to Zabriskie, 4.10.21; M.D. to Hecker, 12.10.21; P.C.B.: memo by Zabriskie, 29.4.25.
44. P.C.B.: Hecker to R.C.B., 1.9.23.
45. M.D. to Hecker, 21.2.21.
46. Knight to Zabriskie, 16.1.26.
47. M.D. to P.C.B., 2.10.19.
48. Hightower, J. V.: *Chem. Eng.* 58, No. 5, 162–3 (1951).

15

Steve Mather goes to Washington – the National Park Service is created

Following the acquisition of the Sterling Borax Company by Pacific Coast Borax and with a long-term contract for the supply of ore, Mather and Thorkildsen were able to concentrate on refining and marketing, and their business continued to prosper. They were now both wealthy men, but had quite different ideas about the way their time and money should be spent. Thomas Thorkildsen was a confirmed playboy, and only needed the necessary cash to cut loose, while Steve Mather was preoccupied with the idea of putting his wealth and position to some different purpose, and thoughout his working life in Chicago had been much involved with charities and community affairs. By 1914 Mather was in a restless mood, and it was at this point that he decided to leave the running of their business to Thorkildsen and to embark on a remarkable second career. This career, although not strictly part of the story of borax, is so interwoven with it that its inclusion here seems justified.

As part of his recovery from illness, Steve Mather and his wife had spent three months in the summer of 1904 visiting Europe – they walked and climbed in the Swiss Alps, and made an expedition to Italy to see the borax industry in Tuscany, and also visited many other European countries. The splendour of the Alpine landscape and the arrangements which enabled so many who loved natural beauty to get about so easily and to enjoy it in unspoilt surroundings sowed the seed of an idea which on Mather's return to America never left him. He saw in California, and particularly the Yosemite Valley, a land comparable to the Swiss Alps. He joined the Sierra Club, and in 1912 he organized his own first big mountain trip taking his wife, daughter and nine others out to California, where he met the great conservationist John Muir. In the autumn of 1914 Franklin K. Lane, who had been appointed Secretary of the Interior by President Wilson, received a letter; it was one of complaint – not unusual – describing the National Parks as a 'federal disgrace'. It was signed by Stephen T. Mather. Lane had known him at the University of California, Berkeley, thirty years earlier, and discovered he was now a self-made millionaire in the borax industry with a reputation as a philanthropist and a mountain-climber. The Secretary replied, suggesting that if he didn't like the way the National Parks were run, would he come to Washington and run them himself?

Mather was now forty-seven. On 20 January 1915 numerous friends in Chicago gave the Mathers a farewell dinner, presented Mather with a writing-desk and wished him well in his new assignment as Assistant to the Secretary of the Interior. He went to Washington with the idea of doing what he had to do in a year. He stayed for fourteen years, and brought into being the National Park system on a stable and permanent basis. His work on the promotion of the use of borax in the home on a national scale, together with his previous experience as a reporter on the *New York Sun*, fitted him well for his new task, in which he proved himself to be a born and inspired communicator. To be effective, there were two areas where he had to become known and accepted: the first was the Congress and the second was the Press. He set out at a tremendous pace to achieve this in the first nine months. When Mather had arrived in Washington Lane had asked him to meet a young man aged twenty-four called Horace Albright, a graduate lawyer from the same campus at Berkeley, who was working in another part of the Department of the Interior. Horace Albright happened to be the son of the undertaker in Candelaria, Nevada, mentioned earlier (p.46), and of course he knew well the world of 'Borax' Smith, of Mather's father, of Zabriskie and of the '20 Mule Team Borax'. They talked, and thus began a working partnership that lasted the rest of Mather's life. Whereas Mather detached himself from the borax industry to deal with the National Parks, Horace Albright (as will be seen later) was to leave the Parks in order to manage an important development in which Borax Consolidated were to be involved.

Mather's personality was described by Albert Atwood, a well-known writer for the *Saturday Evening Post*, who met him when he was writing about the forests and parks of the West. Atwood said:

> Mather had a wonderful incandescent enthusiasm for his subject – persistent, never discouraged, never gave up, but always tempered by good nature and good taste. Did he ever get tired? I often wondered. I certainly couldn't keep up with him and I was ten or fifteen years younger. He was always entertaining people and did it supremely well. He had the graceful touch. Of course, there was his charm and good looks – those things do help. He used tact and diplomacy in dealing with people. Knew how to listen, didn't argue too much, but put in a word here and there. He combined the zeal of an agitator with a charm and graciousness I've rarely seen equalled.

Mather was confident he knew how to put the plan for a National Parks Service across, but finance was another question. He soon found there was only one way to defeat the restrictions of his budget, which was by use of his own money. His first of many similar moves was to hire his old friend Robert Yard, Sunday Editor of the *New York Herald*, to join him as the National Parks publicity man at five thousand dollars a year. Another of his early moves was to organize a National Parks conference and he chose for this the chemistry building of his old university at Berkeley in California, accompanied by a fund-raising luncheon in the Bohemian Club at San Francisco. Thomas Thorkildsen, who was in the city, having just completed the first leg of a 12,000-mile cruise in his recently acquired 135-ft sailing yacht, came along, and was among many $1,000 contributors.

Yellowstone had become a National Park nearly thirty years earlier, but the cause of conservation only really started up when President Theodore Roosevelt took office, and The Antiquities Act which he signed in 1906 gave him the right to set aside areas of historic and scientific interest as national monuments. In the next few years parks and monuments increased in number, but the whole problem of maintenance, access and proper management was a haphazard situation, divided between several Federal agencies with no central plan for decisions about policy and expenditure. This was what the Department of Interior wanted to rectify, but the political obstacles which faced Mather and Albright were considerable.

Until Congress felt the impact of public opinion concerning the National Parks, money would not be available to do the job, so Mather and Albright had a 'chicken and egg' problem. The whole concept of withdrawal of lands by the government and the use of taxpayers' money on recreation areas was not exactly in keeping with popular sentiment in much of the United States of that era. However, Mather's theme was enthusiastically supported by national journals such as the *National Geographic Magazine* and the *Saturday Evening Post*, and the Press as a whole found his charm and eloquence very hard to resist. Mather did a lot of entertaining and organizing trips out West for Congressmen, business-men and the Press, and he never forgot the importance of the wives in the process of their conversion; all this was mostly at his own expense. As a result by the time the 64th Congress assembled in December 1915 it was decided to introduce a Bill to create a National Park Service without any further delay. Hearings were completed by April, but since it was election year the outcome was unpredictable. By 1 July the Bill was passed in the House of Representatives, but its prospects in the Senate were more precarious. Mather was committed to leave for the high Sierras on a camping trip with important supporters, and he left Albright to do his best with the Senate, but actually fearing the worst. On the 5th of August Senator Smart of Utah got the Bill passed, but not for the last time it was a different version of the measure passed by the House. Horace Albright's anxieties continued until a Conference Committee finally reconciled the differences between House and Senate on the 22nd of August, and his diligence not only managed to get the Bill signed by the President three days later but also got the pen with which the deed was done.

Mather was named first Director of the National Park Service, and Albright became his assistant. His appointment was accompanied by some political infighting, and unpleasant things were said. Sadly, the strains of political life and hyperactivity brought Mather another breakdown, and he was away for eighteen months. Albright took over as acting Director, and started to organize the new bureau and to defend its first request for funds. His report called for the termination of administration by the Cavalry of the U.S. Army in Yellowstone Park, the establishment of the Grand Canyon National Park, the enlargement of the Sequoia National Park and the annexation of the Jackson Hole area to Yellowstone Park. This last proposal started a battle with cattlemen and landowners which continued for over thirty years.

Without Albright's loyalty and experience it would have been extremely difficult, if not virtually impossible, for Mather to contend with the illness which periodically afflicted him, and to be able each time to return to his job. The many improvements and additions to the National Park System and the political battles surrounding them are well described in Robert Shankland's biography of Steve Mather. His continuing interest in the borax industry always kept him in contact with Pacific Coast Borax. Yosemite and Yellowstone National Parks and Death Valley were all places in which he met periodically with Zabriskie, Baker and others to discuss their contractual arrangement for the supply of borax ore and to discuss environmental matters.

In January 1927, when Pacific Coast Borax Company were moving operations from Death Valley to the Mojave Desert, Mather and Albright met Baker and Zabriskie to discuss the potential of the Death Valley area for recreational purposes. The borax men no doubt felt a nostalgic interest in inviting back the individual who thirty-five years earlier had conceived the '20 Mule Team' trademark, but their main purpose was to ask him whether Death Valley – and in particular that area of beauty surrounding Furnace Creek Inn, where much of the land was owned by the company – could be incorporated in the National Parks system. Because of his own borax background, and because California now had four national parks and soon perhaps a fifth, a Redwood National Park, Mather felt for political reasons it was not possible for him to initiate this idea. Those in favour of National Parks wanted to see them spread around, those against wanted to see them go elsewhere, and California had had too much attention. The subject was pursued further when Lord Leven, Chairman of Borax Consolidated, came out in April and met Mather in Yosemite National Park, but nothing happened till after Mather's time, when Death Valley did enter the system as a National Monument in 1933.

In the early years in Washington Mather left his borax affairs to Thorkildsen and the business prospered. Mather's main role was to keep good relations with Pacific Coast Borax, and specifically with Zabriskie. Prosperity found Thomas Thorkildsen enjoying the proximity of the capital of the newly created movie industry to the borax-mining area. He acquired a Hollywood mansion complete with swimming-pool and the necessary amenities to hold his own in the fast-moving society of the time. Some of his wilder and more scandalous escapades hit the headlines, and when they continued to refer to him as 'Partner of National Parks' Chief', Mather – who would otherwise probably have ignored the whole thing – felt it was time for the parting of the ways. Horace Albright still remembers advising Mather not to send the cable to Thorkildsen which suggested he should either buy out Mather's share or sell his share to Mather. However, the cable was sent, and Mather got the business, and as a result he paid too much at a time when profits from borax had disappeared. From about 1924 Thorkildsen and Mather Co., even more than Pacific Coast Borax, had begun to feel the impact of competition from the new plant of American Trona at Searles Lake (see p.160) and the depressed borax prices which followed. Later Thorkildsen is said to have lost or spent everything he had, but it seems he was kept from destitution by a pension from some part of the Pacific Coast Borax's operations in the ongoing borax industry.

At the time of the 1912 Presidential Election, Mather, a Republican, had supported Teddy Roosevelt and his 'Bull Moose' Party when they split from the official Republican Party, whose candidate Taft was seeking re-election. Both were opposed by the Democratic candidate Woodrow Wilson, who as a result of the split Republican vote was elected President. Thereafter Mather could be described as first a conservationist and second a Republican. By 1928 he had served under several Presidents – Wilson, Harding and Coolidge – and the prospect of Herbert Hoover in the White House filled him with excitement. No one since Teddy Roosevelt could have been a happier choice for him and the cause of conservation. Mather had taken Hoover on expeditions in the Parks, and Hoover knew and admired him and his work.

Mather planned to hear the election results with friends in his old stamping-ground Chicago, and having arrived the day before, he spent the morning on borax matters with 'Olie' Mitchell, the manager of his Chicago office, and then

The National Park Service is created

called on his attorney, Harold White, to discuss election prospects. Then, settled in an armchair and smoking a cigar, Mather stopped in mid-sentence: he had had a stroke, and was paralysed on one side, and was without his voice.

The following month Mather realized that his career was finished, and he asked that Horace Albright, at that time Superintendent of Yellowstone National Park, should succeed him. This took place in January 1929. After months in hospital he made something of a recovery and learnt to walk and talk again, but following a further stroke he died in January 1930.

Tributes came from everywhere, from Hoover (now in the White House) downwards. While his name is almost unknown outside America, the traveller there finds it much remembered and with increasing emphasis, as the creation of the National Park Service becomes an ever more significant part of the history of the United States. In Alaska there is a peak named Mount Mather, in Mount Rainer National Park there is a Mather memorial highway, at the University of California at Strawberry Canyon there is a Mather arboretum of redwoods and at Lake George in New York State there is a memorial Mather Forest. A Congressional Bill passed in 1931 authorized a Mather memorial for the City of Washington or near by, but it was not until 1969 that the scenic gorge of the Potomac river below Great Falls was renamed in Mather's honour. The bronze plaque placed there and in each of the twenty-three national parks and thirty-three national monuments shows Mather's profile in bas-relief with an inscription which reads thus:

> HE LAID THE FOUNDATION OF THE NATIONAL PARK SERVICE DEFINING AND ESTABLISHING THE POLICIES UNDER WHICH ITS AREAS SHALL BE DEVELOPED AND CONSERVED UNIMPAIRED FOR FUTURE GENERATIONS. THERE WILL NEVER COME AN END TO THE GOOD THAT HE HAS DONE.

REFERENCES

Shankland, R.: *Steve Mather of the National Parks* (Alfred A. Knopf, New York, 1951).

Department of Interior, National Park Service: 'A Special Publication to Commemorate the 65th Anniversary of the National Park Act August 25, 1916.' 25 August 1981.

Sevain, D. C.: *Wilderness Defender – Horace M. Albright and Conservationism* (University of Chicago Press, 1970).

Horner, C. B.: *Preservation Comes of Age* (University Press of Virginia).

Travis, N. J.: Recollections by Horace M. Albright at a meeting at his house in Sherman Oaks, California. September 1981.

U.S. Borax Archives, Los Angeles, California.

16

1914-26: the end of an era in Death Valley and a new discovery

By 1914 the Lila C. mine in the mountains above Death Valley was believed to be nearing exhaustion. Since 1908 John Ryan, the mines supremo, had been studying other remnants of the former Coleman empire to find a suitable successor. Financial considerations and the obstacles presented by the mountainous terrain were limiting factors in the choice among the six available mining claims, but by the end of December 1914 the new Biddy McCarthy mine was ready to send ore by the Death Valley railroad down to the new roaster at Death Valley Junction, and soon 80–90 tons a day were moving along this route. In April 1915 this narrow-gauge railway was reported to be in excellent condition and giving no more trouble through sliding rock, and prospects were hopeful for mining and transporting any quantity of ore that might be required from the new deposits.[1]

At the six Ryan mines – Played Out, Upper Biddy (McCarthy), Lower Biddy (ditto), Grandview, Oakley and Widow – the reserves of ore available had all been increased by development work and mining, the first five moderately so, while the reserves at the Widow Mine, which presented the poorest chance of all from surface indications, were increased enormously to well over a million tons.[2] Thus in this group of mines using one common mining camp at New Ryan there were reserves of colemanite far in excess of the world's needs at the time. In addition, although Lila C. had closed in April 1915 as worked out, Fred Corkill (the second) reported in June 1919 that a large body of ore was still there. This expensive lapse appears to have been the result of deliberate non-cooperation resulting from a foreman's intense dislike of the temporary mines superintendent, and hasty plans were made to put in a narrow-gauge spur to retrieve at least 100,000 tons of this good ore.[3] It was not until November 1925 that the last shipment from Lila C. was made, and the camp and the railroad as far as Horton Junction was dismantled. Not unnaturally, Baker wrote that 'it makes one rather jumpy about borate'; for F. M. Smith, who in spite of his financial recklessness had a reputation for having a good nose for a borax deposit, had left the company; Fred Corkill (the first) had retired; and John Ryan, the trusted veteran of Death Valley borate-mining, had died in May 1918.

Baker in the years after 1918 was thus left without experienced management at the mines, and he did nothing much to remedy it except to become involved himself to such an extent that when he was not in the United States he became absorbed in a voluminous and almost daily correspondence about every detail of matters at the mines, mill, refineries and every part of the organization of which he was now President. After Smith, it was a marked change in style and method, and while Zabriskie and his staff responded, it would be difficult to imagine that they were enthusiastic. Today this method of management by letter and coded cables – with no telephones, of course – appears cumbersome, but Baker was a hard-working and meticulous man who did not avoid decisions, and nothing escaped his attention.

The exposed ore at Ryan was colemanite, and a new mill had been built at Death Valley Junction – a calciner (or roaster). However, the ore bodies at Ryan are a mixture of colemanite and ulexite impregnated with lime and shale, and at depth the ulexite begins to appear in increasing quantities, with resulting milling problems. Ulexite does not disintegrate in a calciner to a fine flour-like product as does colemanite, and instead of passing through the screens it stayed with the shale and ended as waste on the tailings dump. Until this problem was solved ulexite had to be separated as much as possible at the mine, and large dumps accumulated.

When Herbert Faulkner, a very tall – six-foot-six – English mining engineer (his height brought him the nickname 'Highpockets'), arrived at the Ryan mines the ulexite problem was coming to the fore; either mine production dropped or there were loud complaints about the ulexite content from the mill at Death Valley Junction. He set about trying to solve the difficulty, with little assistance and a good deal of cynicism from those at the mill. It was a problem that needed time and technical experiment, and fortunately someone recognized that outside help was needed. Tom Cramer has described a visit to the Alameda refinery by Professor Richards of Harvard,[4] a leading authority in America on ore-treatment, who carried out trials on ulexite. As a result, a wet concentrator plant was built at Death Valley Junction which separated light and heavy particles (using a process known technically as jigging). After fairly prolonged teething troubles this plant began to reduce the piles of ulexite, but when it was burnt down in 1924 it seems that no one wanted to go back to the wet process. Rasor suggested calling in a consultant, Dr Stebbins of Los Angeles,[5] who developed a dry process based on grinding the ore to fine particles and separating the ulexite from the shale by using an air separator and vibrating tables. Fred Beik, an engineer from Wilmington, was sent to Death Valley Junction to build a new plant,[6] which successfully disposed of both ulexite stocks and tailing dumps, and this meant that the Widow mine – which was the most inexpensive to work but which contained ulexite – remained the main producer.

Faulkner, who followed the popular Billy Smitheran, had a tough time, not the least of his problems being ex-employees who jumped company mining claims and defended them with guns, and he was no doubt ready to go when he left in 1921 to take charge of the Borax Consolidated mine in Turkey.[7]

Another side of mining life is revealed in a letter written by Baker in August 1923, which gives an original view of Ryan and Death Valley Junction.[8] The background was the attempt to find a job in mining for the son of Colonel Reid, the vice-chairman of Borax Consolidated, an attempt which did not meet with the gratitude it deserved:

The camp at the Junction is certainly not the same as Ryan, which is the Ritz of American mining camps and the best in the States and, as it would appear, spoilt your son for the average life at American mines. I have never seen at the Junction the general untidiness and litter he mentions although, as the railroad passes through the camp and the passengers are of a mining type a good deal of stuff may at times be thrown about by them. . . . Also a great number of motors pass through that district and the people use our camp and, generally, are not of the most tidy type. The so-called hotel, in which your son has been accommodated, was recently bought by us for the purpose of getting rid of a saloon which attracted men with bootlegging liquor and for gambling, and also it is our intention to use this place for more accommodation providing more conveniences for the men. . . . The lavatory accommodation is necessarily not like that at Ryan where the camp, being on the side of the mountains, allows for splendid drainage. Our camp at the Junction is on the flat and nothing but earth closets are possible. . . . I have been glad to use one of these places myself when at the Junction, until recently, when bungalow accommodation has been provided for the management.

Improvements continued in all aspects of personal comfort at these camps. In November 1924 a civic centre was completed at the Junction, and Baker wrote:[9] 'Although rather expensive, [it] is a very good advertisement for the Company. . . . I am glad to see that the men have started a baseball team and I hope with you that they will give a good account of themselves.'

In fact the civic centre cost $300,000, and everyone thought that the mines and mill in Death Valley would be the centre of the borax industry for generations to come. In addition to the feeding and sleeping accommodation for two hundred men, the civic centre included a general store, company offices, a hospital unit, theatre, recreation hall, pool-room, and guest-house. The *Inyo Register* described it as 'a city under one roof'. For opening night at the recreation hall – named Corkill Hall – Baker crossed the Atlantic, Zabriskie came from New York, and all Ryan and Wilmington were there. All who could wore dinner-jackets and starched shirts, ladies were in long dresses, and an abundance of fresh flowers were delivered from Los Angeles by the Tonopah and Tidewater railroad. Thereafter nothing escaped celebration at the Junction – Easter, May Day, Thanksgiving, and Christmas were big affairs, and of course on St Patrick's Day the whole town burnt green – but on Sunday Corkill Hall became a church.

In the midst of all this there still stood on 'the other side of the tracks' the one institution that had existed at the Junction since the railroad was built, and before all this civilization arrived. Bob Tubbs's saloon and desert store were combined with what was sometimes called an hotel and sometimes a 'travellers' rest', but was in fact a brothel. To Tubbs's credit, he applied to the state school system for the first teacher to come to the Junction, and as a result 4 ft 8 in, eighteen-year-old Bess Davis duly arrived with her suitcase at the Junction and reported to the school trustee – namely, Tubbs. She was puzzled by the fact that she was not allowed to board with the Tubbses but was happy in the Corkill home. She stayed for two years, was a great success, married a Pacific Coast Borax man, Frank Grace, and 'lived happily ever after'.[10] Some probably thought the Junction had become too civilized when Tubbs was asked to move, but a good solution was found, and the organization moved a few miles away to a 'ranch' at a remote and romantic spot called Ash Meadows, where the Tubbs tradition continued to flourish until the house was destroyed by fire in the 1960s.[11]

These communities were remarkable examples of the resourceful pioneering spirit which pervaded the western United States during these times and earlier. Harry Gower – who joined Pacific Coast Borax in 1910 and retired half a century later – planned, built and developed a township at Ryan which included a store, a recreation hall, a school, a church, a brass band and a way of life far above that of the usual rough mining-camps in the West. Homes, wives and children all had a place, and before the days of home refrigeration or air conditioning it was a real test of self-reliance and courage. Since it was adjacent to one of the hottest places on earth, it needed people of character, and there were many such who worked there. Pauline Gower, Harry's wife, taught the children, and was always ready to play on the piano whatever anyone wanted. Amateur theatricals were a feature of life at both Ryan and the Junction. But these mines and this mill were becoming expensive indeed, and their lifeline, the Tonopah and Tidewater railroad (about which much has been said earlier) continued profitless and a drain on Borax Consolidated's finances which no amount of optimism about possible new sources of revenue along the route could diminish.[12]

This was a disquieting feature at a time when competition from the American Trona Company was somehow refusing to subside as expected, and when there were other factors in the situation to make Borax Consolidated 'jumpy about borate'. Among these the rise in labour costs during the 1914–18 War has already been mentioned. It continued throughout the first post-war years until 1921, when reductions of wages in many industries became possible without fear of strikes, owing to world-wide reduction in demand. Consequently both the Death Valley mines and the mill at the Junction were shut down for seven months during 1921, and restarted on reduced wages. After that things rapidly went the other way, and a boom and the Bolshevik scare – which deterred the US Government from admitting enough immigrant labour – forced wages up rapidly.[13] Overseas, the Turkish situation did not help. It had been hoped to resume supplying British and European works from the Turkish mines soon after the War, but Mustapha Kemal's nationalist rising cut off supplies from 1920 until 1927, and production of colemanite had to be boosted at the American mines to fill the gap.

In the USA itself the 1914–18 War had made life very difficult for the New Jersey works at Bayonne. To an abnormally high level of wages had been added the cost of moving raw material from Death Valley mines in the West to the east coast. After the War advantage was therefore taken of the vastly shortened sea route to the East offered by the opening of the Panama Canal in 1914 and all borax-refining became based on the west coast. For this the 'Chandler' site was acquired at San Pedro for $50,000 in July 1921,[14] and since it was on the Wilmington side of Los Angeles harbour it was renamed 'Wilmington'. Although the refinery was not fully completed until nearly 1930, the US headquarters of Pacific Coast Borax moved there in August 1924.[15] Boric-acid production remained at Bayonne until 1929,[16] after which the deserted site awaited a purchaser until it was sold to the rival Stauffer Chemical Company for $100,000 in 1945.[17]

Wilmington was not the only new works to be established by Borax Consolidated at this time. Inevitably, the War had altered the European sales scene, and in particular the collapse of the Austro-Hungarian Empire had left the company's works at Stadlau near Vienna serving the needs of only a small republic, instead of a vast multiracial territory.[18] More especially, the majority of the enamellers and glassmakers were now included in the new state of Czechoslovakia, and it became imperative, particularly owing to tariff and currency considerations, to supply them from somewhere within this new state.[19] Accordingly the Aussig works (which formerly belonged to the refiner Billwaeder of Hamburg) was acquired for £9,000 from one Max Ullman in September 1920.[20]

There was also talk of establishing factories in the enlarged post-war Poland and in Germany, but nothing came of these ideas.[21] An attempt to go back to pre-war days, and to gather the German refiners again under the wings of Borax Consolidated, was stifled by the dozen or so refiners who brought suits against the company for non-fulfilment of raw-material contracts made just before the War broke out.[22] Meanwhile the European market became a happy hunting-ground for competitive sources of raw material, including F. M. Smith, American Trona, Stauffer Chemical Company, and other United States producers, who were now determined to move into Borax Consolidated's overseas markets.

In 1922, following the imposition of an import tariff in Spain, Borax Consolidated decided to build a new refinery there at Badelona near Barcelona and the Spanish subsidiary company, La Productora de Borates, was formed.[23] After three years, in order to carry the overheads involved, an additional plant was built to make tartaric acid, which was an appropriate project for a wine-growing country.[24]

In the decade following the War sales of industrial borates grew steadily. Enamelware was the principal industrial use, with increasing interest from the glass industry. The possibility of producing sodium perborate and ferro-boron was being studied, and the use of sodium pentaborate, particularly for optical and heat-resistant glass, was under investigation. Attempts were also being made to extend the uses in agriculture, particularly for citrus-fruit preservation from 'blue rot'. '20 Mule Team' packaged borax for household use, and 'Boraxo' hand-cleanser continued to make progress, but Smith's cherished innovation of toilet and laundry soaps containing borax had disappeared.

But for those who find sales reports dull, there is the story of a certain Harry Dumont, of whom Baker wrote to Pacific Coast Borax:[25]

> I see that it has occurred to you to put Dumont at the head of the Sales Promotion Department, and that he is delighted to have the opportunity to take charge of this. He should do well because that is, so to speak, literary work that he is more capable of doing efficiently than some others. Certainly it will be necessary for you or Mr Shannon to keep a good watch on his effusions as we do not want him to launch into poetry over a business matter.

When Shannon quit for another job Dumont moved up to a position of general sales manager of Pacific Coast Borax, but seems to have disappeared into thin air soon afterwards. According to one account, Mrs Dumont woke one morning in 1925 to find a note on the bureau. It read: 'Bye-bye, Baby Bunting, Daddy's gone a-hunting.' Maybe he realized that sales work was wrecking his poetic genius.

The atmosphere of post-War uncertainty in the market and the rising tide of competition were reflected in the profit record of the leading producer, Borax Consolidated, in these years. Gross profits showed an abnormal peak of over £500,000 in 1916, and settled down to something just under £400,000 a year from 1918 until 1925, except for the brief post-War depression of 1921; but after 1925 they went into a sharp decline following a collapse of borax prices, which put the industry into a constant state of jitters.[26] The beginnings of the Searles Lake challenge have already been outlined, and also there was always the possibility that others might discover borax deposits which were better placed than the remote Death Valley mines, and the certainty that mining rivals of pre-War years, together with one or two new interlopers, would rush in to exploit any chance that thus offered itself.

The continuing pre-eminence of Borax Consolidated in these conditions owed a great deal to the acquisition of the services of an outstanding man, Clarence M. Rasor. Born in 1874, he was raised in Clay Township outside Dayton, Ohio, and after leaving school took a course in civil engineering and worked in mining at Aspen, Colorado and Yellow Jacket, Idaho; but he soon sought adventure in the Spanish–American War, serving with the US Cavalry Regiment known as 'Torreys Rough Riders'.[27] Recuperation from malaria brought him to San Bernardino in California, where he joined his elder brother Edwin, who was working as a consulting engineer. The firm of Rasor Brothers succeeded, and moved to Los Angeles in 1907.[28] Among other jobs Clarence worked as a Deputy Mineral Surveyor in California, a position which was that of a private contractor approved by the US Surveyor General of Mineral Claims (in San Francisco), and he was employed and paid by the parties filing their mining claims. This gave him an opportunity to obtain unique knowledge of mining-law and claim procedure, and of what was going on in California.

Rasor was ideally suited, physically and in temperament, to the scramble for new sources of raw material which took place in the borax industry during his thirty years of association with it, which started in 1905 when he was asked to join Corkill and Ryan to plan and build the Tonopah and Tidewater Railroad. At some time after 1910 he left Rasor Brothers to become the Railroad's Chief Engineer, and when Corkill retired in 1915 he succeeded him as head of Pacific Coast Borax's Land and Exploration Department.

Rasor's priceless value to his company in this era was something less easily defined than professional skill. He had contacts in the world of mineral prospectors throughout the West, and travelled widely. He combined a shrewd understanding of the importance of a little financial ground bait, even when information was of no particular value, with a personality which inspired trust and confidence. Thus Rasor often came first to the mind of those who had something to tell and to sell in an age when legal agreements were not part of the armoury of the mineral prospector and treachery was a hazard of the game. As will be seen later, it was no coincidence that, on two big occasions in the history of Borax Consolidated, Rasor was there when it mattered.

For Borax Consolidated had powerful rivals in this epoch. The Stauffer Chemical Company remained the foremost threat, even when partially allied to it by agreement. Their approach differed from that of American Trona, who made it clear that they intended to wipe Borax Consolidated off the map by flooding the market with borax produced as an inescapable by-product of potash-production from Searles Lake. The Stauffer Chemical Company had endeavoured to advance into borax as yet another line, from the secure basis of the profits made by other chemicals which this company was already producing, and to follow the conventional course of finding a cheap source of borate raw material from which to manufacture. Mention has already been made of the borate mines in which Stauffer had an interest in the hills of Ventura County, and of how low prices had driven this company to dispense with them, so that by 1914 they had adopted the policy of negotiating for cheap raw material for their refinery from other producers, particularly F. M. Smith. After Smith's departure from Borax Consolidated Baker took the view that the best way to stop Stauffer encroaching further into the borax business was to make a reasonable contract with them for 'non-participation and supply' and, after lengthy negotiations a contract was concluded in April 1915.[29] In the circumstances the terms had to be 'generous to

Stauffer' and indeed they were, as Stauffer obtained a steady supply of borax ore at well below market price, which was more profitable than mining their own. There was regret at the time that Stauffer would agree to accept only a three-year contract, with an option for a two-year extension, but soon Borax Consolidated started regretting that they had ever made the contract at all. Stauffer was looking round for a really good mine, while using the contract to compete for Borax Consolidated's trade wherever the chance arose.

Stauffer's opportunity was provided by a certain Dr John Suckow, a man who was destined to harass Borax Consolidated for over twenty years.[30] Suckow was born in Germany, and after some training as a pharmacist he emigrated to the United States and hitch-hiked to California. While he was working in a Los Angeles pharmacy he studied medicine at the University of Southern California and graduated in 1905, specializing in the broad area of aches and pains known as rheumatism, for which medical dogma prescribed the dry warmth of the high desert.

Suckow sought to establish his own sanatorium and in 1913 set about acquiring land in the Kramer district in the Mojave Desert under the Federal Homestead Act and, as part of the work necessary to 'prove up' a homestead, he proceeded to drill for water. The driller, Mr Leslie Griffin, advised the doctor by telegram that he had struck shale and that this indicated water was unlikely. However, he was instructed to go deeper, although all that was obtained were some pieces of white rock, a sample of which later adorned the mantelpiece of Suckow's consulting room in the Bryson Building in Los Angeles. It was there that a patient called Lew Rasor, none other than Clarence Rasor's brother, saw it and offered to identify the mineral, which was colemanite. Perhaps not surprisingly Clarence Rasor appeared on the scene and Pacific Coast Borax later bought the Doctor's mineral rights known as the Otelia Suckow Placer Claim (named after his wife) and part of his land on Section 22*. A number of holes were drilled and exploration shafts sunk in the vicinity but nothing of merit was discovered. However, cash had stimulated Suckow's interest in borates and in 1917 he located colemanite in the eastern half of Section 22 of quality sufficient to interest Stauffer.

Besides this new discovery, Borax Consolidated had also to contend with discoveries by several other parties in the same area. Baker considered this 'rather disturbing as it points to the possibility of this [Kramer] deposit extending under a large portion of that flat [land] and where it may begin and end we cannot even surmise.'[31] However, Zabriskie concluded that Pacific Coast Borax could not keep control of the situation in the area without going into joint ownership with Suckow, and by October 1918 Pacific Coast Borax had 40 per cent of the equity of the new Suckow Chemical Company, while Stauffer held 35 per cent and Suckow 25 per cent.[32] Stauffer gave Suckow the voting rights of their share for a period of ten years, thus he had control of the Corporation and its management. Suckow's need for cash had caused him to share his discovery, but he was in a position largely to ignore his partners and proceeded to do so.

Meanwhile, every few months saw a new scare in the colemanite world, and Rasor was constantly on the move between Nevada and Lower California.[33] Early one morning in March 1920 Rasor set out from the township of St Thomas, Nevada, in a farm wagon with his camping and prospecting kit and accompanied

---

*See map VIII, pp.176–7.

by a prospector called Perkins, who guided him to a remote spot at the southern end of the Muddy Mountains.[34] Perkins had discovered the visible signs of a colemanite deposit, and as a result Pacific Coast Borax paid him five thousand dollars and filed a group of claims in an area which Rasor christened White Basin, about forty-five miles east of Las Vegas, and near what later became the Hoover Dam.

Perkins liked to recount his success story in the local saloon, and two attentive cowboys, Lovell and Hartman, soon made their own colemanite discovery at a place some thirteen miles from White Basin, called Calville Wash. They invited both Rasor and F. M. Smith to inspect their find, and asked for bids. Rasor described how they all met at this remote spot in the Nevada Desert and standing apart the rival parties scribbled out their bids and then handed them over. Rasor's $60,000 was swamped by Smith's $250,000, and Smith had seized the opportunity to be back in borax-mining for which he had been waiting so long.

The usual borax rush followed these discoveries, and as might be expected the ubiquitous Blumenberg arrived on the scene. He began staking placer claims at White Basin on behalf of the American Borax Company in the same area as Pacific Coast Borax's lode claims, and even started extracting colemanite. After lengthy litigation Pacific Coast Borax established their rights. However, little colemanite was ever mined at White Basin, and much to Rasor's annoyance Calville Wash proved a much better prospect.

The discovery at Calville Wash was made on the 23 January 1921, the day of Smith's wedding anniversary, and he christened it the Anniversary Mine. Production started in 1922, and with the Stauffer Chemical Company as his sales agent, Smith at the age of seventy-five was back in the borax business, fired with the determination of a fighter to show those who earlier had witnessed his downfall that he could still make a comeback.

In a manner designed to cause maximum irritation in London, Smith's company then started offering its 'Twenty Aeroplane' brand in both America and Europe. About this choice of names Baker remarked to Zabriskie:[35] 'These people are, as you say, a good deal "up in the air" and probably the definition is a very accurate one. At the same time we cannot but look on it as an infringement of our trademark. . . .' But it appears Smith himself did not think much of this brand-name and soon reverted to calling it 'F. M. Smith Borax',[36] a situation reminiscent of Smith's first reaction some thirty years earlier to Steve Mather's suggested 'Twenty Mule Team' trademark. The Anniversary Mine required a substantial investment (which in the course of time totalled two million dollars) and Baker had no doubt that this would raise something of a hornet's nest among Smith's creditors.[37] Later the collapse of borax prices in 1924 hit colemanite-mining badly, and by 1928 Smith had closed the mine, relying on his Searles Lake operation for borax-production.

In 1922 Rasor made up for losing out to Smith at Muddy Mountain by discovering colemanite at Shoshone, which became the Gerstley mine.[38] No one stole any further march of importance on Rasor's exploration efforts during this period. In fact, it looked as though the Ryan mines would be the unchallenged centre of borax-mining for years to come, and that the big unresolved question was how well would they compete with the challenge from Searles Lake.

Efforts continued to tame the pirates in the business. In July 1924, when the Suckow mine was producing about 150 tons per month of saleable colemanite, Rasor met 'Uncle John' Stauffer.[39] Each wanted news of their partner Dr

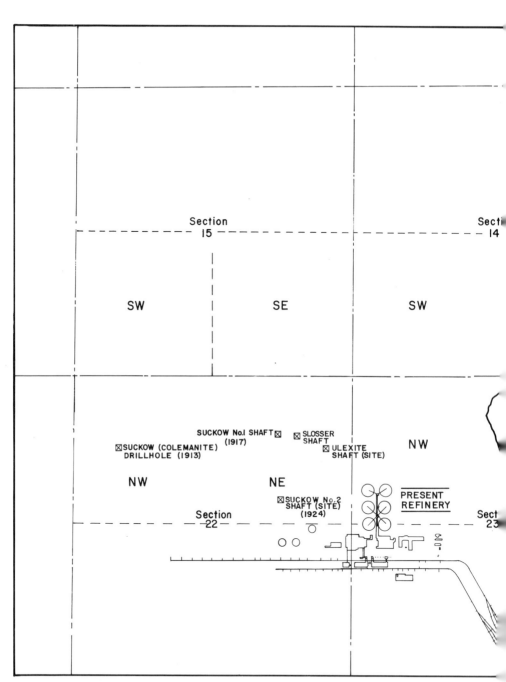

Map VIII

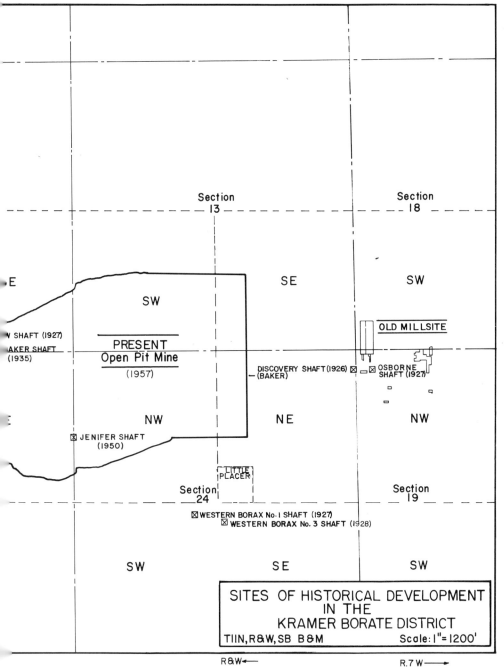

Suckow, and each blamed the other for creating a situation which enabled the Doctor to enter the business at all. They also discussed Blumenberg, who since Daggett days had managed to lay claim to colemanite with sufficient success to tease both Pacific Coast Borax and Stauffer. Rasor chided Stauffer that he had bought out Blumenberg just as he had been about to run out of ore, and predicted – as it happened correctly – that it would merely enable him to show up again in borax-mining after a suitable interval. No admirer of Blumenberg, Rasor commented that he 'was cached away somewhere in Los Angeles. I am advised he devotes most of his time to young ladies of the movie class.'

As for Suckow, Rasor was unable to trace him, but he saw his agent Kleiner, who was 'sore as a boil', complaining 'that Suckow seldom comes to the office and when he does he spends most of his time "calling up chickens [chicks]".' Writing to Zabriskie, Rasor went on: 'I think I have intimated to you heretofore that Suckow seemed to be troubled with, or rather affected by some kind of sex complex. It has become very much apparent in the last year, and is getting much worse.' Rasor also reported that the Suckow mine contained considerable ore reserves, and it was not long before the financial demands of the Doctor's life-style brought an offer from him to sell, which was reluctantly accepted by Pacific Coast Borax in 1925.[40] However, Stauffer showed no interest in sharing Suckow's holding with them, and agreed to exchange his interest in the mine for Suckow's borax-refinery in Los Angeles, which ceased to produce in about 1927.

At this time Stauffer was endeavouring to organize a consortium of borax and potash interests to acquire American Trona, and a number of meetings with Dr Teeple took place which Stauffer and Zabriskie attended.[41] Baker was unwilling to consider taking such a step in harness with Stauffer, and also about the same time he turned down the proposal presented by a discontented shareholder called Gallois and a banker-creditor of Smith called Carlson, to try to acquire control of Smith's West End Chemical and Mining companies.[42] Smith had managed, in some seven years only, to enmesh both these companies in a web of unsound finance and long-term low-price contracts, and Baker and his colleagues had steadfastly refused to become embroiled again in his affairs, either to hinder or to help him. But even Baker allowed himself an uncharacteristic outburst when he heard Smith was offering borax in early 1925 at $6 a ton below Borax Consolidated price, and 'a fine specimen of the low-down, unscrupulous, lying Hun' was his uninhibited reference to Smith's ancestry.[3]

In his later years Smith ran into antagonism from a section of West End's shareholders in the form of a shareholder's protective committee, and in 1926 he resigned as President of the mining company.[44] He continued as President of the chemical company until he fell seriously ill in 1928; and then, on 27 August 1931 this old warrior of the West died at the age of eighty-five. Smith was succeeded by his wife Evelyn as President, assisted by her brother George Ellis, who in turn became President of West End Chemical Co. in 1940. This ran successfully and profitably, and in 1956 was acquired by Stauffer Chemical Company.

Meanwhile, the Pacific Coast Borax Company had by 1925 invested about $350,000 in getting title or contracts covering seven sections in the Kramer district, and portions of several more. This area had been familiar territory to Rasor ever since Suckow's colemanite discovery in 1913. Rasor's men had sunk about forty drill-holes but no mining had resulted, except to the small extent of the Suckow No. 2 shaft on Section 22. Drill-holes and exploration shafts at Kramer showed colemanite and ulexite averaging about 10 per cent boric oxide

(B_2O_3), while at Ryan there were vast reserves averaging 30 per cent (B_2O_3).[45] Thus after twelve years' exploration Kramer did not seem to be an area of much importance. Moreover, the Death Valley mines represented a substantial investment for Borax Consolidated, and unless there was some competitive or very significant cost reason for establishing new production facilities elsewhere, it was about the last thing they wanted to contemplate.

However, Rasor kept a watchful eye on the Kramer area and Corum – who drilled for Rasor, but also drilled on his own behalf and for others – reported to Rasor late in 1920 that at the eastern end of the Kramer district they had drilled down to basalt bedrock without finding anything.[46] Little of interest occurred in the area until October 1924, when Zabriskie reported to Baker that considerable activity had again commenced there, to which Baker replied:[47] 'I am glad to see that people are drilling wells in the latter district [i.e., Kramer] where there is not any prospect of their finding borate ore and that they are more likely to find water'[48] – this on the eve of a great discovery!

Another prospector, Widdess, who had also been working there for over five years, and had established claims, saw Rasor just before the end of 1924: he told him he had struck some blue shale at 440 ft and a green-flame test showed the presence of borate, adding that he would come back if he struck colemanite.[49] Rasor was nervous, but expected nothing.[50] However, by 11 March 1925 Zabriskie was cabling Baker[51] that Rasor recommended negotiation with Widdess for Section 18, as he had struck 13 ft of shale at 580 ft showing 11½ per cent boric oxide (B_2O_3) in the form of a soft white material like ulexite, and no colemanite. Rasor was already checking the ownership of other sections of land in the area, and got the Southern Pacific Railroad to agree not to sell Section 19 to anyone for the time being.[52] By the end of March Widdess was offering a quarter of Section 18 for $60,000, and was also approaching Stauffer and others.[53] At that point Baker pointed out that they had a good deal of that kind of stuff at Daggett and 'we require to keep our powder dry for other purposes'.[54]

A sample from a core sent to the Wilmington laboratory confirmed the 11½ per cent B_2O_3, and that the borate was largely, if not entirely, ulexite.[55] At the end of April Widdess started further drilling on Section 18 when Suckow appeared, claimed the mineral rights and put some men to work. On 24 April Widdess called the Sheriff and had Suckow arrested and taken to Atolia, and both then retreated to Los Angeles 'to begin suit'.[56]

By the end of May Corum's independent activities on Section 24 also began to occupy Rasor's attention. His first hole had struck blue shale some 260 ft nearer the surface than the Widdess strike, and showed about 67 ft of borate-bearing shale of the same type. Rasor concluded that borate was likely to be in Section 19 on the Southern Pacific Railroad's ground, and he agreed with Zabriskie that they must buy the north half of Section 19 and the west half of Section 17 without delay.[57] These were Railroad land grants from the US Government which carried with them all mineral rights, and these came with the land.

Corum had four partners: Hannan, an experienced mining prospector and Dowsing, an ex-vaudeville actor specializing in hypnotic stunts (these two worked together on the drill); Corum's brother-in-law Edenburg; and Knean, an Englishman, president of the Santa Monica Paving Company, and also a man of means and strong financial connections.[58] In August Corum brought samples to Rasor, which Zabriskie described as 'native or crude borax, different from any tincal I have ever seen'.[59] Rasor sent four samples from the drill core to the laboratory at Wilmington. They were small samples from which to draw any conclusions. Cramer reported[60] that most of the borate was soluble sodium borate with clay and shale impurities indicating tincal ore, and that one small crystal sample was gypsum only. In September Zabriskie wrote to Baker that Section 24 was a great surprise, 'and as it is tincal [it] opens an entirely new feature as to what the costs would be of refining'.[61]

From November onward things began to move fast. Rasor was convinced that although on the evidence available it was something of a gamble, the existence of a new large borate ore body was more than a possibility, and that Pacific Coast Borax needed to move quickly if it were to consolidate the position before too much was known by others who would be interested in picking up separate pieces.[62] Zabriskie's letters to Baker reflected Rasor's views, but the problem was to get Baker to agree to an expenditure of a possible $500,000 for the Corum and Widdess claims. This required testing powers of tact and persuasion because Baker had been sceptical about the Kramer district all along. Sitting in London, he was for good reasons preoccupied with the prevention of any unnecessary expenditure and the ever-present shadow of Trona competition from Searles Lake, and he began to ask for the kind of hard facts about the geology of the area which were simply not available, and which if they had been would have meant that the price and the number of people interested would almost certainly have put it out of reach of Borax Consolidated. On 20 November Baker was cabling that further information was necessary for him to form any opinion, that he could not understand how any assessment could be based on two drill-holes. He suggested that an option should be negotiated, and hoped to be in California in the New Year.[63] To this Rasor replied with spirit: 'Fat party is such a windjammer, I do not know how to credit him Stop Will see them all next week believe delay dangerous.' The 'fat party' was the talkative Dowsing.[64]

Meanwhile Knean was trying hard to persuade his partners not to sell and to develop the property themselves, but they declined and wanted cash.[65] Quite surprisingly, the property was offered to Rasor, and on 26 November he reported that Corum estimated their claim contained three million tons of ore between 15 and 40 per cent borate, and that there was no question of getting an option. Rasor felt there could well be a million tons of ore of grade 20 per cent or above, and he was against buying unless the Widdess property was bought as well.[66]

Baker continued to show scepticism about estimates based on two drill-holes and wrote: 'We have shown too much anxiety to deal with these people and their ideas have become too inflated. . . . A douche is needed – we cannot buy the earth.'[67] Ten days later, on 18 December, Baker felt that $250,000 was about the limit to which Borax Consolidated could go. And so, without any understanding having been reached, events moved to a board meeting of Borax Consolidated on 23 December at which a decision would be made. On the previous day news had reached Baker of a third drill-hole that supported the other results, together with an assessment of Borax Consolidated's position by Rasor and his firm recommendation to buy both the Corum and Widdess properties. It would be interesting to know what was said at the board meeting, particularly by Baker, but no record of the discussion survives. A nervous Zabriskie and an elated Rasor received a cable agreeing to negotiation for both properties for up to $450,000, subject to clean claims.[69]

It is greatly to the credit and judgment of Rasor that he maintained such a positive outlook while negotiating with two sets of difficult people and dealing with the complex legal background he found in trying to establish clean claims.[70] He was certainly getting no encouragement from London, and his own description, after attending a hearing of the Suckow case in Bakersfield, that he was carrying water on three shoulders, describes the situation well, as does his modest comment that 'strange to say, [I] do not believe I spilled any'.[71] Apart from the legal action which Suckow eventually lost, only just avoiding a charge of perjury,[72] Pacific Coast Borax had the difficult task of seeing that the claims for these properties were properly filed, and of assessing the problem that certain properties containing sodium salts like Searles Lake were now only eligible for US Government leases under the Sodium Leasing Act of 1920 and not for the preferable patented lode and placer-mining claims applicable to other minerals discovered on public lands. Practically nothing was known about the sodium borate in these deposits or its possible extent. It was vital that the discovery should be kept secret to avoid a 'borax rush' in the area, and it appears that Corum and partners co-operated and followed the advice of Pacific Coast Borax and Frank Wehe, their lawyer in San Francisco, which enabled these claims to be registered on a sound legal basis. A lot had to be done over Christmas 1925 and the New Year to enable final negotiations to start. The claim on the Widdess Section 18 presented no problem, except for Suckow's unsuccessful intervention, as sodium borate was not then in evidence and the claim could be registered as a normal colemanite and ulexite discovery.

On 14 January 1926 Rasor closed the deal to acquire from the Corum syndicate their interest in Section 24 for $375,000 and from Widdess the south-west quarter of Section 18 for $75,000.[73]

Baker must have decided to recommend the proposition to Borax Consolidated, Limited's Board – no one else in London could have known enough about the situation to say much, other than to ask questions. He certainly got precious little additional geological information to help him. In his final assessment, recommending that they buy both Kramer properties, Rasor had written:

> The thought of Searles Lake of course always gives one a chill. No telling what will happen there in the end. With what we know now of Kramer District, I believe we would have in the new strike a formidable weapon with which to go after Trona. Whether we go into the Kramer District or Searles Lake District, the Ryan mines should be closed. The costs are about double what they should be; this is not the fault of the mines. It is the system that prevails at that place. Gerstley (mine) could furnish the tonnage for a year or two at half the Ryan cost.[74]

The difficult and expensive possibility of having to fight American Trona with a new Searles Lake operation had been haunting Baker for some time, and this, with the frightening thought of what competitors might be able to do to Borax Consolidated if Kramer was what Rasor thought it could be, were probably the arguments he used to convince his colleagues. However, to commit his company to pay a sum equal to its likely profits for the next two or three years, on the basis of such limited geological information, was a highly entrepreneurial decision of immense importance. It was unlike Baker not to have been in California during the previous six months, but after the deal was done he wrote explaining that 'but for being in the doctor's hands' he would have been there, and with his usual candour he was telling Zabriskie and Rasor he was still very dubious and they had better be right![75] Although this was not understood in London, Rasor's agreement with Corum and his partners was to acquire only the east half of Section 24. As was cabled in response to Baker's query,[76] the north-west quarter was already theirs, and had been purchased in 1920, but, said Rasor, the 'ownership of the

south-west quarter [is] unknown. It is on a granite hill and can be of no value.'
This was an assessment which later was to cause him embarrassment.

Even before negotiations were concluded, Rasor recommended that they should sink a small shaft as soon as possible on Section 24.[78] Baker was doubtful, for it was *not* just a question of developing 'as soon as possible'. American Trona were in the middle of yet another expansion of their plant, and it was not yet possible to compare their new production costs with the cost of producing borax from a new mine at Kramer. Who could say what the Trona borax price might be if the linked potash price stayed at a reasonably high level? Borax Consolidated were in fact still thinking about their own production at Searles Lake, and possibly terminating altogether the mining of borate in the United States. Zabriskie asked Rasor to put his case for sinking a shaft,[79] with one of those ever-persuasive sentences at the end specially for Baker's consumption: 'I want you to give me your recommendations with your detailed reasons for making the same, all of which I can place before Mr Baker so he can have our views, whether they coincide or not, which would assist him in coming to his own decision in the matter, which of course is the one which rules.'

By February the Borax Consolidated board had sanctioned the expenditure.[80] Rasor had of course been making plans in anticipation of this decision, and, according to L. D. (Roy) Osborne (formerly Superintendent at the Lang Mine), he had already been assigned the task in December 1925 of making plans to sink 'the cheapest shaft possible in a shale formation'. The secretive Rasor had volunteered no further information, and Osborne knew better than to inquire. So, Osborne says,[81] 'I proceeded with these meagre instructions to design a modest wooden headframe, select an old 30-horse power hoist from the then closed Lila C. Mine, select a small new portable compressor, plan to remove a small wooden building from Lang for a hoist-house and sink a $4' \times 7'$ wood cribbed shaft "somewhere in soft shale".' And then in February Rasor drove Osborne to a point in the Mojave desert and said 'This is it.' An Italian crew was assembled, and excavation began about 40 ft west of the dividing line between sections 24 and 29. Early in August 'thin borate of lime' was struck at 328 feet. The drill core had indicated sodium borate about 380 ft, and as this depth was reached and passed tension grew. In Osborne's words:

> Finally on a certain morning in August [actually the 15th] the foreman wakened me about 5.30 a.m. and said he guessed that night shift had struck ore as the dining room table was covered with white rocks. I hurriedly dressed and we roused the hoist engineer and the foreman and I descended the shaft to examine the discovery. The entire bottom of the shaft was a white mass of the purest borate ore that I had ever seen. . . .
>
> We all felt a very personal interest in the discovery of such a find on the barren Mojave Desert where we had sweated it out for the last few feet. You see, the ore occurred 10 ft deeper in the shaft than in the nearby drill hole. . . . Mr. Rasor had been at the mine two days previous and had left in a very subdued mood. . . . I drove furiously to Lancaster, so as not to send a telegram from the nearby Kramer office: "I have something of interest to show you". Mr Rasor came out at once and examined the strike which was then 10 ft. in the ore, and hurriedly left to cable Mr. Zabriskie, in code, of the good news.

On 18 August Zabriskie sent Rasor's report by cable to Baker:

> Shaft entered deposit 7 feet below point shown by drillers' log. Hanging wall sharply defined. Material is hard compact crystal which will be ideal for milling. Seems to be higher grade than cores from wells [holes]. Should be through deposit four days. Generally gratifying. Cramer reports as follows: 'General Rasor sample average 12 tons

runs 35.6% B.A. [B_2O_3]. Picked sample clear crystal 51.5%. Believe latter is new form borax containing four molecules water crystallisation.' Believe large part values in general sample is new form. Full investigation and report to follow.

Thus the revelation of a new and immense ore body of unique character was about to begin, and the prospect of a mine – as opposed to an exploration gamble – was becoming a reality.

REFERENCES

1. M.D. to Pacific Coast Borax, 22.12.14 and 14.1.15.
2. Gower, H. P.: '50 Years in Death Valley – Memoirs of a Borax Man', in Publication No. 9, *Death Valley '49ers* (San Bernardino, 1969), pp.31–2.
3. M.D. to Boyd, 16.4.20; Corkill to Zabriskie, 12.6.19; P.C.B.: Zabriskie to Rasor, 3.9.19.
4. Tape-recorded recollections of T. Cramer, 17 April 1972, Carlsbad, New Mexico, in U.S. Borax archives.
5. M.D. to Zabriskie, 2.10.24.
6. Beik, F. A.: 'My Connections with Pacific Coast Borax Coy', 10.3.47, p.2. in U.S. Borax archives.
7. Gower, H. P., *op. cit.* Ch. IX, p.39.
8. M.D. to Reid, 28.8.23.
9. M.D. to Zabriskie, 1.4.25.
10. Glasscock, C. B.: *Here's Death Valley* (1940), p.266, and Gower, H. P., *op. cit.*, p.63.
11. Visit by N. J. Travis, Nov. 1959.
12. M.D. to Ryan, 19.1.16.
13. M.D. to Zabriskie, 24.3.23 and 8.5.23.
14. M.D. to Zabriskie, 7.7.21 *et al.*, also Minute Book, 6.7.21.
15. Letter-headings and correspondence.
16. M.D. to Zabriskie, 25.4.29.
17. Minute Book, 13.8.45.
18. M.D. to Zabriskie, 1.10.19.
19. Hecker to R.C.B., 5.3.20 and 23.9.20.
20. Minute Book, 2.3.20.
21. M.D. to Hecker, 14.1.20 and 9.10.25; Hecker to R.C.B., 9.1.20.
22. M.D. to Sir F. Crisp (solicitor), 3.3.19.
23. Minute Book, 12.12.22.
24. R.C.B.: Anderson of Pfizer to R.C.B. *et al.*, 5.7.27; also 9.4.26.
25. M.D. to Pacific Coast Borax, 18.3.21.
26. Published Annual Reports.
27. Ingersoll, L. A.: *Ingersoll's Century Annals of San Bernardino County 1769 to 1904*, (*Embellished with Views of Historic Subjects and Portraits of Many of its Representative People*) (Los Angeles, 1904), pp.876 and 877.
28. Private communication from Robert Rasor Jr., Westerville, Ohio, 23 Dec. 1981.
29. M.D. to Pacific Coast Borax, 22.1.15 and 6.4.15.
30. (a) Sierra Services Bulletin, Nevada, California.
    Electric Corporation, Riverside, California, Vol. XIV, No. 8, Aug. 1939.
    (b) Notes taken in Dec. 1981 by Otis Vary, U.S. Borax, about the recollections of Mrs Anita Bryant (formerly Mrs Griffin and wife of Mr Leslie Griffin deceased who drilled for Dr Suckow).
    (c) University of Southern California, Alumni Association records in Los Angeles.
    (d) Noble, L. F.: 'Borate deposits in the Kramer District, Kern County, California' *US Geological Survey Bull.* 785c, pp.45–61.
31. M.D. to Pacific Coast Borax, 7.11.17.
32. M.D. to Zabriskie, 15.10.18.
33. M.D. to Zabriskie, 6.4.20 and 31.5.20.
34. Rasor's Field Notes, U.S. Borax Archives.
35. M.D. to Zabriskie, 30.8.21.
36. P.C.B.: Knight to Dudley, 30.9.21.

37. Zabriskie to R.C.B., 11.4.21.
38. M.D. to Zabriskie, 17.1.22.
39. P.C.B.: Rasor to Zabriskie, 1.8.24.
40. P.C.B.: Cable from Zabriskie to R.C.B., 25.6.25 and reply from M.D. to Zabriskie, 7.7.25.
41. P.C.B.: Cable from Z. to R.C.B., 7.1.25, and letter of 8.1.25.
42. P.C.B.: Zabriskie to R.C.B., 20.1.25.
43. M.D. to Zabriskie, 27.1.25.
44. Circular of Stockholders Protective Committee, May 1926.
45. P.C.B.: Zabriskie to R.C.B., 28.7.31; Siefke, J. W.: 'Geology of the Kramer Deposit' Jan. 1979, U.S. Borax files.
46. P.C.B.: Rasor to Zabriskie, 19.10.20.
47. M.D. to Zabriskie, 15.10.24, re Zabriskie's letter of 6.10.24.
48. M.D. to Zabriskie, 26.10.24.
49. Rasor to Zabriskie, 22.12.24.
50. Zabriskie to M.D., 1.2.25.
51. Cable from Zabriskie to M.D., 11.3.25.
52. Rasor to Zabriskie, 9.3.25.
53. Cable, Zabriskie to R.C.B., 23.3.25.
54. Zabriskie to Rasor, 2.4.25, quoting R.C.B.
55. P.C.B.: Cramer to Rasor, 10.4.25.
56. Rasor to Zabriskie, 28.4.25.
57. Zabriskie to R.C.B., 23.6.25.
58. Rasor to Zabriskie, 15.6.25; Zabriskie to R.C.B., 28.11.25 (in this letter Edenburg is referred to as Greenleaf).
59. Zabriskie to Baker, 28.11.25 (handwritten letter 20.8.25).
60. P.C.B.: Cramer to Rasor, 20.8.25.
61. P.C.B.: Zabriskie to R.C.B., 17.9.25.
62. P.C.B.: Rasor to Zabriskie, 18.12.25.
63. M.D. to Zabriskie, 20.11.25.
64. P.C.B.: Zabriskie to R.C.B., 23.11.25.
65. P.C.B.: Rasor to Zabriskie, 10.11.25.
66. Cable, P.C.B. to R.C.B., 26.11.25.
67. Cable, M.D. to Zabriskie, 7.12.25.
68. M.D. to Zabriskie, 18.12.25.
69. Cable, M.D. to Zabriskie, 23.12.25.
70. M.D. to P.C.B., 11.1.26.
71. P.C.B.: Rasor to Zabriskie (manuscript letter), 26.12.25.
72. M.D. re cable from Zabriskie, 28.12.25.
73. P.C.B.: Cable from Rasor, 8.1.26.
74. P.C.B.: Rasor to Zabriskie, 18.12.25.
75. M.D. to Zabriskie, 26.1.26.
76. M.D. to Zabriskie, cable 15.2.26.
77. P.C.B.: Rasor to Zabriskie, 31.12.25.
78. M.D. to Zabriskie, 19.1.26.
79. P.C.B.: Zabriskie to Rasor, 6.1.26.
80. M.D. to Zabriskie, 2.2.26.
81. Osborne, L. D.: An account of his experience with the Pacific Coast Borax Company dated March 1947.

17

The aftermath of the discovery of sodium borate at Boron

When the discovery shaft struck the 'crystal borax' mineral Rasor had thought that its appearance indicated a form of colemanite and was somewhat disappointed[1] as unlike the tincal (sodium borate) which the drill-holes had indicated, colemanite would require an additional cost of milling. On 17 August Rasor put a bulk sample in the back of his 'Rickenbacher' and headed for the Wilmington laboratory.[2] It so happened that Cramer, Connell and Knight, the company's trinity of Stamford University graduates, were there at a meeting when he arrived, and indeed they all took a hand in carrying out the analysis of the needle-like crystals taken from one of the large lumps of ore that Rasor had brought. To everyone's astonishment they identified a crystal form of sodium borate never before reported[3] – $Na_2O.2B_2O_3.4H_2O$ – that is, one containing four instead of ten molecules of water of crystallization as in the case of tincal, a result which reminded them that Connell had written a 'fanciful' article some years previously (1922) predicting that such a form of borax could and perhaps should exist.[4]

The new sodium borate mineral was for the time being referred to within Pacific Coast Borax as 'Rasorite', until it was later officially christened kernite by the US Government, thus indicating its place of discovery, Kern County, California. 'Rasorite' was registered as a trademark in honour of Clarence Rasor, and has ever since been the brand-name of some of the products derived from this ore body.

In some published accounts the discovery of kernite at Kramer is (perhaps rightly) attributed to Hannan and Dowsing,[5] who were working on the drill in the north-eastern quarter of Section 24 when it intersected sodium borate in June 1925. The drill cores must have contained kernite, but the logs of the drill-holes refer to 'borax' and 'crystal borax', and this was assumed to be tincal, the only known mineral form of sodium borate. Neither Hannan nor Dowsing nor any of Corum's partners, nor anyone in Pacific Coast Borax, identified kernite in the drill cores, and this only occurred over a year later in August 1926 in the Wilmington laboratory.

A year earlier Zabriskie had called attention to the unusual appearance of the crystal borax in the Corum drill core. Pure tincal contains 36·5 per cent boric oxide (B_2O_3), but Corum claimed the crystal borax he had found had been shown by analysis to contain 40 per cent B_2O_3. A manuscript note in the margin of one of Rasor's letters suggests that the tincal sample had probably been partially dehydrated to give this high borate value[6] and thus another clue that something unusual was present was overlooked (pure kernite contains 51 per cent B_2O_3). The Wilmington laboratory, which might have been expected to identify kernite earlier, seem to have been confined to one meagre set of drill-core samples sent there by Rasor in August 1925, when they identified sodium borate but observed nothing special about it.

It was typical of contemporary exploration to keep samples and information relating to drill cores closely guarded, and Rasor certainly trusted no one, not even those within his own organization, when he was trying to establish mining claims. Twenty-five years in the West had taught him a great deal; Osborne, his assistant, asked few questions and answered none, and the Italians sinking the shaft knew neither English nor what it was all about.

As ore reached the surface in increasing quantities, self-congratulation and euphoria was evident on both sides of the Atlantic. As Zabriskie explained to Baker,[7] the new ore introduced 'an entirely new feature'. It was really a 'concentrated borax of very high quality or borax minus the water, on which we would have to pay freight were we shipping the refined article'. He had discussed with Newman, the superintendent of the Bayonne Works, 'the features of refining this material into borax' and they had come to the conclusion that it would be simple and inexpensive, not requiring the addition of alkalies or other reagents', but there is no evidence that the Bayonne works was asked to process any material in bulk. Tests conducted in England found that the new ore contained almost pure crystal borax, no arsenic or lead, and only a minute trace of chlorides or sulphates.[8] 'It was definitely of the top grade . . . too good for commercial borax.' Baker and Zabriskie therefore concluded that this new mineral would probably require no processing, and could perhaps be ground and then shipped to customers for industrial use without further treatment. But it was not the first time nor the last in the history of mining that the examination of samples did not reflect the real problems to be encountered when mining and processing on an industrial scale.

In the same year, 1926, following their decision to build a new plant, American Trona (which had now become the American Potash & Chemical Corporation) announced[9] that they would soon be producing half the world's requirements, or 50,000 tons of borax a year, and that the reduced prices in the last four years had been due to their efforts. Baker summed up the situation:[10]

> The sooner we can get large quantities of the ore the stronger will be our position. There is every indication that we have to face quite low prices in the meantime. At the meeting of Consolidated Goldfields yesterday, the Chairman referred to a cable he had just received from Count Dru stating that by January 15th [1927] they will be in a position to produce two and a half times as much . . . and it goes without saying they will have to sell it somewhere and somehow.

Therefore nothing, it seemed, could have been more timely for the fortunes of Borax Consolidated than the discovery at Kramer, but nobody realized what an obstinate mineral kernite was going to be. However, before long they were able to mine about 1,500 tons a month through the discovery shaft, now christened the

'Baker' shaft. By November a drill-hole on Section 13 showed a satisfactory extension of the deposit north of Section 24.[11] Another hole on Section 19, some 350 feet east of the discovery shaft, confirmed the thickness of the ore, and this area was chosen for sinking the main ('Osborne') shaft for what was to become the 'Baker Mine'. In December, much to Baker's relief, Section 19 had also been acquired from the Southern Pacific Railroad.[12]

Arrangements were made for the Santa Fe Railroad to build a spur to the mine as soon as the main shaft was completed, and consideration was being given to constructing the sort of economical mining camp that Baker had always dreamed of, but which had just as often eluded his cost-saving supervision[13] – 'We do not wish at the present time to go to a large expense in putting up a camp similar to Ryan but rather, as you say, to run the place on much the same lines as we are working at the lease' (i.e., the Gerstley colemanite mine). More explicitly, if they started with married men and families they would soon have the same sort of expensive upkeep that they had at Ryan, although they must expect that a more elaborate organization would have to be considered eventually.

The same spirit seems, in fact, to have guided the whole financing of the beginnings of this great mine. There is nothing to show that any share issue was made or any loans raised in these early years to provide the necessary capital; and the conclusion remains that it must have been financed from revenue or reserves, in spite of the deteriorating financial position of Borax Consolidated. As already noted, the district in which these new discoveries were located was called Kramer, but as operations developed and a mining centre was established the township was given the name Boron, by which it has now been known for over fifty years.

Early in 1927 Rasor estimated the deposit to contain at least 5 million tons of mineral, but as mining progressed it became apparent that the ore varied, and averaged about 70 per cent kernite and 30 per cent tincal[14] and that the discovery shaft had struck an area of almost pure kernite where the borate content was abnormally high, averaging about 40 per cent B_2O_3. There were increasing references to black shale, and the difficulty of obtaining a uniform product from the mine increased. Hand-picked samples of kernite were tried by the enamel trade in Europe and the United States, which scented a new low-cost raw material. As often happens, the customers, although able to handle an impure product, needed one of constant composition, and variations in iron and other impurities in the mined product soon damped their enthusiasm. It was gradually realized that run-of-the-mill ore would average far less than the first indications from the discovery shaft.

Although shipments of kernite crystals selected by hand were sold in Japan for some years, within six months the whole question of what to do with kernite to make it acceptable for sale was preoccupying almost everyone in the organization, and the Baker–Zabriskie dialogue is loaded with technical discussion and the need for a quick solution to the problem.[15]

The difficulty which kernite posed was that its solubility in water was about one-hundredth that of borax. The whole idea of such an insoluble sodium borate was a surprise, and the simple methods of refining which had been envisaged were not feasible.[16] Before long Wilmington, Bayonne, Death Valley Junction, the Kent and Staffordshire refineries in England and Coudekerque in France were all struggling to find a way to deal with kernite, with Baker at the centre applying a large stick and not so large a carrot in order to try to keep Borax Consolidated in business.

At the beginning of 1927 trials were made at Wilmington using an autoclave (a

50 lb pressure vessel) to produce a temperature of 200°C in order to dissolve the kernite[17] and Cramer started working on the problem of large-scale refining using autoclaves and a batch process. Thoughts then turned to upgrading kernite by roasting it in the calciners at Death Valley Junction,[18] and Corkill, aided by the consultant Stebbins, started what proved to be a lengthy period of process development.

In the event it took two and a half years before Borax Consolidated could base its production on the new ore at Boron and supply its refineries and customers with products which were reasonably satisfactory. The Ryan mines at Death Valley ceased production in June 1927, and stocks and the Gerstley mine provided what additional colemanite was needed.

Today's mining world, with chemical, mechanical, civil and all kinds of specialist engineers on call, and with plant design based on research and pilot plants preceding full-scale operations, is difficult to reconcile with the industry of the 1920s. Pacific Coast Borax was no different from most of the mining industry, which has a long tradition of scepticism about the ability of laboratory-based scientists to answer their problems. 'Borax engineers' were few in number, and they had to deal with exploration, mining, milling, construction of railroads and other facilities; when any problem arose they had to tackle it when the rest of the day's work was under control. They worked under tough conditions financially and climatically, with Baker in London expecting quick and inexpensive answers. Hiring additional qualified staff probably never occurred to Baker, and no one suggested it. Apart from a bout of recruitment to strengthen the technical staff before the First World War, Pacific Coast Borax carried on with the old guard almost unchanged. Baker clearly suspected 'educated' engineers. Not only were they expensive but their ideas were liable to cause further expense, and to infect adversely those who had been trained to practise strict economy. When a mine superintendent was needed for the new mine at Boron the choice was Osborne, who was wholly 'borax' educated, and Baker and Zabriskie went to some lengths to sidetrack the Ryan manager Major Boyd, a graduate mining engineer, into a consultative role. Short bursts of extravagance on consultants were occasionally permitted, but only to solve special technical problems.

However, in fairness to Baker, it must be remembered that Borax Consolidated was afflicted with declining profits and sources of cash, and only a determined and skilled navigator could have steered the company past the Scylla of the Great Depression and the Charybdis of the low-priced by-product borax coming in increasing quantities from Searles Lake.

As the hopes of selling ground ore direct from the mine faded it became clear that the recalcitrant kernite needed further processing to produce an acceptable product. Trials at Death Valley Junction in the calciners had shown that roasting changed the kernite needles to a product rather like popcorn, which could be separated from the shale by an air-separation process; but the product picked up excessive amounts of moisture, and was impossible to handle commercially. Unsuccessful efforts were made to compress this fluffy borax into bricks, using pressure of up to 10,000 lb per square inch. In the end, in order to obtain a product which could be packed and shipped commercially it had to be rehydrated by damping with water and then crushed. This product, which contained about 42 per cent water – i.e., about the same as the original kernite – was satisfactory, and could be refined by simple methods. It also had a substantial freight advantage over ordinary refined borax, which contains 64 per cent of water. With limited financial and human resources Corkill and Stebbins, working in gruelling heat throughout the summer of 1928, had finally arrived at this answer to the problem.[19]

For the process to be economical, the residual calcined shale (which contained a considerable amount of borax) had to be treated on a Stebbins vibrating table, and a fraction which was high in borax was added to the main stream and rehydrated. The end-product was called Calcined 'Rasorite' (C.R.). The whole process was something of a hotchpotch of technology, depending on the extraction of borax from the shale residues and the collection of the borax dust from all parts of the calcining process.

To send kernite to Death Valley for roasting was of course uneconomical, but by September 1928 there was sufficient confidence in the process to start moving the aged calciners to Boron.[20] However, in May 1929 Zabriskie was still explaining why sufficient calcined 'Rasorite' was not yet available for Europe, and that Corkill, now at Boron, was often working till 2 a.m. on the new process, while Cramer at Wilmington was just beginning to get production of refined borax up to the required level from the autoclaves.

Calcined 'Rasorite' (C.R.) just about saved the refineries of Borax Consolidated and those of their customers in Europe; and so the mining of borate of lime in South America and colemanite in Death Valley, after some fifty years of considerable activity, both reached the end of the road.

Meanwhile, Borax Consolidated had to face the uncertain outcome of exploration and mining activity all around its new property. Early in 1927 Baker wrote:[21] 'There is a good deal more publicity being given to the Kramer District than we like, but it is inevitable.' Indeed, general interest shifted from the Death Valley area just over a hundred miles south-west to Boron with some rapidity, and irritations followed Pacific Coast Borax to their new pastures like a swarm of flies after a well-favoured milch-cow. Every effort was made to keep the nature of the discovery quiet,[22] and even schools, museums and learned institutions received replies to their requests for samples saying that the mine was not yet fully developed, and that a perfect sample could not yet be offered. Scientific curiosity from the US Geological Survey found Zabriskie wishing that a prominent member of their staff 'would get interested in aerial explorations of the North Pole and spend most of his time in the Arctic regions'.[23] There was always, of course, the possibility that another sodium borate deposit might exist in some other part of the USA,[24] and there were two unfounded claims made within three years of the discovery of the Kramer deposit; but the real danger was rightly considered to lie in the probability that the deposit itself was extensive, with the consequent risk that some rich part might be left for others to claim.

The presence of interlopers in the vicinity soon made itself felt. In mid-August 1927 a syndicate headed by Buley, an oil-prospector, struck sodium borate at a depth of 890 feet in the south-western quarter of Section 24.[25] 'This is very annoying', wrote Rasor, 'as we fully believed that that part of the section, being on granite formation, was perfectly safe. . . .' This syndicate became The Western Borax Company, and rapidly produced two unpleasant surprises. Stauffer immediately concluded a contract with them for ore supplies;[26] and the realization dawned on the Pacific Coast Borax team that in this case both tincal and kernite had been found below the hard rock layer that they had taken as bedrock, and that the same might be true in other areas where they had drilled and found nothing worth while.[27] Immediate instructions were therefore given to probe deep into relatively unexplored areas – notably to the west of the deposit.

A further menace to Pacific Coast Borax developed almost simultaneously. In September the notorious Dr Suckow, with whom they shared a placer claim on the southern half of Secion 14, had started drilling, and had found tincal.[28] Zabriskie's and Rasor's quick reaction was to hand Suckow a letter[29] 'declaring our joint interest, that we would pay half of the expense and notifying him not to make contracts for the disposal of the ore without our approval', although the lawyers had doubts whether they could insist on this last point.[30] Needless to say, Suckow characteristically proceeded to ignore any such possibility and to exploit this area as though it were entirely his own.

Both the Western Borax Company and Suckow were to cause Borax Consolidated considerable trouble over the next few years, and in the short term they proved as big a menace to them as had the Searles Lake producers. By the end of 1928 'Western' had contracted to supply a group of ten German refiners who were former customers of Borax Consolidated,[31] and by mid-1929 their physical presence was becoming uncomfortable, as they were reported to be mining within eight feet of the Baker mine.[32] The man who had started the mine, Buley, had now been pushed into the background, and Stauffer and Blumenberg had acquired equal shares of a 51 per cent interest in it.[33] During 1930 Western Borax started to put up a refinery at San Francisco in preparation for tackling world markets seriously. However, the Depression was already over a year old, and Western Borax, like everyone else, were feeling the strain of shrinking markets. As usual, Blumenberg was believed to be 'pressed for money',[34] and to Stauffer the borax-mining business had always been an expendable hobby.

It is not surprising, therefore, to find tentative negotiations starting in mid-1931 for Borax Consolidated to take over Western Borax, and Osborne reported that 'the ore body was extensive and on the whole of better grade than ours and a mixture of kernite and tincal'.[35] However, the mine was in a dangerous condition, and shortly after Osborne's visit there was a bad cave-in which started a flow of water that endangered the mines of both companies. It became an urgent matter for Borax Consolidated to conclude the negotiations, but these were still to take many more months owing to an understandable reluctance on their part to assume the obligations of Western's low-price contract to supply the group of German refiners known as the Deutsche Borax Vereinigung. This problem was eventually solved by Borax Consolidated, Limited, agreeing to supply ore at cost, and sharing with them the profit on the sale of refined borax, and so on the 13th of May 1933 Western's mine and mill were acquired for $700,000.

Dr Suckow presented a different type of problem. Technically speaking, he was a co-owner and co-operator with Pacific Coast Borax in Section 14, but he went ahead on his own and proceeded to try to capture their European market. A suit brought to regain sums which should have been shared, and also the enormous salary increase which Suckow had awarded himself, was decided against him in April 1932 in the sum of $334,000 plus interest.[36] His other creditors had already started proceedings against him, and in March 1933 the Suckow Company was adjudged bankrupt.[37] However, the Suckow mining properties at Boron were outside this company, and it was not until April 1934 – when the Referee in Bankruptcy approved a five-year lease of the Suckow properties to Borax Consolidated[38] – that they could feel themselves safe from the manœuvres of the Doctor, which had so easily upset the stability of the whole industry. When the Suckow mine was taken over it was found to be a high-grade tincal ore body, but as in the case of Western Borax, the mine was in a dangerous condition and unfit to work.[39]

The aftermath of the discovery of sodium borate at Boron

This mine was not restarted until 1936, and following a visit from Jenifer (then Vice-President of Pacific Coast Borax) Osborne, now Superintendent at Boron, wrote:[40] 'Curiously enough we encountered the anticipated fault in the winze* the next round after you saw the face. It is a superstition that you will lose the ore if you take a woman into the mine, and it may behave the same for Vice-Presidents, although I had never heard of the circumstance before.' On this Jenifer commented:

> It is hoped that Osborne does not credit me with all the peculiarities and defects of the Doctor and things with which he is connected. If taking a woman into the mine will lose the ore, one would naturally think that Suckow would have long ago lost the whole property, since it is understood that he used to be accompanied by a cabinet of women on his underground inspections.

The Doctor's reputation as a ladies' man is still remembered in Boron till this day.

After a cave-in in mid-1937 the Suckow mine was considered too dangerous to continue production, and was only to be opened again for a short period in 1941. In this pre-war period Borax Consolidated were trying to negotiate the outright purchase of the Suckow properties, and did their best to cover themselves by mining as much ore as possible before their lease expired, in case the purchase failed. Shortly after obtaining the lease in 1934 a new shaft called the West Baker was sunk on the Suckow property, and from 1937 this new mine successfully supplied tincal ore for the next seventeen years.

Thus it was, and thanks perhaps to the traumatic years of the Great Depression, that Borax Consolidated were able to consolidate their position at Boron and to face their competitors at Searles Lake with some greater hope of success.

Competition had produced a steady decline in borax prices from 1925 onward, and the arrival of the Depression in 1929 saw a severe contraction in the market for industrial borax throughout the world. As a result Borax Consolidated's net profits remained at a low level of about £50,000 for each year from 1929 to 1933, during which time no dividends were paid.[41] Although the new mine at Boron had not yet produced prosperity, there is little doubt that without it Borax Consolidated would not have survived these difficult years.

Unfortunately, there was little to relieve the gloom that surrounded the Tonopah and Tidewater Railroad after the closure of the Death Valley mines. This Company had not paid a dividend since its formation in 1907, and in fact had not at any time earned the interest on its debentures. Currently it was not even covering its operating costs, and it was in debt to Borax Consolidated for about $3·0 million. With the transfer of the mining operations to Boron the whole of the borate ore traffic had been diverted from the T. & T. R. to the Santa Fe Railroad, and the shares were quite valueless. The objection to the most obvious solution, closure of the line, was that it must be kept in operation or it would lose its franchise, which would face Borax Consolidated with the burden of repaying the Railroad Debenture holders forthwith.[42] But the only new business that could be obtained was a contract for calcining and carrying a clay found in the locality, for eventual sale to oil-refiners.[43] This soon showed a

---

*A downward-sloping shaft connecting two different levels underground.

loss, but long before it failed new and unsuccessful attempts were being made to dispose of the line as a going concern.[44] However, this hope faded, and Borax Consolidated were left with the problem of the T. & T. R. among other remnants of the old colemanite days.

By contrast, once it was clear that colemanite mining in the traditional area of Death Valley was to be wound up, a novel scheme to exploit the scenic beauties of the area was introduced with surprising speed. Major Julian Boyd, the Australian who succeeded Faulkner as Superintendent of Pacific Coast Borax's mines in 1920, had for some time been discussing with Baker a scheme to attract tourists to the area. At the end of 1924 Baker commented to Zabriskie:[45] 'As to a one million dollar investment in a big hotel at the Ranch, as suggested by Major Boyd, this looks to me visionary. In any case we are not the people to go in for an enterprise of that kind.' It is therefore all the more surprising to find the idea of establishing a tourist hotel at Furnace Creek Ranch appearing in correspondence less than three weeks after the discovery of borates in the Section 24 shaft at Boron.[46] Admittedly Death Valley itself and the Furnace Creek area had been abandoned as mining propositions many years before; but the idea of using Ryan also as a tourist centre was being mooted even before an official decision had been made to close down the colemanite mines in the vicinity.[47]

This decision was finally made, and the consequent resolve to go into the hotel business pursued with a zest which might not have been expected from a mining company. Experienced engineers like Boyd and Gower vied with each other to become hoteliers. By the end of 1926 the bunkhouse erected for miners at Ryan had already been converted into a tourist hotel and was to be used as a 'distributing centre for autobuses and tourists'.[48] In 1927 the twelve-roomed Furnace Creek Inn was opened in Death Valley, publicity for this and the Death Valley area being provided by the Union Pacific Railroad.[49] In spite of all the muddy water that had flowed under the bridge since 1914, one of the first (but unfêted) visitors to the inn was F. M. Smith, accompanied by his second wife Evelyn,[50] whose name had graced one of the stations on Tonopah and Tidewater Railroad from the time it was opened.

It was of course hoped that the tourist trade would provide the near-bankrupt Tonopah and Tidewater Railroad with a few dollars to compensate for the loss of its original *raison d'être*,[51] the carriage of ore from the Death Valley mines, and there was talk at the end of 1926 of procuring Pullman cars to take visitors to Furnace Creek Inn.[52] However, the rapid development of private motor transport between the wars soon demolished this hope, and left yet a third and lesser known 'hotel' created in this period – the Amargosa Hotel at Death Valley Junction, which had been converted out of the expensive civic centre so triumphantly opened in 1925 – rather high and dry, but still managing to find custom.

The rest of the hotel story during this period was not one of success. Boyd reported a rush of tourists in 1928,[28] but the onset of the Depression in 1929 dealt the enterprise a blow from which it was only just starting to recover a year or two before the European war broke out in 1939. As Baker put it in March 1929, Furnace Creek Inn could be run like Palm Springs or alternatively as a cheap holiday resort, and yet it seemed to lose money whichever policy was adopted.[54] In 1930 Ryan had to be closed down owing to 'the poor outlook of the tourist traffic due to present conditions and the weather'.[55] Attempts were being made to sell off the hotels to the professional hotel trade as early as 1929;[56] but all was not unrelieved gloom in Death Valley. Publicity continued, and on 31 March 1932

The aftermath of the discovery of sodium borate at Boron 193

Baker wrote to the US management: 'I am glad to have the news as to the Broadcast Easter Service in Death Valley. I hope that the people who heard it will buy a few extra packets of Borax in consequence. Cleanliness is next to Godliness.'

The nine-hole golf-course was laid out in about 1930 by someone with a flair for design. During the months when Death Valley is habitable water flows through the golf-course and converts the scorched earth of summer to a beautiful green oasis. From the fairways one sees magnificent scenery in all directions, dominated in the west by the snow-capped Telescope Peak (11,000 ft) in the Panamint mountain range. Reflection of the bright light from the rocks provides a beautiful variety of colours, especially in the late afternoon. Where else in the world is it possible to finish a round of golf below sea-level by moonlight?

The climate of Death Valley in winter is as beautiful as its scenery. Those in the 1920s who conceived it as a place for a hotel, relaxation and pleasure made one mistake – they were before their time, as the ever-increasing number of visitors has come to show.

Those who dreamed of the conversion of Death Valley into a tourist centre also had other and less pleasant problems on their minds. In the second half of 1926 suggestions were made in London that Lord Chichester, the Chairman of Borax Consolidated, might meet Lord Brabourne, Chairman of Consolidated Goldfields (owners of American Potash), to discuss the problems of the increasingly unprofitable borax business. But Lord Chichester was not keen to meet Lord Brabourne, with whom Borax Consolidated had exchanged sharp words in the early days of the Trona enterprise at Searles Lake, and it was suggested that Count Dru (now in charge at Searles Lake) should come to London and meet Baker. Shortly afterwards Lord Chichester died and was succeeded by Lord Leven,[58] and there followed a series of meetings suggesting various arrangements by which Borax Consolidated might purchase the borax output of American Potash with a view to marketing the product more profitably.

The scheme, designed to avoid breach of the Anti-Trust Laws, would have necessitated Borax Consolidated buying the whole of the output of American Potash of 48,000 tons per year for ten years at an agreed price. Borax Consolidated considered the risk of loss from possible contraction of their own production, and possible further downward pressure of prices from other producers, to be too great, and negotiations ended.

The next attempt to bring price-stabilization to the borax industry came about early in 1929, after its proposer Bates, a company promoter, had obtained some sort of indications of approval from the US Government regarding any legal implications of a scheme that other producers should wholesale their output to Borax Consolidated, who would distribute it.[59] He made little progress, and it all came to nothing when Count Dru withdrew his interest.[60] The likely real reason for this was probably to be found in an announcement made by the Chairman of Consolidated Goldfields at the beginning of May.[61]

This was to the effect that their potash interests had been sold to an undisclosed purchaser at a large profit. Zabriskie was asked by Baker to try to find out, through 'our Banking friends', who the purchaser might be. Various 'certain' tips were given to him in the following days, including du Pont, Howe (ex-Kali Syndicate sales agent in the USA) and his partner Forbes, and the American part of the German I. G. Farben Company. These rumours were passed on to Baker;[62] but very soon, on 23 May, a cable was received by Zabriskie stating: 'Confidential. We know source through whom sale effected and therefore only possible buyer. Do not make any further enquiry.' From that day until the published announcement by the US Government in 1942, five years after Baker's death, there was never a mention in the Borax group's correspondence as to who the purchasers of the American Potash shares actually were, although from the extensive correspondence on the subject at the time it does not seem possible that Borax Consolidated's directors were left in ignorance of this important fact. Baker's letter to H. M. Albright in January 1933 gives a great deal of information on the subject:[63]

> The stock of the American Potash & Chemical Corporation was, until the spring of 1929, practically entirely owned by Goldfields Consolidated but at that time they sold their holding, or most of it, as stated by themselves at their meeting of shareholders. At this time an issue of 6½% German Potash bonds was made here and placed privately. It was about £2¼ million at 96, the remainder of the total issue of £3 million being issued in Holland and Switzerland. This concurrent transaction raised some suspicion as to where that £2¼ million went. It was stated to me by Count Dru, who was a director of Consolidated Goldfields and also the President of A.P. and C.C., that it was placed through Schroders with certain financial interests in various countries on the Continent. It was never disclosed who held these bonds or on whose behalf they were held.

He continues:

> The issue of £2¼ million was equal to $11 million and I knew that that was the amount asked for the A.P. & C.C. because we were negotiating for the purchase ourselves at one time until advised that to go on would be to bring ourselves within the Sherman Anti-Trust Law on the ground that we would be aiming at a monopoly in borax. This appeared to us to be rather stretching a point as we had all the business prior to the operating of A.P. & C.C. and would be only getting back our own business. But we had very competent legal advice and therefore dropped the negotiations. If the deal had gone through it was agreed that we would also have taken over Count Dru and his contract with Goldfields Consolidated. This is noteworthy as, after the sale of the stock of the A.P. & C.C. by Goldfields Consolidated, Count Dru retired from the G.C. Board and was appointed as a manager of the North European Oil Company, being quartered in Berlin, . . . The North European Oil Company is that which works the oil property of one of the leading German potash producers. You can draw your own conclusions as to the probable owners of the A.P. & C.C. . . . If the ownership of the A.P. & C.C. does lie in that direction, it may be surmised who dictates their policies. That these have been directed against us, there is plenty of evidence. . . .
>
> Their position as potash producers is handicapped by the production of one ton of borax to each two tons of potash, whereas normal potash salts tonnage is at least thirty times that of borax. They must, if they recover their borax, in order to keep down the cost price of their potash, produce more borax than they can sell and ultimately must throw back the borax liquors into the Lake which will increase the cost of their potash but they would sell the borax at less than cost and indeed I think they are doing so now . . . they are using their potash profits to support selling borax at an unremunerative price and are trying to put out of business the legitimate borax interests.

When, therefore, Zabriskie wrote in June 1929, only a few weeks after the purchase, that 'By inference I gathered that Mr. Wells [President of Solvay] considered that the Kali people were the purchasers',[64] he was giving the correct answer to the riddle. He had said,[65] that if any European potash interests had made the purchase they would have had good reasons for keeping it covered; and indeed American Potash & Chemical Corporation continued to operate under that name with few internal changes, except that, confusingly, another 'Baker' – F. C. Baker – succeeded Count Dru as President.

Kali Syndicate now controlled the biggest source of American home-produced potash, as well as the major share of potash imports into the USA. The reasons why they immediately decided to double the capacity of the Searles Lake plant are a matter for surmise, but it could have been partly due to the insecure financial position of Germany at the time, which required a good return in dollars on capital invested in the USA. It could also have been that greatly increased American home production of potash might have been used as a justification if foreign control of Searles Lake ever came to be challenged by the isolationist-inclined administration of President Hoover. Whatever the reason, the proposed plant-expansion faced Borax Consolidated with the possibility of their established borax market being wholly captured by A.P. & C.C.'s by-product borax, and forced them into a second major development – to fight back on the potash front in order to defend their borax trade.

Baker made this clear to Diehn of Kali Syndicate in June 1929, but whether he was writing in real or feigned ignorance that they now controlled A.P. & C.C. cannot be deduced from the wording:[66]

> As you are aware, we have been attacked in our industry by the A.P. & C.C., which has been able to do this through the profits they are making on Potash; they have claimed that we have an inferior position, as we have only one product, Borax, whereas they have two and can make their profit on Potash and sell the Borax at a low price to realize it. This being the position and there evidently being a large profit on Potash produced at Searles Lake, the only course left to us was to utilize our own property at the Lake and produce both products, much as has been our repugnance to this step, as our desire has been to confine ourselves to our own business. As I explained to you, we have now arrived at the parting of the ways and we cannot wait longer; we must do one of two things: make an arrangement by which all the substantial producers of Borax will benefit, or take the other road and attack the A.P. & C.C. with not only Borax but Potash, and we are well equipped to follow out this plan.

The question of how well equipped Borax Consolidated were to follow out this plan is not, of course, set out in detail in this letter to Diehn, and the difficulties of starting from scratch in this new line were their main preoccupation in the next few years.

REFERENCES

1. M.D. to Zabriskie, 18.8.26.
2. L. D. Osborne's Notes on his work with Pacific Coast Borax, March 1947, U.S. Borax Archives.
3. Cable: Zabriskie to R.C.B., 18.8.26; also *The Story of Pacific Coast Borax Co.*, Division of Borax Consolidated, Limited (1951), p.47.
4. U.S. Borax Research Corpn. Archives, Connell's unpublished paper d. 1922.
5. Tucker, W. B.: 'Salines and Borax 1929', *Los Angeles Field Div. Mining in California* Vol. 25, pp.71–81.
6. Rasor to Zabriskie, 16.12.25.
7. P.C.B.: Zabriskie to R.C.B., 23.9.26.
8. M.D. to Zabriskie, 21.12.26.
9. *Ibid.*, 14.1.27.
10. *Ibid.*, 3.12.26.
11. *Ibid.*, 26 and 30.11.26.
12. *Ibid.*, 14.12.26.
13. *Ibid.*, 6.9.26 and 9.8.27.

14. *Ibid.*, 17.2.27.
15. M.D. letters, 1927–8.
16. M.D. to P.C.B., 30.12.26.
17. M.D. to Zabriskie, 8.2.27.
18. *Ibid.*, 24.2.27.
19. *Ibid.*, 7.5.28 and 29.5.28.
20. *Ibid.*, 12.9.28.
21. *Ibid.*, 17.2.27.
22. *Ibid.*, 4.8.27.
23. Zabriskie to Rasor, 16.12.26.
24. M.D. to Zabriskie, 28.2.27.
25. *Ibid.*, 16.8.27.
26. *Ibid.*, 12.9.27.
27. *Ibid.*, 30.11.27.
28. Cables: Zabriskie to R.C.B., 7.9.27 and 9.7.27.
29. Cable: Zabriskie to R.C.B., 16.9.27.
30. M.D. to Zabriskie, 20.9.27.
31. P.C.B.: Zabriskie to R.C.B., 2.11.28. Cable from Rasor, 8.11.28; Zabriskie to Cramer *et al.*, 15.11.28. Cable from M.D. to Zabriskie, 12.11.28.
32. M.D. to Pacific Coast Borax, 20.8.29.
33. M.D. to Zabriskie, 30.5.30.
34. *Ibid.*, 7.8.31.
35. P.C.B.: Cable from Zabriskie to R.C.B., 8.8.31.
36. *Ibid.*, 18.4.32.
37. *Ibid.*, 8.3.33.
38. *Ibid.*, 17.4.34.
39. M.D.: Gerstley to Lesser, 21.10.37.
40. P.C.B.: Osborne to Jenifer, 21.2.36, and Jenifer to R.C.B., 27.2.36.
41. Borax Consolidated, Ltd – Annual Reports.
42. M.D. to Crisp, 7.2.35.
43. M.D. to Zabriskie, 27.1.28.
44. M.D. to Jenifer, 6.1.32.
45. *Ibid.*, 15.12.24.
46. *Ibid.*, 6.9.26.
47. P.C.B.: Zabriskie to R.C.B., 19.8.26. Also M.D. to Zabriskie, 13.9.26.
48. M.D. to Zabriskie, 17.12.26.
49. Union Pacific to R.C.B., 26.12.27, *et al.*
50. P.C.B.: Boyd to Zabriskie, 11.3.27.
51. M.D. to Pacific Coast Borax, 2.7.28.
52. M.D. to Zabriskie, 14.12.26.
53. M.D. to Pacific Coast Borax, 19.4.28.
54. *Ibid.*, 28.3.29.
55. M.D. to Jenifer, 14.2.30.
56. M.D. to Pacific Coast Borax, 4.7.29.
57. M.D. to Zabriskie, 21.7.26.
58. Minute Book, 21.8.27.
59. M.D. to Diehn, 25.6.29. Zabriskie to Jenifer, 14.3.29.
60. M.D. to Pacific Coast Borax, 29.4.29.
61. M.D. Cable to Zabriskie, 3.5.29, and reply from Zabriskie.
62. *Ibid.*, 23.5.29.
63. M.D. to Albright, 24.1.33.
64. P.C.B.: Zabriskie to R.C.B., 24.6.29.
65. *Ibid.*, 13.5.29.
66. M.D. to Diehn, 25.6.29.

18

The start of potash-mining in America, and borax in the years before the Second World War

Borax Consolidated intended originally, as indicated by Baker's letter (25 June 1929) to Diehn, to fight A.P. & C.C.'s policy of low borax prices and profitable potash prices by also producing borax and potash from their own base on Searles Lake, in spite of the previous Borosolvay venture there.

As early as June 1928 Knight, the Pacific Coast Borax 'watchman' at Searles Lake, had started production experiments there, urged on by Baker stressing that time was of the greatest importance.[1] This approach might, however, have proved an inadequate defence against A.P. & C.C. competition if it had had to be taken further. The legal but dormant connection with Solvay in Borosolvay had first to be unscrambled if Borax Consolidated were to be free to make full use of the Jones-Grover patents covering the carbonation process, and this had to be done without rousing the German Solvay Company, who were said to be 'interested in Kalisyndikat'. By early 1929 Solvay were prepared to assign to Pacific Coast Borax their share in the patents,[2] but Wells of Solvay indicated that, while they had no present interest in making a joint arrangement for production of chemicals from Searles Lake, the soda there 'might have a future interest',[3] and only three years later they were discussing going into soda-production jointly with the West End Chemical Co. Furthermore, Dr Hecker, on a visit to Searles Lake in October 1928, had reported that A.P. & C.C.'s process was probably infringing the Jones-Grover patents,[4] and it was decided to sue them; a suit which Borax Consolidated eventually lost.[5] Also about this time two newcomers, Howe (the former sales agent of Kalisyndikat) and his assistant Forbes, were preparing to start a scheme of their own on Searles Lake in rivalry to American Potash and all other producers. Taking all these circumstances into account, the chances of a Searles Lake scheme proving the saviour of Borax Consolidated's crumbling fortunes could not have been rated very high.

Other ways were sought. At the end of 1928 Rasor had begun investigating potash possibilities in places other than Searles Lake when he met a Mr Dwyer, who had volunteered the information that potash fields in New Mexico recently discovered would probably control the potash situation in future.[6] 'I could not draw much definite information on this subject,' wrote Rasor, 'but he told me that a New York firm had completed a $250,000 drilling campaign and were well pleased with the results.' Rasor did not hurry to New Mexico. In February 1929 he was investigating a hole bored somewhere in Utah by 'two men' called Mulvey and Means, which was alleged to contain carnallite (potash) but turned out to be a dud.[7] Meanwhile he had been making routine inquiries about New Mexico at the Bureau of Mines Geological Survey Department, and there he came across a great deal of information about the exploration work of a company called Snowden and McSweeney.[8]

V. H. McNutt, technical director of this company – who were oil operators – had obtained in 1925 a Federal Oil and Gas Permit to drill for oil on Federal land in Eddy County, New Mexico.[9] Permit-holders on Federal lands were required to send a portion of each drill core to the Geological Survey, and confirmation came back from the Survey that potash was present in the core of the first well drilled. Accordingly McNutt applied for and obtained a potash-prospecting permit on these same lands, and a series of holes were drilled between 1926 and 1930 to determine the extent and depth of the potash beds in the area.[10] Altogether about 3½ miles of 3 in. cores were obtained, which showed the existence of four beds of potash in the form of sylvite (potassium chloride ore). This was the most soluble and economic form of potash ore for conversion into refined muriate of potash (potassium chloride), or for sale in its crude form as 'manure salts'. The four beds discovered between 762 ft and 867 ft depth contained an estimated total of 231 million tons of potash, but the fourth or bottom bed, at 29·86 per cent K_2O, was almost twice as rich as the other three, and a fifth bed was appearing below it.

So Rasor went off, in September 1929, to the small township of Carlsbad in New Mexico, having promised Baker that 'the expense will not be very much'.[11] By December he delivered his report,[12] showing that the drill cores contained good potash values and claiming that the deposit contained enough ore to supply the needs of the United States for a hundred years.[13] The deposit was owned by 'the United States Potash Company', which was dominated by two elderly men, Snowden and McSweeney.

Baker, during one of his periodical visits to the States, met Snowden, and the meeting and Rasor's report had convinced him that this could be a major potash proposition,[14] and that even the Germans had not got a better one on which to work. On the other hand, the Snowden and McSweeney company, after abortive negotiations with American Cyanamid, decided to take a look at Borax Consolidated. McNutt visited Wilmington, and appears to have been suitably impressed with the company's technical resources, as he recommended to his board that an interest in the potash venture be offered to the company.[15]

Apart from the considerations dictated by the struggle with A.P. & C.C., there were more normal reasons which made the proposition an attractive one for the Borax Consolidated board. In 1929 the US potash market was substantial, and domestic production supplied only a small part of it.[16] At existing prices and estimated costs at Carlsbad the investment would show an attractive return.

The aspect of the negotiations which proved troublesome, however, was the proposed financing of this venture. At first all seemed most promising, even to the cautious Baker, who was confident that Snowden would appreciate that Borax Consolidated's knowledge of the mining and chemical sides would be useful, as soon as it was decided what financial arrangements were required.[17] He thought that the venture would best be financed in the joint interests of both parties by raising a part or the whole of the capital required from the public, as Kalisyndikat had recently done.[18] Preliminary negotiations were conducted principally with Snowden, but he died shortly after they began, leaving it to his younger partner McSweeney (then aged seventy-five), to allow Borax Consolidated the precious extra 1 per cent over the 49 per cent interest in the U.S. Potash Company which Snowden had offered, and also the management of the venture.[19] The price eventually agreed for the 50 per cent interest was two million dollars, and the board of Borax Consolidated approved the deal on 8 September 1930. In a subsequent letter to J. P. Morgan[20] this was alluded to as 'a large sum' and it was paid in the form of notes maturing at 12, 18 and 24 months. This was in itself a heavy burden, which Lord Leven (Chairman of BCL), Baker and Gerstley eased by purchasing a total of 125 of the 2500 USP shares.[21] But there were further sums to be found. USP itself required an estimated $1,761,000 for the overall cost of the proposed mine and refinery, and a further $3m for working capital, and these were the figures put to J. P. Morgan of Wall Street, with the suggestion that Morgan take up an issue of 5 per cent preference stock or bonds.[22] This approach soon proved to have been a great mistake. Morgan stated that they could not take up the business, but would be pleased to give introductions to other finance houses[23] who then tended to take the attitude that, Morgan having considered the matter, they could not touch it. Morgan had in fact turned the idea down because they wanted a good share of the equity of United States Potash for a number of years, and Baker was not prepared to concede this, considering that the prospects for the project were too good to be given away for the sake of ready cash.[24] It was alleged that Morgan tipped off the other banks that this was the line they were going to take, and certainly Baker and his colleagues, after approaching the Continental Bank, Chase National Bank, and others, soon came to the same conclusion. A basic reason for the whole difficulty was, of course, that the effects of the Depression had reached all corners of the business world by 1931, and loan money was hard to obtain.

The method eventually adopted was to rely largely on financing the erection of the plant by deferred payments to contractors consisting of a certain part in cash and the rest in USP notes of the same type as those offered for the original purchase of the 50 per cent interest.[25] This meant that some financing would have to be done at once, though on a more restricted scale than that previously anticipated, and first thoughts were that it would have to be handled in Britain. Unfortunately, when Britain went off the gold standard on 20 September 1931 it put an end to any hope of raising money on the London market, owing to the potential exchange losses involved. It was therefore not until December 1931 that a deal was eventually made with a contractor in the USA to erect the refinery.[26] The Moore Company would put up the plant and would arrange the finance, and Borax Consolidated would pay 15 per cent of actual cost and bank interest, the total sum involved being about $800,000, which was to be paid over 17 months. In the absence of an adequate loan from the banks the monthly payments to Moore would probably have to be financed by USP drawing on Borax Consolidated in London, and this emphasized a weak side of this potash enterprise – the strain that it imposed on Borax Consolidated's finances throughout the 1930s. 'We are in the position', wrote Baker,[27] 'of having to provide money for the Potash Company as well as our own business in these most difficult times and we have advanced in money and plant a minimum of $600,000 up to date. . . . All this is depleting our reserves.' So perhaps Snowden and McSweeney's main motive for letting Borax Consolidated into the enterprise in the first place was not so much their admiration of their technical efficiency and probity as their intention to profit from the financial support afforded by a large mining group. In April 1932 Baker described in more detail the serious financial problem in which his company had become enmeshed and added an illuminating personal touch:[28]

> The burden is very heavy and I am feeling my share of it rather much just now. Our partners in the potash enterprise are not of any assistance to us and I am rather surprised at the attitude of Mr McSweeney; he is, as you say, a rich man and, if so, he can surely do what Lord Leven and I contemplate doing – raising some money against his securities and taking some of the notes.

There were, of course, other factors exacerbating the situation for United States Potash during these years. Buyers now had a choice of several potash-suppliers – American Potash, USP, the new Potash Company of America (also in Carlsbad), Spain, Russia and the Germans and French – where only German and French imports had existed before, and such buyers as elected for USP were proving extremely slow in providing cash by paying for their consignment stocks at the due date.

Eventually, after arguments with McSweeney, the USP board agreed, in July 1932, to a public issue of $2,500,000 in two-year 10 per cent notes,[29] which was partly intended to relieve the strain on Borax Consolidated, and consequent criticism from their shareholders. These notes were convertible into preferred stock. This proved an important point, as notes were not popular on the stock market in the mid-'thirties and it was really only by the strength of its own production and sales efforts that USP started to pull rapidly out of its financial difficulties from 1932 onward. Sales of 'manure salts' began early in 1930, and by July the custom of the 'Big Six' fertilizer producers had been secured, sales then running at a level of 80,000 tons a year.[30] The refinery (located twenty miles from the mine) for production of refined potash did not start up until September 1932, but rapidly established a satisfactory margin of profit.[31] By the end of that year the company was making a profit, without having raised any capital other than that provided by Borax Consolidated.[32] Only nine months later, in September 1933, USP could claim to have paid off the balance due on its plant from profits and was already engaged on the extension of the plant.[33] It had sold all its output during this period.

In the autumn of 1933 the board of the U.S. Potash Company invited Horace Albright to become vice-president and general manager, and his association with this company, with Borax Consolidated and later with U.S. Borax continued for over thirty years. Zabriskie had been associated with Albright's father, and Baker and Zabriskie had met many times with Mather and Albright in National Park areas like Yosemite and Yellowstone. In suggesting Albright, Baker was proposing a man whose personality and accomplishments as an administrator he knew well. His experience with both Federal and state governments would be a valuable asset in establishing a new industrial enterprise related to the powerful agricultural lobby and located in the state of New Mexico, where Albright was already well known, having recently established the Carlsbad Caverns as the state's first National Park.

The election of Franklin Roosevelt as President in 1932 and the return of the Democrats had been an anxious time for those who had served in the Department of the Interior under a succession of Republican administrations. Albright was determined to stay long enough to protect the Park Service against any meddling politicians; but fortunately the appointment of Ickes as Secretary of the Department of the Interior and the passage in June 1933 of the Executive Order in which Public Buildings, Reservations, National Parks, National Monuments and National Cemeteries were consolidated into a single office (which set up the policies of the Park System as it exists today), all enabled Albright to feel that the future of his and Mather's work was reasonably secure. He could now pursue a new industrial career with people he knew, and who would be in harmony with his own dedication to conservation. Almost his last action as director of the Park System was to obtain admission of Death Valley as a national monument. On the death of McSweeney, Albright succeeded him as president of the U.S. Potash Company and in his ninetieth year, 1980, he was awarded the Medal of Freedom – the highest civilian award in the United States.

The potash project at Carlsbad was highly successful, and on 29 July 1937 Gerstley, in a letter to USP's general manager, was able to write: 'From a political point of view it would be unquestionably a mistake to attract special attention to the affluent position of the company. . . .' Surprisingly enough, however, all this technical and productive effort at the superb Carlsbad deposit, and the plant increases and potash price cutting to which it provoked A.P. & C.C. between about 1932 and 1936, do not appear to have had a great deal of effect on the import of potash into the United States. The statistics show that, in spite of the support of President Hoover between 1929 and 1932, and probably a genuine desire on the part of the US fertilizer industry to 'buy American',[34] imports of muriate of potash into the USA rose from 27,000 tons in 1933 to over 129,000 tons in 1935.[35] Statistics in the *Chemical Trade Journal* show an increase of German potash exports from 514,000 tons in 1933 to over 704,000 tons in 1934,[36] and in May 1935 the *Financial News* claimed that the Germans were taking about 83·55 per cent of the world's potash business of about two million tons a year.[37] Figures sent by Lesser to Albright in July 1937[38] show that USP's percentage share of the fertilizer season sales in the American market increased from 17·30 to 20·54 between 1936 and 1937, while A.P. & C.C.'s share declined from 19·86 to 15·73, so that at least Borax Consolidated (through USP) had succeeded in getting on terms with A.P. & C.C. over the potash menace to its borax business. In fact, it appeared that the productive capacity of the two companies was practically the same, approximately 180,000 tons per annum.[39]

Clearly, the dream of the United States becoming self-sufficient in potash was still far from fulfilment; but the narrower question of whether Borax Consolidated, by bringing into operation the first large-scale potash mine in the country to compete against A.P. & C.C.'s Searles Lake operations, had secured its borax business against attack by the latter's by-product borax is more difficult to answer. United States Potash made good profits in a rapidly expanding market, and 50 per cent of these went to Borax Consolidated, whose own profits showed some measure of recovery from 1934 to 1939.[40] Nevertheless, only in one year, 1937, did these profits reach the level that had been sustained in the period between 1918 and 1926. Meanwhile A.P. & C.C., in spite of a temporary setback and closure in 1931, doubled the capacity of their potash and borax plants between 1929 and 1934, made a significant reduction in the borax price in 1933, and followed the Germans by introducing a potash price reduction in 1934.[41] The two leading Spanish potash-producers also started exporting, and, at one period in the mid-thirties Spain became the third largest potash-producer in the world,[42] while increasing Russian production also became a disturbing factor. But by 1936 the pace of the conflict eased off, as Spain became embroiled in the Civil War and Nazi Germany became preoccupied with its massive rearmament programme.

In spite of competitive pressures, the potash investment was thus highly successful in these early years. U.S. Potash were pioneers, and established the first potash underground mining operation on the North American continent. Having shown that the extensive Carlsbad potash deposit in New Mexico could compete with European sources, other companies were not slow to follow the example. Carlsbad became famous as a potash-producing centre, whereas before it had only been famous as a tourist centre on account of the 'Caverns'.

Borax before the Second World War

The year 1933 was marked by a revival in the American economy, and the Great Depression began to bottom out. Borax sales started on the path to recovery, although profit margins were poor owing to over-capacity and price competition. In 1934 A.P. & C.C. added to the existing worries of Borax Consolidated by introducing a new form of borax from which all water had been removed. This product, anhydrous borax, gave a considerable freight-saving over normal refined borax (which contained 62 per cent water) and was particularly competitive in the glass and porcelain enamel industries, and in the overseas markets of Europe and Japan.

At this time with the financial problem of the T. and T. Railroad Company on top of everything else, the shortage of cash in Borax Consolidated for capital expenditure or any other ambition was about as bad as it is possible to imagine. However, the resourceful Fred Beik was given the task of designing and building a furnace at Boron to produce anhydrous borax, and by 1935 the product 'Rasorite concentrates anhydrous' (RCA) was on the market as an answer to the competition.

In the late 1920s the sale of small packages of 20 Mule Team borax and soap chips for use in the home in the United States was still dependent on the romance of the 20 Mule Team, much as Smith and Mather had promoted it in the early days. The mules and wagons were moved from place to place, and created much interest with the public, wholesalers and retailers, but by the mid-1930s sales promotion needed more modern methods. Death Valley literature had taken over from the live mules, and in August 1934 Jenifer wrote somewhat pathetically to Baker:[43] 'It is a pity that the song-book offer was not more of a success but the number of people who sing, or who think they can sing, is very much less than those who like to read stories'; and added somewhat cynically about the publication proposed for the following January: 'Mrs Cornwall will have been West before then and have had an opportunity to acquire a few more – "stories", shall we call them?'

However, just at that time of uncertainty Jenifer selected a professional advertising agency, McCann-Ericson – quite a new departure – and they developed the highly successful *Death Valley Days* radio programme, based on the timeless popularity of the Wild West, for which a freelance writer, Ruth Woodman, wrote a series of engaging stories based on history and legend. As a result of 'going on the air', the consumer products part of the business did much to restore the stricken financial fortunes of Borax Consolidated as the world emerged from the Depression years and paved the way for greater things to come in the television era.

Chris Zabriskie had succeeded Smith as head of operations in the USA in 1914, although Baker had taken over the office of president. From 1922 Zabriskie was an increasingly sick man, and in 1933 he was succeeded by Frank Jenifer, who had originally joined Smith to be general manager of the Tonopah and Tidewater Railroad. Zabriskie died in 1936, and his memory is honoured by Zabriskie Point in Death Valley.

The next pillar of the old order to disappear was Clarence Rasor, to whom the company owed its possession of the Boron borax deposit, and who was also a prime instigator of participation in the Carlsbad potash project. He was also the untiring defender of Borax Consolidated's numerous mining claims against the encroachments of others up and down the USA. He retired in February 1936, and Baker wrote to Jenifer:[44] 'The replacement of Rasor will be quite a difficult matter as he has been such an exceptional man.' He died on 22 May 1946.

Finally, on 12 January 1937, the end of the road arrived for Richard Charles Baker, founder, with F. M. Smith, of Borax Consolidated, and managing director since its incorporation in 1899. His letters reveal that although he had been an athletic type in his early years, he suffered much from poor health with advancing age (he reached the age of seventy in 1928). Heart trouble plagued him throughout the years of the Depression and the long and frustrating negotiations to raise money for the potash project. His last illness began in November 1936, but he did not let the control of the company's affairs slip from his hands until just before Christmas, only three weeks before he died. 'As you know,' wrote Gerstley to Jenifer, 'he put up a wonderful fight, even joking with the nurses and doctors two or three days prior to his final collapse.'[45]

After wading through over forty thousand typescript pages of Baker's letters, it is impossible to avoid the conclusion that here is the man to whom Borax Consolidated, Limited owed its place in the world. This impression is borne out by reading also what others said about him; and the reason why F. M. Smith has stolen the limelight in so many tales about the early days of borax is – apart from Smith's bold, reckless and colourful personality – to be found perhaps in what Baker once wrote to Zabriskie:

> I note the letter from 'Drug Markets' to me, asking for particulars so that they may include them in their 'Who's Who'. I very much object to personal advertisement of any kind and should they write to you again I shall be glad if you will say I am on the Continent, not well, and unable to deal with any enquiries of this kind at the present time.

In Horace Albright's words,[47] 'Baker knew the West and understood its people as well as any American in his day.' Before the aeroplane or transatlantic telephone, he was a relentless traveller to the Far West, and got to know all the people in that part of the organization extremely well. Much of his success can be attributed to this understanding of the borax world's affairs in its most important area. Baker was undoubtedly a phlegmatic character. He went with Zabriskie to the opera to hear Caruso on the evening before the San Francisco earthquake, and then returned to the Palace Hotel – where, incidentally, both he and the great singer were staying. Baker never seems to have recorded or discussed his impressions of this dramatic event, but unlike Caruso he returned to San Francisco many times thereafter. Pacific Coast Borax's office at 101 Sansom Street was totally destroyed by fire, and was thereafter reopened in Oakland.

From all accounts Baker lost some of his English reserve when faced with the breezy manners of the Far West, but never of course to the extent of using Christian names. Those who worked there referred to him affectionately as 'Lord' Baker, and to them he seemed perhaps more made for the part than the real Lords Chichester and Leven who descended on them from time to time in their capacity as successive chairmen of Borax Consolidated. On one occasion 'Wash' Cahill, the Irish superintendent of the Tonopah and Tidewater Railroad, found himself riding with Baker in the observation car. Having heard of Baker's interest in violins, and wishing no doubt to put him at his ease, he broke the silence: 'Mr Baker,' he said, 'I hear you fiddle a bit.' The response, if any, sadly goes unreported.

Baker was in his eightieth year when he died, and he had been managing director of Borax Consolidated since the beginning. Gerstley, his successor, reached seventy only a few weeks afterwards, and the problem of succession was a serious one, as Lesser – who had been on the board since 1926 – was of the same age group as Gerstley. Writing to Jenifer, Gerstley told him what had been decided:[48]

> Lord Leven [i.e., the Chairman] has discussed the question of the succession with me, indicating his wish that I should carry on, but, for reasons I have already laid before you, I propose that Lesser and self be appointed Joint Managing Directors [quite a common practice in large companies] and thus provide for my retiring in, I hope, a couple of years when conditions are such that the business is organised with another suitable candidate to take my place as joint managing director.

At this time Lesser's son, Fred, rejoined Borax Consolidated in London and Gerstley's son, J. M. (Jim) Gerstley, who had joined the company in 1933, was working in Los Angeles with Jenifer.

The world that now had to be faced was one already changing rapidly from the world of Baker's heyday, and the three events that were principally responsible for bringing this about were Japan's invasion of Manchuria in 1931, and the accession of Hitler to the Chancellorship of Germany and of Franklin Roosevelt to the Presidency of the United States, both in early 1933. Roosevelt's accession brought the National Recovery Administration and its codes for many industries, including the borax and potash industries.[49] As described in a recent history of the United States: 'Written by committees representing industry, labour and the government, NRA codes set up minimum wage and hour standards for hundreds of different industries. They also recognized labour's right to organise, in exchange for which business itself got the right to stabilise prices and production without fear of anti-trust suits.' Unfortunately, the codes brought no joy to the borax and potash industries as once more A.P. & C.C. showed no signs of wishing to join in any price-stabilization agreements.[50] As a result, Borax Consolidated found themselves squeezed between the effects of compulsorily increased wages and reduced hours on the one hand and continuing price competition on the other.

Even after the Supreme Court, in May 1935, declared the NRA unconstitutional, the ferment that it had engendered continued and increased with the Works Progress Administration, the Social Security Act, the National Labor Relations Act and other aspects of the New Deal; and the effects of it within Borax Consolidated became noticeable with more intransigent demands by labour and the unions, particularly for increased wages and the closed shop. One reference to these by Gerstley senior[51] is of significance as showing the ideas which still prevailed in the 1930s and which would soon become out of date:

> I thought it right to put before you the alternative that we should be compelled to resort to if a strike occurs at the Mines, i.e. to immediately re-open South American

production – which, as you know, has lain dormant for years – in order to fill the export trade demands, especially Europe, Australia and India, as our position would not permit of our letting the trade down if the present American source of production forced this issue. But we should very much regret this (and the miners would) . . . as, once set going again, the South American governments concerned would insist on our maintaining our activities there and the consequence would be material reduction in the Californian output, displaced by South American.

Indeed, the international aspect of its affairs was starting to worry Borax Consolidated even more in other directions. By August 1934 the effects of Hitler coming to power the previous year were already starting to be felt and German buyers were reporting difficulty in obtaining permission to buy foreign currency.[52] Within a few months, the situation was obviously developing into something much more complicated and alarming.

From 1936 onward there was a marked accentuation of the difficulty of trading in Germany and Japan.[53] Customers in Germany who required raw material for their refineries found themselves unable to obtain American dollars to pay for it, while they could instead obtain limited facilities to pay for borate ore from Turkey or Chile, where the Germans had an exchange arrangement. Borax Consolidated did not want to lay up marks in Germany or invest there, so that in effect they could not obtain payment from that country, and by the mid-thirties they were owed £50,000, long overdue, for ore supplied and their trade there had fallen off considerably.

In May 1936 came indications that Germany was deliberately looking elsewhere for borate material, and that efforts were being made by the German financial authorities to encourage the shipment of borate of lime from Argentina.[54] The German minister there had been instructed to inquire into a scheme whereby Germany would supply machinery etc. against shipments of borate, and by the end of 1937 Argentinian borate was reaching Germany in fairly considerable quantities in competition with such American material as could get through. As Gerstley senior put it:[55] 'We are up against loaded dice to some extent as the German Government are specially encouraging channels where they can develop a reciprocative barter trade which, as you know, is becoming more and more difficult between that country and the U.S.A.' By February 1938 production at Boron had had to be reduced to four days a week owing to German refusal of import permits.[56]

Meanwhile Japan – which had become the third largest consumer of borax products in the world, and had taken in 1937 well over a third of its requirements from Borax Consolidated – had started to show the same tendency.[57] In that year it also imported some 4,000 tons of crude material from Argentina, and Gerstley commented that neither American nor English sources of supply were viewed any longer with particular favour. Finally, by the end of July 1939 – only five weeks before war broke out in Europe – it had become impossible to import refined borax into Japan, and only raw material for Japanese manufacture was now permitted.[58] The breach between Japan and the Western democracies which had begun with the Japanese invasion of Manchuria in 1931 had now become wide enough to be added to the German menace as a barrier to the normal channels of international trade.

These comparatively minor commercial setbacks were the rumblings that foreshadowed the outbreak of the Second World War, and the coming years would show that it would take a world-wide disaster of tragic proportions before trade and industry could return to something remotely resembling the freedom and prosperity of earlier days.

REFERENCES

1. M.D. to Pacific Coast Borax, 8.6.28.
2. M.D. to Zabriskie, 4.2.29.
3. M.D. to Pacific Coast Borax, 30.1.29.
4. M.D. to Zabriskie, 24.10.28.
5. M.D., Cable from Zabriskie, 16.2.33.
6. P.C.B.: Rasor to Zabriskie, 13.12.28.
7. M.D. to Pacific Coast Borax, 5.2.29. Also M.D. to Zabriskie, 13.6.29.
8. Gertrude Stiehler (former secretary to Snowden and McSweeney) letter to J.M.G., 25.10.75.
9. *Ibid.*
10. M.D. to J. P. Morgan, 3.3.31.
11. M.D. to Pacific Coast Borax, 7.10.29.
12. M.D. to Zabriskie, 6.12.29, and Zabriskie to R.C.B., 26.12.29.
13. *Ibid.*, 11.6.30.
14. *Ibid.*, 17.6.30.
15. Gertrude Stiehler, *op. cit.*
16. M.D. to J. P. Morgan, 3.3.31.
17. M.D. to Zabriskie, 17.6.30.
18. *Ibid.*, 23.6.30.
19. *Ibid.*, 5.8.30, and Gerstley to Jenifer, 1.3.37.
20. M.D. to J. P. Morgan, 3.3.31.
21. Minute Book, 8.9.30.
22. M.D. to J. P. Morgan, 3.3.31.
23. M.D. to Jenifer, 1.7.31 (Private).
24. M.D. to Zabriskie, 15.7.31.
25. M.D. to Jenifer, 1.9.31.
26. M.D. to Paul Speer, 10.12.31.
27. *Ibid.*, 10.12.31.
28. M.D. to Zabriskie, 11.4.32.
29. Minute Book and cable from Jenifer, 12.7.32.
30. P.C.B.: Zabriskie to R.C.B., 10.3.31 and M.D. to Secretary, International Investment, Ltd., 21.7.31.
31. M.D. to Jenifer, 27.9.32.
32. M.D. to Albright, 29.12.32.
33. M.D. to Rossetti, 5.9.33.
34. M.D. to Zabriskie, 13.2.31.
35. M.D. to Albright, 24.7.35.
36. *Ibid.*, 8.2.35.
37. M.D. to Jenifer, 27.5.35.
38. P.C.B.: Lesser to Albright, 16.7.37.
39. *Ibid.*, 10.5.38.
40. Annual Reports and J. Gerstley's 'Notes'.
41. M.D. to H. M. Albright, 5.6.34.
42. *Ibid.*, 25.8.33.
43. Jenifer to R.C.B., 20.8.34.
44. M.D. to Jenifer, 11.2.36.
45. R.C.B.: Gerstley to Jenifer, 20.1.37.
46. M.D. to Zabriskie, 8.6.26.
47. Travis, N. J.: Recollections of Horace M. Albright at a meeting at his house, Sherman Oaks, California (November 1981).
48. M.D. to Jenifer, 8.2.37.
49. Miller, Wm.: *A New History of the United States* (Paladin, 1970).
50. M.D. to Jenifer, 20.2.34.

51. M.D. to Jenifer, 4.3.37.
52. P.C.B.: Jenifer to R.C.B., 20.8.34.
53. M.D. to Albright, 2.10.34.
54. M.D. to Jenifer, 26.5.36.
55. M.D. to Jenifer, 10.12.37.
56. P.C.B.: Jenifer to Lesser, 18.2.38.
57. M.D. to Jenifer, 12.1.38.
58. *Ibid.*, 24.7.39.

19

The Second World War

Politically and commercially, the Second World War broke out long before hostilities actually began; and some of its antecedents which affected Borax Consolidated have been outlined at the close of the preceding chapter. In fact, from the point of view of the Allies and neutrals, an account of the direct effects of the Second World War on the borax industry must mean almost entirely an account of its effects on the Borax Consolidated group, which alone had a substantial network of factories and agents in Europe in 1939. The indirect consequences for American Potash and other rivals will become apparent later in this chapter.

The worsening political situation began to be felt more directly by Borax Consolidated from 12 March 1938, when the Nazis marched into Austria. On the 15th of that month James Gerstley reminded the American headquarters that the group's factory in Stadlau – until recently in Austria, but now under German tutelage – might henceforth require German permits for import of raw material, and if there were any doubt in regard to payment to Borax Consolidated it would be necessary to require payment in dollars in advance before shipment.[1] All the same, Gerstley told Rosenstern (the Borax group's agent in Hamburg) a few days later that a substantial shipment had gone forward to Deutsche Borax Verein that day; but he took the precaution of sending Norman Pearson (who had joined Borax Consolidated after Baker's death) to investigate the situation in Germany, Austria, Poland and Czechoslovakia.[2]

Gerstley seems, like so many people at that time, to have remained remarkably relaxed about the international situation. On 16 August 1938 he wrote to Albright,[3] then on holiday in Europe:

> If you should glance at the foreign press while abroad, you may notice considerable malaise existing about the present mobilisation on a grand scale in Germany. Personally, I see no reason to anticipate any trouble at present, although possibly there may be some eye to effect and political pressure but no justification for the gold hoarding generally taking place throughout Europe.

About the future of the Stadlau factory, on 1 September 1938, just as the crisis over the Sudetenland was warming up, Gerstley wrote to Dr Hecker at Baden-Baden and asked him to reassure Dr Seitz (who was in charge of Stadlau) that there was no question of disposing of the Borax group's interests there, in spite of a visit paid to Stadlau some time previously by a party of German refiners.[4] No suggestion of the kind had arisen, especially as it was understood that a new German law had been promulgated immediately Austria was absorbed into Greater Germany, prohibiting any German individual or firm from acquiring an Austrian factory or interests up to a certain date. This was intended to show the Austrians that the Germans did not wish to swallow up their individuality.

The Munich crisis a few days later naturally shook the Western democracies out of this relaxed attitude. By this time Gerstley had come round to the view that, in face of Hitler's most recent announcement, war was inevitable.[5] Writing to Jenifer in Los Angeles, he mentioned Stadlau's predicament in such an eventuality, and went on to talk about the Borax group's other endangered factory, at Aussig. This place was situated in the Sudeten area of Czechoslovakia, within about forty miles of the German frontier and in a territory which would probably be invaded at once. It was in the nominal charge of Dr Tullner (appointed in 1923), but by the time that Hecker managed to visit it about 20 September 1938 Tullner had fled to Berlin, whence he sent a remarkable letter to London 'explaining' his conduct. Gerstley said that it simply conveyed the impression that he was suffering from a fit of nerves. He mentioned no word about Brettschneider or Proksch or any other German employees at Aussig, and it was incomprehensible that the German manager should abandon them, as they would surely be exposed to the same risks. If Dr Tullner was not in a condition to return – and Hecker was of the opinion that he would be exposed to danger in returning – then they would have to appoint a temporary manager.

The Sudetenland crisis completely changed Gerstley's mind about future prospects. On 4 October he wrote to Jenifer in the USA that he had very little confidence in the permanent nature of these European peace efforts as long as the existing regime and system prevailed in Germany.[6] All shipments were now to be covered against war risks, all stocks of refined material in store were transferred from Hamburg to Rotterdam, and it was decided to transfer the French office from Paris to Coudekerque.[7]

Lesser, however, continued to take an optimistic view of the future. 'Regarding the situation in Europe,' he wrote to Albright, head of U.S. Potash, as late as February 1939, 'I am one of the few who do not believe in war in Europe. . . . Of course, there are many others, amongst them some of my colleagues, who have different views of the European situation. . . .' This may account for the fact that throughout the last twelve months before war broke out in Europe his optimistic efforts continued to try to persuade the German Control to change its mind about not importing from the United States.

It was obvious, however, that hostility to the Nazis was growing in the USA, particularly after the final rape of Czechoslovakia in March 1939, and that this was likely first of all to divert fertilizer orders from German to American producers. The US government also started to take note of this feeling. Lesser had to concede that although they had thought that they had at last succeeded, through the help of influential friends in Germany, in proposing to the competent German Control authorities the possibility of permitting exports to

Germany under barter or compensation agreement, the fact that the US government had suddenly announced a 25 per cent countervailing duty on all German products which paid import duties in the USA had rendered the position 'now not clear again'.[8]

The position became very clear again, however, on 3 September 1939 when Britain declared war on Germany. The USA did not come into the war until over two years later – December 1941, after Pearl Harbour – but the duality of the Borax Consolidated group's position, as a British company with the major part of its assets in the USA, lined up the whole of the group against Britain's enemy. On 7 September James Gerstley wrote to Jenifer: 'Of course, all shipments to the enemy will cease' and went on to define that this meant exports to the Deutsche Borax Verein, Schott & Genossen and various other famous customers, as well as their own Boraxwerke, Stadlau. It might still be possible to supply two small buyers in Hungary via Trieste while Italy remained neutral, but exports to Poland through Gdynia would have to be suspended until the end of the war. Gerstley had no doubts on the hostile policy to be pursued:[9] 'You will, like ourselves, we trust, read with equanimity the news that the s.s. Olinda, which was sunk by the British, had on board 750 tons of Argentine borate destined for the Deutsche Borax Verein, and 203 tons for Aussig Works (Czechoslovakia), now both enemy destinations.'

Nevertheless, although not on the scale apparent in the First World War, some attempts at evasion of this policy by other traders show up in the records. Just before war broke out, Borax Consolidated were offered large orders for borax and boric acid, requiring rapid transport to German ports, with immediate payment in cash in London. Although the orders came from bona fide customers, the circumstances were suspicious as they would normally have been for shipment from the United States, so Borax Consolidated simply replied by temporizing, pleading heavy bookings, and the business fell through as they had intended. A year later, in September 1940, a request came from the Industrial Surveys Corporation, Detroit, asking for a price for 600 tons of powdered borax 'intended for a friendly neutral country'. Gerstley commented that there was no friendly neutral country capable of consuming such quantities. He thought that it would probably be shipped to a neutral port for diversion to France or Germany, and asked to be informed in future about any proposed shipments to Spain or Portugal as they might mean a leak to the enemy.

With the opening of the German blitzkrieg on the West on 10 May 1940, Borax Consolidated lost most of its interests in Western Europe almost immediately. There is, naturally enough, an almost complete break in recorded information about all the European works, except Badalona in Spain, between the end of 1940 and mid-1945; but the picture can be pieced together from news and rumour in the first year of the War, combined with reports on the state of the various factories when liberation came in 1945.

In the Western blitzkrieg the first factory to fall to the enemy was the CIB works at Loth in Belgium.[10] According to a report coming through Spain in August 1940, this factory had suffered relatively little damage. All the staff were well with the exception of two – one reported a prisoner of war in Germany, and another believed to be in France. Two directors, Van der Smissen and Jasper, were also missing and believed to be in France at that time, but by mid-September they were reported to have returned to their homes and their work.[11]

Coudekerque fell to the enemy a few weeks later, and of course could not escape severe damage by reason of its close proximity to Dunkirk. A letter from Gerstley[12] mentions M. Arthur van Batten, manager of Coudekerque Works, who had sent his wife off to the South of France and then stuck to his post. Nothing further had been heard of him, and Albright was asked whether he could try to get a letter delivered to him through the American Red Cross.[13] Eventually news came back in November through a different source, a contact in the Vatican, that van Batten was well and at work at Coudekerque. Meanwhile a 'pleasing cable' had been received from 'the same secret source' early in October,[14] 'passing us on the information that the works we are interested in, in France, have only been slightly damaged . . . and they are carrying on their working to 30 per cent capacity' but if this referred to Coudekerque, its accuracy is questionable. When Johnson, the Borax sales manager, visited Coudekerque in August 1945[15] he found that there was neither water nor sanitation available, no roof over the building, no food in the town (you had to bring it with you) and, on 5 October 1945 Gerstley's son, J. M. Gerstley, reported that the Dunkirk factory was not operating, and probably could not start to operate for another six months at best.[16]

The next loss was the Paris office of Borax Français. The Germans entered Paris on 14 June 1940, by which time the Borax office was supposed to have moved to Nantes, although judging from a letter written by Gerstley on 5 July there was doubt as to whether it had actually done so. Nothing had been heard of 'Mr Thomas' and Maurice Lebreton, the senior clerk, since several days before the fall of Paris and it was not until December that news was received that Thomas had been interned in the military prison of St Denis in Paris.[17] He had been replaced at the office by someone called Van der Donk. Meanwhile, on 10 June 1940, Italy had declared war against the Allies, which ended any further hope of importing Tuscan boric acid to help Borax Consolidated in times of production shortages, and made it unlikely that they could count on any more boracite ore being shipped through the Mediterranean from Turkey, which was a further blow to boric-acid production.[18]

Then in the autumn of the year (1940) came the blitz on Britain. Belvedere Works, on the Thames estuary near Gravesend, was bound to receive a great deal of passing attention from the Luftwaffe, and an unsigned letter (almost certainly from Lesser) dated 20 September to Harding, the works manager, reads:

> Meantime I trust Belvedere Works still stands where it stood and that the bomb remains unexploded in the crater you showed me yesterday or has it been removed by the Bomb Disposal crowd to a safe distance? About the holiday, I should think the men would be very glad to take advantage of this next week to all enjoy a nerve rest. . . .

A temporary headquarters was acquired at Oxshott in Surrey at the outbreak of the War, for use in case of need, but most of the administration remained on in London. A letter from Gerstley dated 25 September stated that[19] 'our premises were rendered temporarily untenable (on 9th September at 3.30 am) but, everything considered, we were extremely fortunate as, apart from slight damage to the furniture and records, we have removed everything necessary to this address [i.e., Oxshott]. . . .' A few weeks later there were further visitations:[20] 'Unfortunately at our works we have suffered further damage from enemy attentions (a high explosive bomb on 15–16th October) and our town residence has also again been visited. . . . Our builders assure us that all the damage will be repaired within 8 weeks. . . . Nobody was hurt, all the men being in our shelters. . . . A goodly portion of the roof was blown away which pro-tem prevented night work. . . .' Nevertheless, the final mention of the subject was reassuring when Gerstley wrote to the American office at the end of 1941: 'It may interest you to learn that at our works, which, as you know, are situated in the Estuary, and therefore infinitely more vulnerable, the damage suffered had proved inconsiderable thus far and, we hope, we may be spared anything worse.'

Britain (and with it the UK side of the Borax Consolidated group) were bombed, but hit back. On 2 September 1941 Gerstley wrote to Jenifer:

> You will recollect in our Chairman's speech the proposal to donate a Spitfire, which proposal was carried with acclamation by the shareholders present. This Spitfire is now ready, christened 'Boron'. . . . We trust it will make a deep and lasting impression on the enemy.

The following March Gerstley reported the sequel:[21]

> This year, instead of contributing to another Spitfire (the first, we are sorry to say, recently came to grief over enemy territory but the pilot, a French Canadian who had established an excellent record, escaped capture), we propose to distribute the sum of £3,000 to be divided equally between the benevolent fund for the R.A.F. and a similar fund for the merchant seamen, to both of which we owe so much for helping to keep British trade routes open.

The Spitfire pilot eventually reached Britain again, and was awarded the DFM.[22]

As regards production and sale conditions affecting Borax Consolidated operations in Britain in the Second World War, there is mention of labour being attracted away from the Belvedere works by the high wages of a near-by munitions works,[23] although labour difficulties might have been much worse if borax and boric acid had not been classed as 'Essential Industries' at the beginning of the war. One reads also of attempts to modify the boric acid plant to use rasorite being terminated by unexpected arrivals of Turkish boracite;[24] of package boric acid increasing in popularity in multiple stores such as Woolworth's and finding its way once more, via food parcels, into the boots of troops on active service; and of an 'Order in Council' at the end of 1944[25] once more allowing the sale of 'margarine which contains borax' and 'bacon which contains borax', whereas under the regulations then applying in the USA the use of practically all food-preservatives was forbidden.[26]

The last stages and immediate aftermath of the War brought the fate of Stadlau and Aussig, the two factories in Eastern Europe, once more into prominence, as both had been engulfed in Russian-occupied zones and no one yet knew what had happened to them. The ubiquitous and evergreen Dr Hecker, who had left the German Army and managed to return to Borax Consolidated service in the USA during the First World War, had not been so successful during the Second when he found himself on an inspection tour of the works in Eastern Europe at the outbreak of hostilities. At some stage, however, he got back to America 'after all sorts of troubles', as he expressed it, and after much difficulty he obtained permission to stay there.[27] He turns up again as the recipient of correspondence from Gerstley from August 1945, when he was obviously anxious to revisit Stadlau and Aussig. Gerstley had to point out to him[28] that it was preferable in the circumstances of the time for an Englishman to visit Czechoslovakia and Austria and to come into personal contact with the officials of both places, considering the natural and intense hatred of Germans prevailing in Eastern Europe immediately after the War. Accordingly, they were keeping in regular touch with one of their key men,[29] Norman Pearson (at that time a colonel and in Greece), who had been on active service throughout the War and expected to be allowed home for 'demobbing' early in September. They had arranged with him, particularly as he had the advantage of still being in military uniform, to try to come back via Vienna (visiting Dr Seitz at Stadlau) and thence via Prague and Aussig, seeing Dr Pollak and Dr Brettschneider.

This was obviously going to be a difficult mission as Pearson had telegraphed to say that the problems of getting to Vienna were likely to be insuperable. A more junior officer, who had also been a member of the Borax staff, had already tried to get across the Russian zone in Vienna to visit Dr Seitz and had failed.[30] However, only a week later a cable was received from Pearson saying that he had managed to reach Dr Seitz. Unfortunately, he had to report that Dr Seitz's home had been badly bombed and the building was a shambles, as it had been in the thick of the fighting close to the Sud and Est Bahnhof. Pearson traced him to his next address, Mayerhofgasse 12, Vienna IV, but reported that the family had lost practically everything and were living with borrowed furniture in two rooms. They were in the Russian sector, where conditions were bad and food practically nil. However, they were bearing up most pluckily. Both Dr Seitz and his wife were well and remarkably cheerful in spite of all that they had endured. Since the occupation Dr Seitz had worked desperately hard to clear up and keep the factory going. He had been at the factory three times a week, including a three-mile walk each way, and, for a man of seventy-five this was pretty good. Vienna, Pearson reported, represented a tragic picture, as the damage was as widespread as the starvation and the wreckage of battle which was still uncleared.

Pearson also visited the Stadlau factory where, on the outbreak of the War, Hummer, a member of the Nazi Party had been made *Verwalter* (Trustee) by the Reich Office and, after a short spell in the army as a chemist, had returned to Stadlau and stayed until February 1944.[31] During this time he doubled his own salary despite Seitz's protests. He spent a week to a fortnight each month with his family at Linz, installed central heating at his house at the company's expense, and when the raids became too frequent disappeared with his family to Linz, leaving Seitz to hold the fort. Now Seitz was once more reinstalled and Hummer – who reappeared and attempted to assert himself – was sent packing by Pearson. The rest of the staff were in good shape. It appeared that Stadlau was well furnished with various kinds of raw material and had carried on steadily all the time. The works continued to produce sufficient only to pay running expenses in order to satisfy the Russians that it remained actively in operation. When the economic and political situations became clearer it would be possible to form a better picture. The office and the laboratory had been destroyed and the house looted.

Aussig suffered a worse fate. According to information subsequently received in 1945 through the Borax group's lawyer, Dr Richard Pollak of Prague,[32] the Aussig works had been expropriated on 14 April 1943 by the German Reich Commissioner and purchased by Schaffer and Budenberg, who used part of the building to store spares for the German Air Force. Later the plant was sold to the Aussig Verein by order of the Reichstelle für Chemie, and the Aussig building became a shell. Meanwhile the manager, Dr Brettschneider – although forced to pretend Nazi sympathies, which enabled him to continue in charge of the Borax group's interests in Aussig – had done everything he could to protect these interests.[33] By September 1945, however, Czechoslovakia had fallen under Russian control, and it was impossible as yet for an Englishman to visit the Aussig works. So although Pearson, who managed to reach Prague as well as Vienna in September 1945, saw Dr Pollak and Dr Brettschneider and 'helped to establish our claims',[34] Aussig never reopened as a Borax Consolidated works, and was the one real loss of the War in terms of the group's spheres of activity.

While these tragic events were taking place in Europe there were indirect repercussions on the borax industry in the United States, although, even after the actual entry of that country into the Second World War in December 1941 these were very far from being as severe as had been expected. Gerstley had written in October 1939,[35] on the occasion of a threatened strike at Boron, that the general position had greatly changed. There was the European war, which directly affected the production of the American mines, as it automatically closed (as the men well knew) the largest individual market in Europe – namely Germany, including Austria and Czechoslovakia, which were now under German sway. It had also suspended all shipments to Poland, and also to central Europe – Hungary, Yugoslavia etc. – as the Baltic was in practice closed. Gerstley added that, indirectly, trade with the rest of Europe was also being upset by shipping shortages, but nevertheless he had to admit in conclusion that it would appear from prices in the USA that Europe's emergency was being regarded as America's opportunity. Indeed, as the months went by it soon became obvious from difficulties encountered in finding sufficient production capacity, and from sales results, that the industry's business was booming, a trend accentuated by the wartime rise in prices. An almost embarrassing stimulus to Borax Consolidated prosperity came from an unexpected quarter. In March 1940 a strike closed American Potash and Chemical's works[36] at Searles Lake, and this strike was not settled until July 1941.[37] During this period Pacific Coast Borax had sufficient borax capacity to meet requirements, and there was no overall shortage.

The Japanese attack on Pearl Harbour was of course a decisive event in the War, for the borax industry as well as for the United States as a whole. First reactions in the States tended to be based on the previous British experience. Pacific Coast Borax cabled: 'Investigating possibility cost insuring refineries against bombardment bombing, sabotage'; but Gerstley replied, bearing in mind America's remoteness from the enemy, that in London's view there was no necessity to include sabotage. The only risk would be one of enemy action, which they could not help thinking and hoping was extremely remote.

But again, as after the outbreak of the European war in 1939, there were other events which did not follow a predictable course. The most sensational of these was that, some months after the entry of the United States into the War, the American Potash and Chemical Corporation fell under the scrutiny of the Alien Property Custodian, and its German ownership (which as already mentioned in this book had been guessed at by its rival Borax Consolidated for at least a decade) became officially published fact. On 20 October 1942 90·79 per cent of the capital stock of the American Potash and Chemical Corporation was vested in the Custodian by a Vesting Order published in the Federal Register.[38] This of course was not the end of the American Potash and Chemical Corporation, which was soon reconstituted to compete again with renewed vigour; but some of the details of the changeover, as contained in a letter from Lesser to Jenifer of 16 November 1942, are interesting:

We have read with interest the publication made in the Federal Register containing the order of the Alien Property Custodian and were surprised to find that the shares in question actually stood in the register of the A.P. & C.C. in the names* of Consolidated Goldfields, Goldfields American Development Company Ltd, and New Consolidated Goldfields. Quite a small number were registered in the name of a Dutch concern while the Hope Syndicate of Amsterdam had an option to repurchase 5,000 shares held by Goldfields American Development Company. Otherwise the order of the Custodian confirms the suspicions you and we had for some time. . . .

If and when the impounded shares are disposed of, we would also like to know about it and would, of course, be particularly interested to learn what interests in the U.S.A. would take over these shares. . . . Whoever the new owners may be, we also hope that you and we shall in the future, as in the past, be able to hold our own against this competition. . . . P.S. Since dictating the above we have learned from friends here that the chairman of the A.P. & C.C. has resigned not long ago and that a Mr. W. J. Frodish has been appointed by the Custodian of the A.P. & C.C. . . . The latter Company are opening their own office and have taken over practically the whole staff, including the President, all of whom have severed their relationship with Goldfields American Development Company.

An encounter with Germans in another but less important sphere was ultimately rather more profitable. Dr Suckow's former mine had been acquired, as already related, before the War, but negotiations to acquire his 50 per cent share of the important Section 14 had so far secured only a lease for Borax Consolidated and not the freehold. Dr Suckow was an old man by the time the War broke out in 1939, and the negotiations fell under the control of Otto Tobeler,[39] a new director of Suckow's set-up, whom Gerstley described as 'a direct purchasing agent for the German Government and hence an individual to be given as wide a berth as possible'. Lesser had no doubts about the relations between Tobeler and representatives of the Nazi Party in California, including Dr Hohne, who had been arrested, and had warned of possible interference by Tobeler in the proposed sale of the leased property to the Borax group. While the negotiations dragged on Suckow died, in February 1942, which added a further complication, for as Lesser put it: 'As far as the various ex-wives, sisters, and daughters of old Dr Suckow are concerned we regret we are quite unable to find our way through the muddle that exists. . . .' However, some way must eventually have been found, or Tobeler must have given up the struggle when his country's fortunes in the War started to wane, and the acquisition was completed successfully in January 1943 for the sum of $350,000.[40]

The industry's encounters with the US Government were not as successful as those with expatriate Germans during these war years. As in the First World War, difficulties over securing adequate shipping space to the United Kingdom and Europe were ever present. Gerstley summarized one aspect of this problem: 'As one official put it, "if household linens are not so white during the war through shortage of Perborate, or other cleansers . . . the issue is insignificant compared with winning the war," with which we agree.'[41]

Then there was Borax Consolidated's perennial problem of what to do with the

---

*Vesting Order 249 on page 8757 of the Federal Register of 20 October 1942 shows that these various Goldfields companies and the Dutch Company were merely nominee shareholders 'for the benefit of Wintershall AG, Germany and Salzderfurth AG, Germany'.

now useless Tonopah and Tidewater Railroad and its elaborate substrata of undischarged financial obligations. By early 1939 the group had reached the conclusion that the temporary abandonment of the railway service was the only way in which to staunch this running sore, and application was made to the Inter-State Commerce Commission accordingly.[42] The Commission approved the application, and agreement was reached with certain objectors to abandon the line finally on 31 March 1940.[43] But it was not until August 1941 that the board of Borax Consolidated decided to dispose of all the assets of the Railroad and to transfer the proceeds of the sale, less expenses, to a special fund, which would be earmarked for the redemption of the £175,000 redeemable sterling bonds, the gold bonds already being covered by a sinking fund.

In October Gerstley was hoping that it might be found possible to clear up this incubus, preferably as a going concern, in connection with Las Vegas's development.[44] However, the delay proved too long for the march of history. The USA was now at war, and in mid-June of 1942 the US Army stepped in and started taking up the rails and fastenings for shipment to the Persian Gulf, where they were to provide transport to the Russian front. An official notice requisitioning the railway arrived a few days later. Although the Borax group were glad to be rid of the physical assets of the railroad, the abrupt action of the government forestalled their plan to rid themselves simultaneously of the formidable burden of debt that the railway had built up.[45] With the disappearance of the railroad the common stock became valueless, as the funds realized from the sale were insufficient to cover even the sterling bonds.[46] The sequel, therefore, was that after a few months Borax Consolidated had to resort to selling a significant proportion of its shares in United States Potash to pay off the railroad debentures. For the railroad it was a sad ending to what had begun with Smith's triumphal, if dusty, progress down the line in the first official train in 1907.

The War, however, did not bring the hoped-for ending, sad or happy, to another saga of Death Valley. The efforts of the Borax group's hotel business to prosper financially during the thirties have already been recounted. When, a few weeks before the outbreak of the European war, a 'serious enquiry' was received from a big hotel operator in Los Angeles,[47] Gerstley commented that the investment at Furnace Creek Inn was not turning out in the way they had all expected. Naturally enough, after a few optimistic months in early 1940, the whole venture became an even heavier liability, relieved only by periodical bursts of optimism as possible candidates for taking the burden off the books made brief appearances on the scene. They were, in order of appearance, the Park authorities at the end of 1941, a Flying School established in the Valley temporarily in early 1942, and an Army training interest which faded.[48] By the end of 1942 the hope of cashing in on the Park authorities' reviving interest was finally wrecked by the British Treasury's refusal to treat a possible deal as a Lend-Lease transaction. At this stage the management obviously began to feel desperate about this hotel burden, as Lesser, in Britain, had to write to Jenifer, in the USA:[49] 'You may suggest the possibility of dismantling the hotels and selling the furniture and equipment.... We can hardly imagine that they can deteriorate so much as to make this sale a necessity, unless the war should last much longer than is generally anticipated.' The hotel problem, in fact, dragged on until peacetime brought a gradual return of popularity to Death Valley as a tourist resort.

The Lend-Lease Act of 1941 might well, however, have produced a much more serious predicament for Borax Consolidated. The principle behind the agreement between the US and UK governments was that British-owned companies operating in the USA were obliged to transfer their securities to the British Treasury to hold as security for advances from the US Treasury to Britain. However, the memory of a harsh transaction which the US Government had forced on Courtaulds before Lend-Lease came into existence still haunted British companies operating in America. In return for American aid to Britain at that time Courtaulds had had to deposit with the British Treasury 90 per cent of the shares of its Viscose Corporation, and these were later surrendered to the United States Government for sale among the American public. 'It may well be,' wrote Lesser to Jenifer,[50]

> that after the Courtaulds deal is out of the way, deals are contemplated for other large British interests in the U.S.A. and that small affairs – such as our interests – may be left for later, and this may afford the writer the desired opportunity to reach New York to help you and to take part in the discussions which may have to take place.

Jenifer had been empowered to negotiate with Sir Edward Peacock, the British Treasury spokesman in the USA, if negotiation were forced on the group, and if Lesser failed to reach New York in time for such negotiations in his wartime journey there via Lisbon.[51] Lesser reached New York on 22 April but the spectre of a forced takeover of the American side of the Borax group vanished soon afterwards as the negotiations were postponed indefinitely. The reason was that the case of Pacific Coast Borax was not as straightforward as most of the others. The majority of the assets operated by it belonged, in any case, to Borax Consolidated in London, so that the shares of Pacific Coast Borax represented little that would give them much of a sale value.[52]

The spectre of a forced takeover might have vanished, but the War was not over before the United States Government produced another and more aged spectre to take its place. The scope and interpretation of the Sherman and Clayton Anti-Trust Acts had concerned the industry for nearly half a century, although since the end of the First World War it had been felt increasingly that in face of the fierce competition between Borax Consolidated, American Potash & Chemical, Stauffer, West End, Dr Suckow, and other American producers the chances of any contravention of these Acts had become remote. Not unnaturally, therefore, Lesser wrote to Jenifer on 3 December 1943:[53] 'You can well imagine our surprise when we received your cable advising of the commencement of investigations of the borax industry. . . . What the cause of such investigation of our industry can possibly be we at this distance fail to understand.' Jenifer offered a possible explanation, but Lesser replied:

> We are surprised to learn from your letter that the new investigation was a continuation of that made in 1929, which you stated related to Boric Acid prices. . . . It is difficult to understand how any concentration of economic power can be alleged when other producers of boron products have for some time made larger sales than Pacific Coast Borax or ourselves.

There exists extensive documentation about this investigation and the subsequent criminal and civil cases, but none of it answers the surprised question that the Borax Consolidated group continued to ask through the subsequent months of preparation, appearance before a Grand Jury, indictments, and trial: 'Why on earth bring up this accusation now? Is there any political or other motive behind this move?'

As Gerstley wrote to Jenifer on 24 July 1944:

> While we have no idea what is behind the authorities' action one must always bear in mind the possibility of misleading press reports; hence we deemed it advisable to cable you today asking whether it was possible to furnish us with any data as to the subjects of the enquiry before the Courts and, if there was any suggestion of a possible infraction of some law we would like to have fuller information.

Finally, on 5 October 1944, Lesser wrote to Jenifer commenting on 'copies of the criminal indictment which we have read with dismay and surprise' and adding: 'The indictment and the civil suit are practically identical. From whom and how information contained therein has been obtained we do not know.'

Before setting out the principal points in the indictments, Borax Consolidated's defence, and the subsequent Consent Decree to which Borax Consolidated had to agree, it is possible to conjecture that the whole affair may well have developed out of a resolution which a certain Senator Key Pittman from Nevada steered through the US Senate in mid-1936.[54] This attacked the borax and potash industries and revived the idea of a 'foreign-owned Borax trust'. Baker commented to Albright at that time:[55] 'The details you give as to the way in which the Pittman resolution was put through would seem to show trickery and action which could hardly be anticipated.' In a letter to Jenifer two months later he implied that it was instigated by a rival of pre-war days.[56] Be that as it may, it led to the setting up of a Senate Committee to investigate the potash industry, and this provoked the last letter that Baker appears ever to have written as managing director, on 16 November 1936, just over two months before his death.[57] It was characteristically pungent to the last: 'I am glad to see that the imprisonment penalty in the common jail is not less than one month nor more than twelve, and I am wondering which of our competitors may be penalised.' Like so many committees and commissions, however, this committee's deliberations slowed down markedly after a first burst of enthusiasm, and Lesser is found writing to Albright, head of U.S. Potash, as late as 26 August 1938: 'It seems most remarkable to extend the investigation about the potash industry until 1941, particularly when the investigation has been dropped for quite some months.' It is possible, however, that it slumbered on in government offices until the files reached someone who started the anti-trust actions of 1943–5.

Those actions and their results are best summarized by quoting the official statements of both sides after their conclusion.

Gerstley and Lesser received a cable dated 16 August 1945 from San Francisco, which read:

> Attorney General Tom C. Clark announced settlement in San Francisco, California of two anti-trust suits, one criminal and one civil, involving the borax industry. Both suits were filed on September 14, 1944 and charged seven corporations and eleven individuals with conspiracy to violate the Sherman Act in mining, processing, manufacturing, selling and distributing crude borates, refined borax, and boric acid by acquisitions and trade practices which had driven out or prevented competition by American concerns, had allocated among themselves foreign and domestic markets and customers and had agreed upon restrictive selling and distributing methods, terms and conditions, including prices at which such products are sold. Both suits alleged that practically the entire world supply of borax and boric acid is produced from raw materials found and mined within the State of California, that borax is extracted from crude borates and from lake brines, and that boric acid is made from either the crude borate or borax, that virtually the entire world supply of borax and boric acid made from crude borates is made from kernite, all the known deposits of which are located in Kern County, California, that substantially the

entire world production of borax made from lake brine is made from the brine taken from Searles Lake, San Bernardino County, California, that the British Group control all known kernite deposits and the German Group (American Potash & Chemical Corporation and subsidiaries) control approximately 85 per cent of the workable area of Searles Lake, and that the two groups control 95 per cent of the world production of borax from crude borates and approximately 90 per cent of the world production of borax derived from lake brine, that the two Groups of defendants permit two American concerns to produce and sell approximately 5 per cent of the world supply of borax and boric acid but dictate the markets, customers and prices for such sales.

The defendants in the criminal case entered pleas of nolo contendere and fines were imposed totalling dollars 146,000.

Here followed a list of the defendants which included, apart from B.C.L. and A.P. & C.C., the 'Three Elephant Borax Corporation of New York', James Gerstley, Lesser, Jenifer, J. M. Gerstley and Winters of B.C.L. and F. C. Baker, Murphy, Vieweg, Curtis and Hatchley of A.P. & C.C.

The statement continued:

The decree entered in settlement of the Civil Suit required the British Group to divest themselves of two mines located in California, one known as the Western Borax Mine, a kernite mine located in Kern County, and the other known as the Thompson Mining Properties, a colemanite mine located in Death Valley. The decree provides for the sale of these properties under a Receiver to be appointed by the Court.

The decree went on to require the British Group to give up any present or future attempt to acquire the Little Placer Claim,* and continued:

The decree also enjoins the British and American Potash Groups from conspiring with one another or with any other producer or refiner of crude borates, refined borax or boric acid:–
1. to divide or share markets in crude borates, refined borax, or boric acid;
2. to divide customers in specific markets for such products;
3. to withhold from the market crude borates, refined borates, or boric acid;
4. to restrict in any way the types of refined borax or boric acid to be sold to any customers;
5. to impose any restrictions whatever on the sale of dehydrated borax and so on – with prohibitions against restriction of production levels and prohibitions on resale or export or on resale price maintenance or on limiting 'the tonnage of the above three products mined, produced, processed, refined, distributed, or sold by the West End Chemical Co., Stauffer, Pacific Alkali or any successor, assignee, or transferees of such companies. . . .

The statement continued:

Wendell Borge, Asst. Attorney-General in Charge of anti-Trust prosecutions, stated that the decree would create conditions over which new concerns could enter the business of mining and selling the above three products since it provided for the sale of the Western and Thompson mines to independent parties and freed the Little Placer Claim for lease by the Department of the Interior to an independent Group. . . . The elimination of the restrictions hitherto existing on the ability of purchasers to resell and export would enable domestic customers of the British and American Potash Groups to compete with these two Groups in both the domestic and export markets. . . .

---

*This refers to ten acres which had been under dispute for many years (see Section 24 in map VIII, p.176).

220   *The Tincal Trail*

The Borax defendants issued the following statement:

> Borax Consolidated and its subsidiaries are innocent of the charges made against them. A trial of one of the two cases was scheduled for the 11th September and would have required the presence in the U.S.A. for many months of the executive heads of Borax Consolidated located in London whose attention to reconstruction problems is imperative. Borax Consolidated possesses refineries in various countries desolated by the War and has mines in other foreign countries affected by the international turmoil. Some of its plants have been destroyed. The executives are therefore confronted by aggravated problems of post-war reconstruction and must turn their attention to these problems. It is impossible for them to sever themselves from the needs of the business in order to come to the U.S.A. to attend protracted trials. The great expense and disruption of business coupled with the impossibility of assembling all necessary evidence in time for a trial in September have led the defendants to terminate the cases through the familiar route of nolo contendere pleas and the entering of a consent decree. In filing nolo contendere pleas the defendants have not admitted that they have been guilty of any offence whatever. The Consent Decree itself shows that the defendants deny that they have violated any law and that the Court has not determined or decided that the defendants have violated any law. Borax Consolidated and its affiliates intend to carry out the decree and will attempt to make it workable in a practical world but they feel that the decree now entered may mean the sacrifice in favour of foreign buyers of the favourable position now enjoyed by the American consumer of borates.

Gerstley had already explained to Jenifer earlier in the year the predicament in which he found himself[58] There was nothing but a skeleton staff at the Borax Consolidated office with so many away in the Armed Forces; Lesser senior had been taken ill (he died the following year), and documents had been destroyed by enemy bombs. It was indeed quite impossible to prepare an adequate defence in the time available under such conditions, and so ended what had been a harsh burden, especially for Gerstley when added to the turmoil of the last six years of war.

REFERENCES

1. M.D. to Pacific Coast Borax, 12.3.38.
2. M.D. to Rosenstern, 28.3.38.
3. M.D. to Albright, 16.8.38.
4. M.D. to Hecker, 1.9.38.
5. M.D. to Jenifer, 27.9.38.
6. *Ibid.*, 4.10.38.
7. M.D. to Jenifer, 27.9.38., and M.D. to Albright, 29.9.38.
8. M.D. to Jenifer, 23.3.39.
9. M.D. to Jenifer 7.9.39.
10. *Ibid.*, 30.8.40.
11. M.D. to Jenifer, 16.9.40.
12. M.D. to Albright, 13.6.40.
13. P.C.B.: Corkill to Borax Consolidated, 19.11.40.
14. M.D. to Albright, 7.10.40.
15. M.D. to Hecker, 28.8.45.
16. J. M. Gerstley to Jenifer, 5.10.45.
17. Lesser to Gerstley, 15.5.41.
18. M.D. to Hecker, 11.6.40.
19. Minute Book.
20. Minute Book.

21. M.D. to Jenifer, 6.3.42.
22. *Ibid.*, 9.3.42.
23. M.D. to Howard, 24.11.41.
24. M.D. to Pacific Coast Borax, 7.7.43.
25. *Ibid.*, 25.5.43. Also P.C.B.: Winters to Borax Consolidated, 28.6.43.
26. *Ibid.*, 21.2.45.
27. P.C.B.: Hecker to Gerstley, 1.4.40, and Gerstley to Jenifer, 7.9.39.
28. P.C.B.: Gerstley to Hecker, 28.8.45.
29. M.D. to Jenifer, 27.8.45.
30. P.C.B.: Gerstley to Hecker, 15.10.45.
31. *Ibid.*
32. P.C.B.: Gerstley, 20.7.45.
33. P.C.B.: Gerstley to Hecker, 20.9.45.
34. *Ibid.*, 29.10.45.
35. M.D. to Jenifer, 14.9.39.
36. P.C.B.: Lesser to Jenifer, 21.3.40 *et al.*
37. M.D. to Pacific Coast Borax, 9.7.41.
38. P.C.B.: Lesser to Jenifer, 30.11.42 and Corkill to J. M. Gerstley, 30.10.42.
39. M.D. to Jenifer, 5.7.40, and M.D. to Pacific Coast Borax, 29.8.40, and to Jenifer, 19.3.42..
40. Minute Book, 10.1.43.
41. M.D. to Jenifer, 27.3.40.
42. Minute Book, 10.1.39.
43. *Ibid.*, 2.1.40.
44. *Ibid.*, 2.10.41.
45. *Ibid.*, 23.6.42 and 14.7.42, and P.C.B.: Gerstley to Jenifer, 14.7.42.
46. P.C.B.: Lesser to Jenifer, 17.6.43.
47. *Ibid.*, 14.7.39.
48. P.C.B.: A. J. Somers to Pacific Coast Borax, 22.6.45.
49. M.D. to Jenifer, 4.11.42.
50. *Ibid.*, 20.3.41.
51. M.D. to Jenifer, 10.4.41.
52. Information from F. A. Lesser.
53. M.D. to Jenifer, 14.1.44.
54. M.D. to Jenifer, 19.6.36.
55. M.D. to Albright, 30.6.36.
56. M.D. to Jenifer, 30.8.36.
57. M.D. to Albright, 20.7.36.
58. M.D. to Jenifer, 29.3.45.

20

The post-war world

The action of the Alien Property Custodian in 1944 had brought about changes in management at American Potash and soon advancing years and the end of the War brought with them changes in the senior management of Borax Consolidated, Limited. Federico Lesser had been compelled to retire in 1943 because of illness, and died in 1946, so James Gerstley – who had been there since before the merger of 1896 between Redwoods and Pacific Coast Borax, and who had hoped to retire a 'couple of years' after Baker's death in 1937 – found himself holding the fort throughout the War and during the troublesome time of the anti-trust suit. Thus he was past seventy before he was able to retire as managing director in 1946. Frederick Lesser, son of Federico, and Arthur Reid were then appointed joint managing directors. Arthur Reid had been a non-executive director since 1937, and he retired as joint managing director in 1953.

In the USA Frank Jenifer, president of Pacific Coast Borax, was involved in a serious car accident in 1946 from which he took a long time to recover, and meanwhile J. M. (Jim) Gerstley, who had been appointed vice-president and general manager, took charge. Gerstley became president in 1950, the year in which Jenifer died.

The borax world moved out of the wartime atmosphere of restraints and shortages very rapidly after the War ended in 1945, and by 1950 the correspondence of the time was full of products, developments and problems unknown or little known in pre-war days only some ten years before. However, the anti-trust suit and Consent Decree, which loomed so large in the industry's thinking in the closing days of the War, had left some matters of unfinished business. The Western Borax Mine (which was to be removed from Borax Consolidated and sold to outsiders in order to create competition within the industry) duly passed into the control of 'the Receiver appointed by the Court'. It was nevertheless not until early in 1947 that any offer was received for the mine, and then it was so much below the 'upset price' of $491,000 fixed by the court that Lesser was driven to write:[1] 'The sale, which Mr. Allen [the Receiver] claims to have negotiated, leaves us with our mouths open in surprise.' Mr Allen had agreed with Messrs Mudd and partner, the would-be purchasers, a price whereby

Borax Consolidated would scarcely get anything more than the original downpayment of $125,000.[2] The matter dragged on, and final payment was not received until August 1954.[3]

In view of the condition of the mine, this was not surprising. 'We are glad to see that no cave-in has taken place yet at the Western mine although it does look as though one is imminent,' wrote Lesser in September 1948; and only a few months later a consulting engineer reported to the Receiver that conditions at the mine were 'dangerous'.[4] Since the Western mine had been acquired in 1933, no mining had taken place, and the dangerous condition of the mine was well known. However, economic conditions rather than the condition of the mine were the reason why there was no rush to take over Western and to get into the borax business. The United States Government itself was also having difficulty in finding a lessee for its unexploited and allegedly more valuable Little Placer lease, for which Borax Consolidated were debarred from bidding by the terms of the Consent Decree, and it was not until 1954 that it became obvious that Little Placer also had gone to the Mudds, and that they were considering production right up to the refinery stage.[5] At this time there was little commercial interest in the colemanite ore reserves in the Death Valley area, and the third property mentioned in the Decree, the Thompson lands in Death Valley, remained on the Receiver's hands right up to mid-1960, when it was sold to the Kern County Land Company.

The pattern of demand for boron products after 1945 saw some fundamental changes. Production facilities immediately after the War were strained by the inevitable demand for restocking empty factory and refinery warehouses, particularly in Europe. However, after the first year this tailed off and a post-war slump followed, in which the near-bankruptcy of nations outside the USA became obvious. This slump was ended by the artificial boost to the metals and chemicals industries provided by the Korean War between 1950 and 1953. Slump conditions then returned for a short time, until the rapid growth in world population, and new developments in the borax industry, all combined to provide what now seemed an ever-growing demand for its products. The total tonnage sales of Borax Consolidated alone had multiplied sevenfold between 1935 and 1955.[6] This reflected the strong production potential which the industry was able to develop very quickly after the War to meet the needs of new and growing industries. At Boron the original Baker Mine had been joined in 1936 by the Suckow property, renamed West Baker, and in 1950 another shaft was sunk to start the new Jenifer Mine.

The new and developing uses of borax in this post-war world were significantly different from the first decades of the century and the Victorian era. The enamel frit and pottery glaze sector was overtaken by the rapidly expanding use of borax in the glass industry. Between the wars borosilicate heat-resistant glass had become available commercially on a world-wide basis, and there was rapid expansion following the Second World War – in the home as ovenware, in all forms of laboratory glass, in industrial plants, car headlamps and the numerous jobs where the desired 'non-cracking' of glass when subjected to heat enabled it to be used. The largest piece of glass in the free world today – a 200-inch, 20-ton disc forming the mirror in the Hale telescope – is made of borosilicate glass. However, what was to become the biggest single use of borates in the world today started in the development of a new process to produce glass fibres on an industrial scale in the 1930s.

Glass fibre has been used decoratively for centuries; first appearing in Egyptian and later in Venetian objets d'art, while French and German glass-makers produced it commercially for this purpose in the 1700s. In 1892 Edward Drummond Libbey exhibited a dress made of glass fibre and silk at the Columbian Exhibition in Chicago. Then during the First World War the Germans developed a method of producing glass fibres to replace asbestos, although this continued to be a process based on single strands drawn through an orifice, until the 1930s, when a process was developed in the United States whereby a continuous filament was formed by multiple strands of glass extruded from a heated platinum box or bushing with many tiny holes. This product was used as a fibre, now familiar in textiles, electrical work and in a resin bonded form as the structural basis of boat-hulls and many similar engineered products. At the same time a way was found to produce staple fibres from which the rolls of glass fibre are made, which are used for thermal and acoustical insulation.

Sales by the Owens Corning Fibre Glass Corporation, who pioneered the process of manufacture of glass fibre, expanded from about $4 million in 1939 to over $80 million by 1950. In the post-war world this industry has sustained a continuously high growth rate, reinforced in recent years by the compelling need to conserve energy in almost every form of building structure in which energy is used, either for the purpose of heating or for cooling.

The chemical sodium perborate, made from borax and hydrogen peroxide, was first produced as a laundry additive in England in 1907. Mixed with soap-flakes – the 'Persil' oxygen washer – it removed stains and produced laundry whiteness with remarkable success. However it was after the Second World War and the introduction of washing machines and detergents aimed at the housewife, with all the marshalled power of the advertising world, that the use of sodium perborate (particularly in Europe) escalated year by year. This 'washing industry' now rivals the glass industry as borax's largest user in this area.

The 1940s had seen some fundamental changes in methods of production of end-products at Boron. The relatively high cost of processing kernite ore by calcination to make concentrated 'Rasorite' (CR) and its treatment in 'digesters' (autoclaves) under pressure to make refined borax at Wilmington, compared with the cost of processing tincal, put increasing emphasis on the mining of tincal and rejection of kernite. Before the end of the Second World War tincal had become the mined ore which was supplied to the processing plants at both Wilmington and Boron.

A magnetic separation process for removing shale and iron impurities from tincal had been developed at Boron in 1933. However, until the Suckow mine was leased in 1936 there was no adequate supply of tincal, and in the absence of any cost system relating to individual processes, and with the reluctance to spend capital, the calcination of kernite continued well beyond the time when it would have been preferable to switch to tincal ore. Some additional magnetic separators were installed in 1937, but calcination of kernite only ceased in about 1943 and there were substantial additions to the magnetic separation plant in 1946 and 1951. In the post-war period about half of the mined ore (tincal) went to the Wilmington refinery and the other half to the Boron mill for conversion to 'Rasorite' products for shipment overseas. Soon after the War ended plant was installed at Wilmington to produce anhydrous borax.

American Potash and Chemical Corporation were tied to producing potash, soda-ash and sodium sulphate in a fixed ratio to match any increase in borax-

Clarence Rasor, the eyes and ears for some twenty-five years of Borax Consolidated's activities in North America

John Suckow, sometime 'partner' and always tough competitor of Borax Consolidated in the Kramer area

Boron, California: the underground mining of borax before 1957 (above) and (below) the open-pit mining of borax after 1957

Furnace Creek Inn, a popular resort hotel in Death Valley, built by Pacific Coast Borax Company and opened in 1927

A group at the opening of the Pacific Coast Borax site at Wilmington in 1923. Left to right: Richard C. Baker, Managing Director, Borax Consolidated; Chris B. Zabriskie, General Manager, Pacific Coast Borax; Clarence Rasor, Field Engineer, Pacific Coast Borax; Lord Chichester, Chairman, Borax Consolidated; Tom Cramer, Refinery Superintendent, Pacific Coast Borax

Lord Clitheroe, chairman of Borax (Holdings) Ltd (1958–70), speaking on 12 November 1960, at the dedication of the new Visitor Centre and Museum in Death Valley National Monument, built on land donated by Borax (Holdings) Ltd

DEATH VALLEY MUSEUM
1960
INITIATED
BY
DEATH VALLEY '49ers
BUILT UPON LAND DONATED
BY
BORAX (HOLDINGS) LIMITED
CONSTRUCTED WITH FUNDS APPROPRIATED
BY
STATE OF CALIFORNIA
EXHIBITS AND ADMINISTRATION
BY
NATIONAL PARK SERVICE

Right: The familiar face of Ronald Reagan working (1962–4) with U.S. Borax on their television series 'Death Valley Days', used to advertise '20 Mule-Team' Borax and other consumer products

production, whereas the Borax group, with independent production of borax at Boron and potash at Carlsbad, now had more flexibility and could take advantage of each market separately. However, A.P. & C.C. were soon to reorganize as formidable competitors in spite of the disappearance of the German ownership which had held over 90 per cent of the issued stock. The former German holding, represented by 478,194 B class shares, was sold by the Custodian of Enemy Property to a syndicate at $32.29 a share, and resold by the syndicate to the public at $35 a share. In the words of James Gerstley, written shortly after this event,[7]

> the Heyden Chemical Corporation, according to our information, acquired 100,000 shares in the A.P. & C.C., and thus can be regarded as having a controlling interest.... During the first world war they were reported to have German associations, but it is now understood that they are entirely independent, extremely active, and have attained a very high measure of success in their diverse activities.

Three Heyden people joined the board of A.P. & C.C.:[8] one was Bernard Armour, who became chairman, and another Peter Colefax who became president. He was the son of Sir Norman Colefax, the distinguished English patent lawyer, and of Lady (Sybil) Colefax, a prominent figure in the English social scene. It soon became obvious that they were going to be just as virile competitors as the German owners, as they set about expanding production and endeavouring to increase their market share by price competition.[9] The all-important European market was now attracted by A.P. & C.C.'s price for anhydrous borax, which began making inroads into the market for refined borax manufactured by the European refiners from crude borax supplied by Borax Consolidated from California. Lesser wrote to Pacific Coast Borax:[10] 'If prices continue to come down we may be forced to revise our whole policy and, in a market like Belgium for instance, close up the factory and let them act as agents re-selling your production.'

The preceding paragraph, by itself, sounds much like a repetition of the facts of A.P. & C.C. competition in the 1930s and might not have deterred a newcomer from competing from a base at the Western Borax Mine or Little Placer, even though the Stauffer Group, West End Chemical and Pacific Alkali at Owens Lake (later the Columbia Chemical Division of Pittsburgh Plate Glass Company) were also all still in the field and expanding. But the War had brought new obstacles to the export of borate products from the USA which would have made the task of finding a market quite difficult for any newcomer. The first may be summarized in the economic jargon of the time as 'the dollar gap'. The principal export market of the American borate industry was in Europe, and Europe was chronically short of dollars. This encouraged European refiners to look instead to soft-currency areas for raw-material supplies and in fact to consider continuing the trend, started by the Nazis before the War, of looking hopefully again to South American borate of lime for raw material for borate-manufacture.[11]

They found post-war nationalistic governments of Chile, Peru and Argentina only too ready to reactivate their borate industries in order to provide the raw material required, but the economic advantages of using crude sodium borate from California in the European refineries were too great except for Spain, where both borax and boric acid were made from borate of lime imported from Turkey.

Daily work on the American side of the business had of course been less disrupted by the War than operations in Europe, where it took varying periods of time to reach the normalities of the post-war world which were a great deal different from those of the decade before the War. For example, in 1946 Lesser wrote to the secretary of the London Chamber of Commerce[12] protesting that the coal-allocation had been cut by 50 per cent, and that this reduction in fuel would cause a reduction as from 1 February to approximately 35 per cent of the normal production obtained at the factory at Belvedere. A few weeks later Glyn Mills, Borax Consolidated's bankers in England, were asked to send some cancelled bonds to the USA, but the London secretarial department had to report:[13] 'There will be a slight delay in carrying out this work as, on account of the present lighting restrictions imposed through the coal crisis, they are able to see in their strong rooms only at certain hours during the day.'

At much the same time, in a lighter vein, Gerstley (Senior) wrote to Wilmington[14] thanking them for

> the tempting parcel of tit-bits . . . which will be shared today at a buffet lunch with a number of our executives. . . . Thanks to rationing, which works fairly for everyone, the domestic fare, which involves queueing up for the housewife, provides an adequate table at home. The city lunch presents the main difficulty, on account of overcrowded conditions involving serious waste of time. To overcome this we arranged supplies of sandwiches for the staff. However, the stodgy content filled with doubtful 'Ersatz' proves indigestible and unsatisfactory. It is a pity you scientists have not discovered commerical application for the resultant gas. The collective ebullient from all the sandwiches should suffice to charge a jet-propelled aeroplane!

On the Continent the factory at Loth in Belgium had been little damaged, and was operating again at full capacity very soon after the War.[15] Coudekerque in Northern France had suffered a great deal of damage, and owing to the difficulty of obtaining replacement machinery, did not start up again until well on into 1947.[16] In Austria, Stadlau had continued operating after the eviction of the Nazis, but the policy had to be adopted of just keeping production going in sufficient quantities for the works to pay the men, as until the exchange rate question was settled it was considered wise to limit the cash currency there to the smallest amount.[17] In Czechoslovakia, as indicated in the previous chapter, any idea of rebuilding and restarting the shattered Aussig factory was soon abandoned, particularly as nearly all the factories in Czechoslovakia had been nationalized, especially those connected with glass-manufacture and chemicals.[18] The difficult problem then had to be faced of how to keep in touch with this profitable market for borate material, and at a later stage a new problem arose from the US Government regulations forbidding many exports, including borates, to Iron Curtain countries. As regards other European territories where no Borax Consolidated works were located, by degrees contact was resumed with various pre-war agents, and assurances of loyal support were received.[19] The European situation was, however, to undergo many changes in the next twenty years,[20] particularly with the formation of the European Economic Community.

The year 1945 saw the beginning of a period of expansion for the U.S. Potash Company. Increased capacity costing $2·75 million was completed in October 1945; after which some half-million short tons of product were sold in 1946, compared with the initial capacity which enabled 57,000 short tons to be sold in 1932. During the period 1944–5 Borax Consolidated, Limited had disposed of 100,000 shares (the proceeds being $3·36 million), and this reduced its shareholding in the U.S. Potash Company from 50 per cent to 31 per cent. The cash had been used to complete the final act of redemption in the saga of Tonopah and Tidewater Railroad: all outstanding debt had at last been repaid.

By 1946 International Minerals and Chemicals Corporation and the Potash Company of America were also well-established potash-producers in Carlsbad, and of 4·3 million short tons of ore mined in that year the U.S. Potash Company mined just under one-third and the others the remaining two-thirds.

REFERENCES

1. M.D. to Pacific Coast Borax, 25.3.47.
2. M.D. to F. M. Jenifer, 29.4.47.
3. M.D. to Pacific Coast Borax, 31.8.54.
4. P.C.B.: A. H. Reid to F. M. Jenifer, 19.8.49.
5. P.C.B.: J. Gerstley (Senior) to P. O'Brien, 1.7.54.
6. Borax (Holdings) Ltd., Managing Director's Reports.
7. M.D. to Pacific Coast Borax, 29.3.46.
8. M.D. to H. M. Albright, 31.7.46.
9. M.D. to J. M. Gerstley, 21.4.49.
10. *Ibid.*
11. M.D. to Pacific Coast Borax, 15.2.51.
12. M.D. to London Chamber of Commerce, 24.1.47.
13. A. J. Somers to F. M. Jenifer, 27.2.47.
14. J. Gerstley to Hecker, 3.6.46.
15. M.D. to Pacific Coast Borax, 11.2.46.
16. *Ibid.*, 31.3.47.
17. M.D. to Hecker, 13.12.45.
18. *Ibid.*, 3.4.46.
19. J. M. Gerstley to F. M. Jenifer, 19.11.45.
20. M.D. to Pacific Coast Borax, 21.1.49.

21

The end of underground mining at Boron and the beginning of U.S. Borax

Underground mining methods at Boron were similar to those used at other mines. A shaft was sunk into or close by the ore, and from it extending through the ore body at different levels were driven what the layman calls tunnels, which mining men call drifts and cross cuts, winzes and stopes, depending on their character. The two main mining methods used at Boron were the room and pillar and the shrinkage stope methods. By the early 1950s continuous mechanical miners had been introduced in the Jenifer mine to cut the ore at the face, instead of drilling and blasting, and mechanical moving belts had largely replaced the ore trains.

The Second World War brought about significant developments in the design and capacity of earth-moving equipment and vehicles. The economics of open-pit mining were revolutionized by these changes, and the Boron mine, with ore at a depth of a few hundred feet, had an unusually good potential for development by this method. However, a decision to change from underground mining to an open pit involved substantial capital expenditure, and in the case of the Boron mine it also gave an opportunity to redesign completely the refining process with a new plant alongside the mine instead of shipping raw material to Wilmington, where refined borax had hitherto been made.

In spite of these economic advantages Borax Consolidated, in the post-war years, was not yet in a financial position to meet the heavy capital costs involved. In April 1947 Fred Lesser wrote: 'We are not discarding the Open Pit method: we are only postponing it.' He added that the postponement would enable them to see the probable future trend of the demand for borax products, and this question was still an open one as late as February 1951, even during the Korean War boom. The serious drop in demand after the Korean War lasted from early 1952 until the autumn of 1953, but when demand then started showing every sign of being firmly on the upgrade again plans for an open pit and concentration of production at Boron were developed rapidly over the following two years.

Meanwhile the US Government had developed an unexpected interest in borates. J. M. Gerstley remembers that some time before Jenifer's death in May 1950 an inquiry about availability of borates was received from the Office of Chief Ordnance at Washington, which indicated a military programme which was 'top secret', and clearly a matter of major importance. Pacific Coast Borax co-operated with the US Government by providing experienced staff to assist in a basic study of the availability of borate mineral resources, particularly in the North and South American continents, which was conducted by the US Geological Survey.

Following the Second World War the development of improved-performance aircraft fuels became an important military objective. The high energy in boron compounds was well known, and in 1947 the British had made an evaluation of their use as ramjet fuels, but decided to take the matter no further. However, in 1952 the US Defense Department initiated Project ZIP, aimed at developing a superior fuel to the hydrocarbon jet fuel JP-4, and as it transpired later ZIP was focused on the use of these compounds of boron and hydrogen called boron hydrides, or boranes. However, for several years little was known about the project, as quite properly it remained a classified top-secret matter. The programme acted as something of a spur to the borax industry, and in U.S. Borax it accelerated the decision on open-pit mining, which permitted a much higher recovery of ore, and therefore greatly increased the ore reserves available for production. Also, many companies increased their research effort to develop new boron chemicals, and not only those who were the prime producers of the raw material borax.

Plans for the open pit and new refinery at Boron were completed by the beginning of 1954. The capital cost of the project of about $20 million was approved by the board of Borax Consolidated, Limited, subject to satisfactory financing plans being developed. This was by far the largest single project undertaken by the company since its formation in 1899. The creation of the open pit involved moving about 10 million tons of earth overburden over a period of two years at a cost of $2 million; however, the financial advantages arising from higher recoveries of ore and lower mining costs, together with a refinery alongside the mine, were ample to support the project.

The raw materials for the new refinery would be crushed and ground in the open-pit mine, and tincal containing about 24 to 25 per cent of the essential boric oxide (B_2O_3) would be converted to three main products: 'Borax 10 mol', 'Borax 5 mol', and 'Rasorite 46'. 'Mol' denotes the proportion of water (or number of molecules of water in crystallization), and Rasorite 46 was a crude product, similar to the Rasorite SCD which had been made at Boron by the magnetic separation process, and which now contained about 46 per cent of B_2O_3. New furnaces for the manufacture of anhydrous borax were also included in the plan.

Borax Consolidated's operations in the United States in the post-war period had enabled about $6 million to be conserved from internally generated funds which was now available to finance the new project, but the international dollar situation prevented the balance of the capital being raised in Britain, and this needed to be done in the United States.

Discussions with the Farmers and Merchants Bank in Los Angeles and the Chase in New York were well advanced in the autumn of 1954, when an unexpected development postponed the financial negotiations and caused some delay of the project, and also had a far-reaching effect on the whole structure and future of Borax Consolidated. A syndicate led by Coleman Morton, a financier in Los Angeles, and the New York finance house of Model, Roland and Stone made a bid in London to acquire the shares of Borax Consolidated, Limited. This was what would be described today as an unwelcome approach, and it started with an unusual event. When a list of those forming the American syndicate was obtained in London a surprising name was that of the chairman of the Chase Manhattan Bank, the company's bankers in the United States, who had received confidential details of the company's expansion plans in order to assist in financing it. The chairman was called to the telephone while on a duck-shooting expedition, only to express total ignorance of Borax Consolidated and the fact that his bank represented them. However, he withdrew from the syndicate, to whom no information had been passed.

Coleman Morton engaged Kleinwort Benson, the merchant bankers, to act as financial advisers to his syndicate, and a series of meetings took place in London. The acquisition would have required approval of the British Government, and it was always possible that this would be withheld because of the importance to Britain of the dollars being earned by Borax Consolidated through Pacific Coast Borax's operations and the U.S. Potash Company. However, the directors of Borax Consolidated felt that if shareholders were aware of the investment programme and the potential increase in profits which would result from it (and which had yet to be announced) there was a strong possibility they would elect to retain their interest in the company. The board of directors of Borax Consolidated, Limited therefore decided to issue a statement to its shareholders about the approach, also outlining the company's plan, and at the same time recommending that the offer should be rejected.

One of the reasons which had prompted the syndicate to make its bid was a tax advantage that resulted from a law enacted in the United States some time earlier that enabled American mining companies to claim depletion allowances; these were not available to Borax Consolidated, Limited as long as Pacific Coast Borax was a division of a British company and not an American corporation. This was a situation which could be remedied by a change in structure of Borax Consolidated. In the event the syndicate withdrew its bid, and the problems of financing the Boron project and reorganizing to obtain the benefit of the depletion allowances in the United States were tackled without delay.

Consultations with Borax Consolidated's financial advisers in New York, Lee Higginson & Co., brought the company's representatives together with Lazard Frères and F. Eberstadt & Co. and also with financial representatives of the Rockefellers. The potential of both borax and potash businesses was soon understood, and the necessary financial support came from these sources, who were ready to join in what was financially an attractive situation for them. The plan involved the creation of a parent company in London, to be called Borax (Holdings) Ltd, which was to own all the shares in a company called Borax Consolidated Ltd. The latter would own and be responsible for all Borax (Holdings) Ltd's assets outside the United States. The reorganization in the United States involved the formation of a new corporation called U.S. Borax and Chemical Corporation, and after a series of negotiations and legal formalities its assets consisted of a merger of those belonging to the Pacific Coast Borax Company and those of the U.S. Potash Company. As a result Borax (Holdings) Ltd owned 3·1 million shares (i.e., 77·3 per cent) in the new corporation, U.S. Borax, and the remaining shares were allocated to the former shareholders (other than Borax Consolidated Ltd) in the U.S. Potash Company, and to the outside financial interests who had supported the scheme.

The final step was to obtain the consent of the British Treasury for the reorganization: this was obtained on 27 April 1956.

In the meantime the plan to proceed with the Boron project, although delayed, had got under way, and the necessary long-term loans were negotiated in the United States to support the capital programme of the newly created U.S. Borax. This all coincided with the increasing effort and expenditure by the US Government on the ZIP project for boron high-energy fuels, still 'top secret' and now creating much speculation in Wall Street and elsewhere. This again is a story to be recounted later.

Construction at the open pit and refinery at Boron was completed in mid-1957, and a dedication ceremony took place in November. In addition to members of the Borax (Holdings) Ltd. Board from London, guests included the British Ambassador to Washington, Sir Harold Caccia, and the Commanding General from Edwards Air Force Base, a close neighbour to Boron. The whole ceremony indicated the strong military interest that existed at that time in this new capacity for borate-production.

In mid-1957 new Research Laboratories were opened at Anaheim, and the US Government's interest underlined the need to form a separate organization called U.S. Borax Research Corporation, as a subsidiary which had only American directors. The corporation was thus able to receive classified information, and also to undertake government contract work involving such matters, from which it had been formerly excluded owing to the parent company's British identity.

The theory behind the military interest in boron high-energy fuels was well known. On combustion a boron hydride would provide about twice the amount of energy obtainable from traditional hydrocarbon fuels. In theory substantial increases in aircraft range could be achieved, or if the range were held constant a comparable reduction in aircraft size could be realized. Both were highly important factors in performance possibilities, at a time when targets could not be reached in the homeland of a potential enemy without in-flight refuelling and aircraft were the only assured means of reaching such targets.

It appeared that considerable progress had been made when the American Government placed firm contracts for plants to be built and operated by industrial firms for the manufacture of these boron hydride fuels. An article in the *New York Times* of 22 September 1957 reported that aircraft could achieve a 40 per cent increase in speed and 40 per cent increase in pay-load by the use of boron hydride fuels in place of hydrocarbons. The *Wall Street Journal* of 8 October 1957 carried the headlines 'Ramjet Test Missile, Using Boron Fuel, Flies 3 Times the Speed of Sound.'

This was all based on a Press release and public demonstration by the National Advisory Committee for Aeronautics of a $2·5 million rocket engine burning a boron hydride fuel. Speeds above 2,000 m.p.h. were claimed. Until this time the Defense Department had of course protected the 'top secret' classification of their research by scrupulously avoiding confirmation or denial of Press statements on this subject for some years. Now the whole programme seemed at last to be taking shape, thus contributing to a restoration of confidence in the USA after its celebrated setback when the Russians were first into outer space.

Borax, the raw material for these high-energy fuels, had suddenly and seemingly – after about 1,200 years of modest use to humanity – become a glamorous product, perhaps second only to uranium in strategic importance. This was an exciting but in many ways an alarming prospect for those involved in the present and future of the industry. Speculation in financial circles was rife and rumours were abroad that these fuels might also power space missiles. It was suggested in the Press in the United States that the first Russian space missile in 1957 was probably powered by a boron hydride fuel – a speculation that proved incorrect, and which was seemingly based on a report that the rocket surged into the sky with a green flame in its wake.

As might be expected, the stock of companies involved in this programme soared on Wall Steet, including that of U.S. Borax, which was listed for trading on the New York Stock Exchange in May 1957. However, the euphoria and promotional antics of the financial community surrounding these developments were brought to an abrupt halt when the US Government announced the cancellation of the programme in mid-1959. This unexpected decision caused shock and surprise in many circles, and a Congressional Enquiry was demanded. The House of Representatives' Committee on Science and Astronautics held 'open hearings' on the nature and effect of the decisions of the Navy and the Air Force not to open the 5-ton per day boron high-energy fuel plants recently completed by the Callery Chemical Company at Muskogee, Oklahoma, and the Olin Mathieson Chemical Corporation at Model City, New York. 'The Committee was interested in determining why it took the Services five years and over $200 million to decide they had no requirements for the fuel . . .' – so commenced the Committee's Report on 'Boron High Energy Fuels' issued on 13 October 1959.

The report covers much ground, and shows that the Navy and Air Force independently each spent an equal share of about $247 million. The Committee concluded it was right to terminate the programme due to changes in requirements, which resulted from technological advances attendant on the missile and space age. The Defense Department was considered slow in recognizing changing requirements, and its decision-making process was criticized. The cost of production of boron fuels was over five times that originally estimated, and engine tests showed nothing like the 50 per cent improvement in performance which the fuel properties had indicated.

The report shows that research teams of great skill and competence were built up to master the production and handling of these dangerous chemicals. Boron hydrides are toxic, and highly and explosively inflammable. Some idea of the problems of designing an engine in which they could be burnt can be envisaged by the fact that the products of combustion (boric oxide) are in liquid form below a temperature of 3000°F. As the temperature decreases this liquid becomes like a glue, which adheres to various parts of the engine, and at about 850°F becomes a glass-like solid. The formation of this solid on parts of an aircraft or engine would be like an unmeltable ice on the wing of an aircraft. Indications were that this problem of boron fuels in a turbojet had not been overcome. The problems in afterburners and ramjets were easier to solve, and sufficient progress had been made to show that this was possible, but the continous ejection of molten glass would have had its own specific environmental problems.

However, the main factors which in the end made this whole programme abortive were the shift of emphasis from manned aircraft to ballistic missiles, which as a new development in the history of weapons of war happened with remarkable speed, and also the increased performance of the B-70 bomber using hydrocarbon fuels which occurred in early 1959, and which was achieved by structural and aerodynamic design features. As a result the Air Force cancelled the J–93–5 engine, which could have used boron fuel in the afterburner.

The borax industry, based on limited sources of raw material, was relieved to get a definite decision on this matter. Working close to an area of classified information, and not knowing whether their raw material was about to become a strategic material, created much uncertainty and speculation for all concerned.

During the period of high-energy euphoria André Meyer of Lazard Frères and Ferd Eberstadt of Eberstadt and Co. – the latter by now a Director of the newly created U.S. Borax – put considerable pressure on the Board of U.S. Borax to agree to a joint venture with the Olin Mathieson Company, in which all research and development efforts of U.S. Borax would be shared with Olin Mathieson. The proposal was turned down by the Board of Directors of U.S. Borax, and this caused disappointment to the sponsors, but when the boron high-energy fuel programme was virtually terminated about two years later U.S. Borax was pleased to be free to control its own destiny in research and development, and thus its own future.

Not for the first time in the history of new mines, plants and industrial complexes, the dedication at Boron in November 1957 did not mark the start of a period of steady and trouble-free production. Teething troubles came, as is often the case, from unexpected quarters. 1958 was a difficult year for the management of U.S. Borax, and in December the Board had to suspend quarterly dividends on the Common Stock, as all the financial resources available were needed to carry out modifications and improvements to the new surface plant at Boron. However, the open-pit mining operation came into production smoothly in accordance with plans, and by 1960 the other difficulties at Boron were well behind the management, and the financial benefits of these new methods of mining and production began to be realized.

PART TWO
South America

South America

Descriptions from early letters (*c.* 1900)

Among the papers produced in court in 1902, in a case brought by an opportunist to challenge the title of Borax Consolidated to the extensive Ascotan deposit was a deposition, made in 1880,[1] to the military governor of Antofagasta. This claimed right of discovery of the deposit 'at a place called "El Borax" at the foot of the Ascotan Post Stage on the cart road from Calama to the interior of Bolivia to the North of parallel 23 but within the territory occupied and held by the Chilian forces under your command'. The length and breadth were stated as some twenty leagues by two leagues, and one of the unusual landmarks stated for identification on the eastern boundary was 'a volcano in eruption'.

Arthur Robottom, whose *Travels in search of New Trade Products* has been quoted earlier in this book, went out to South America in 1890 to seek borate of lime deposits, and made an extended tour which took in Ascotan. This he described as 'the most remarkable mountain I have seen in the Andes. Looking at it from the western side it appears as if the interior had been completely scooped out, and the interior painted with many colours.' He also asserted that 'Ascotan is an Indian word, the interpretation of which is "a place where a dog does not like to live",' and that the range of mountains of apparently irregular formation which bounded it on the west side were called by another Indian name, 'Cebollar' or 'String of Onions'. He also touched on two more places, Carcote and Pintados, which were also acquired on behalf of Borax Consolidated before it came into official existence in January 1899. But of the parts of Arequipa (Salinas), Chillicolpa, Cosapilla and the deposit near Iquique, which were also acquired before 1899, he makes no mention; so we shall probably never know now whether these names also provided similarly expressive Indian descriptions of locality.

The manner of the acquisition of all these concessions by Borax Consolidated has already been described in Chapter 7. However, some ten years after Robottom's visit, and soon after the mergers that produced the Borax Consolidated company in 1899, joint managing director Baker, back in London, decided to send out one of his most liked and trusted associates, W. R. Locke – ostensibly, as he wrote to Lesser,[2] 'to establish, if possible, undisputed title to the various locations in South

America that are at present giving you so much annoyance'. Obviously, however, Locke was intended to make a general report on what was going on out there. It is therefore a rare and fortunate chance that this American ex-mariner, W. R. Locke, possessed a prose style so vivid that it is unnecessary to do more than quote from his letters in order to give a lifelike impression of the borate areas of South America and of Borax Consolidated's activities there way back at the beginning of this century:

Valparaiso, 17th September 1901

DEAR MR. BAKER,

In order that you may get some idea of the nature of the country and people during the short time I was there, starting first of all with Valparaiso [i.e., the seat of the Borax Consolidated Head Office in South America].

This city is situated in a half-mooned bay which may have been formed by a landslide, for if the bluff, which is back of Valparaiso, originally came out flush with the Coast and a portion two miles long by, say, 9 miles wide, should slip into the ocean it would produce a condition now in evidence. . . .

The bay on the land side is separated from the city by a wharf consisting of piles, made from old railway iron, in which stone has been filled and on this wharf goods are landed from lighters at high tide.

The city itself is composed of native brick or 'adobe' buildings, stuccoed and painted . . . The heat in summer is not excessive and there is hardly sufficient cold in winter to necessitate the use of an overcoat, notwithstanding which the rains, which are not too frequent in Winter, so chill and dampen the air as to make the absence of stoves and other heating appliances very noticeable. Deaths from Pneumonia are very numerous and when coupled with other causes of disease, such as Smallpox, typhoid, and scarlet fever, the burial grounds are kept pretty well crowded. The water supply is very poor and it is hardly safe to drink water unless it be boiled first. . . . The streets are narrow and many of the business houses are private houses converted. Our office for instance in Calle Prat, is well located and is the upper floor of an old residence and consists of 3 front rooms, a kitchen and a liberal hall way. We have a telephone and we are convenient to the Banks, telegraph offices and principal business houses. Our office employees, aside from Mr. Lesser, consist of Mr. John, a book keeper, stenographer and also an office boy. . . . A large percentage of the population of Valparaiso speak English, for there are a great many English houses, such as Duncan Fox & Co, Graham Rowe, Balfour Williamson, Grace & Co. & others, and there are a number of shops, where English goods are sold and the language spoken. Electricity is used for lighting, but, as yet, there are no electric cars. From the upper level of the city a most beautiful view is presented. A line of snow mountains in the Cordillera is distinctly visible within 20 to 50 miles and the highest point in all of Chile, namely Aconcagua, is plainly seen, nearly 24,000 ft.

The hotels are remarkable for their inferiority. The butter and eggs are strong and the coffee is weak, the meat is tough and it is practically impossible to get a refreshing bath, for the hotels, like some of the private houses, have bath rooms but they are all almost full of trunks and rubbish, and their use not generally understood.

SANTIAGO is a city of some 250,000 population, situated about 4 or 5 hours journey South East of Valparaiso about 1,600 feet above the sea level in a level pocket in the mountains, surrounded entirely by snow mountain caps. It has a general slope from East to West and abundance of water from the ever melting snow in the surrounding range. . . . During Summer, when disease is most prevalent in Santiago, death averages from 9 to 10%. Typhoid Hospitals are full continually and the death rate on account of pneumonia and other similar diseases in Winter, is probably greater than in Valparaiso, for after sundown the atmosphere becomes very cold, though it may have been very warm during the day. . . .

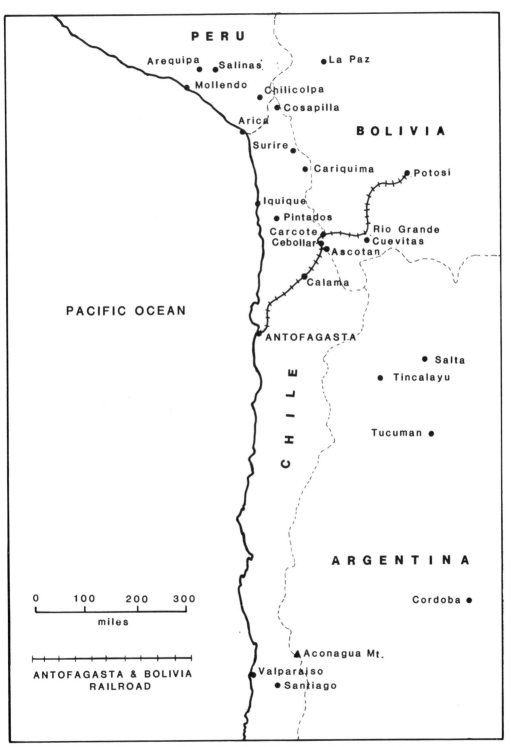

THE BORATE AREA OF SOUTH AMERICA

From an eminence just near the City one can count the number of chimneys visible on one hand: in other words the houses are not heated at all. Most of the houses occupy a very great area and, like Valparaiso, are in the shape of a hollow square, some private houses being nearly as large as the Bank of England, consisting of numerous rooms; The gentlemen's offices, picture galleries, ball rooms and many of them are as well furnished as anything I have ever seen. Pictures are frequently seen, from Paris and Rome, on the walls and beautiful tapestries.

I called upon a number of business men in their private houses and generally found them in the office portion of the house with their overcoats on and with rugs wrapped round their legs and feet, in order to keep them passably comfortable.

Santiago, being the capital, is of course the business and political centre. 46% of the population are lawyers and I should say that the balance of the public are wire pullers and shop keepers. The custom of the country is to retire at midnight, or soon after, and get up anywhere from 10 to 12 in the morning, and sometimes later. Frequently the head of the house wakes up at 10 or 11 and has his coffee and then his work is brought to him and laid on the bed, where he writes, smokes and attends to business, until breakfast time from 10 to 1. He does this because his bed is warmer than any place that he can find to work in.

In business, it is practically impossible to do anything at once, no matter what the subject of your business may be. A Chilian will say that he will always attend to it 'tomorrow', never today, and tomorrow he will probably postpone it until the next day and so on, if you allow him to do so, until it becomes interminable. The people are truthful as long as it is more profitable to be so, and are honest under proper scrutiny. They are intelligent and quick to resent an affront or injury, remarkably polite well dressed, even better dressed than people in London, are easy to talk to and negotiate with, but are very slow to act in conducting any transaction. . . .

Yours very truly,

W. R. LOCKE

Locke then journeyed on to see the various properties and borate concessions of the Company.

Antofagasta 26/9/191

DEAR MR. BAKER,

. . . After a pleasant journey on board the S.S. 'Santiago', we arrived at the above port, safe and well.

ANTOFAGASTA   This little Chilian seaport is two or three days trip by steamer north of Valparaiso. It has not a proper harbour, for there appear to be no harbours in Chile, – anyway north of Valparaiso. It is a half moon shaped bight in the land, with a rocky projection on the south side. The Railway Company have a sort of Wharf or pier, which runs out a little way, but Coast steamers and ships generally rest at anchor outside, except in case of storms, when they are compelled, in many instances, to go to sea for there is no shelter.

At this point we receive our borate of lime from Ascotan lake, some 388 kilometers inland and connected by a little narrow gauge rail road. The town is composed of principally one story buildings, which are much spread out. It has a population of about 18,000 and is a typical Chilian city. Our property is on the Southern outskirts of the Town, and consists of a dwelling for our manager, Mr. Rescalli, having on one side the calcining ovens and on the other the offices of the Company.

The land, owned by the Company, consists of about 3 blocks and not a half of this portion is, at this time, covered by the buildings in use. The little railway leaving this point passes directly across Chile on a very heavy incline and rises to about 14,000 feet at the eastern boundary line between Chile and Bolivia, where our Ascotan property is situated. The

THE WALL STREET JOURNAL,
Tuesday, October 8, 1957

Ramjet Test Missile, Using Boron Fuel, Flies 3 Times Speed of Sound

Top Federal Air Research Agency Shows New Unit, Tells of Rocket Gains

CLEVELAND.—(AP)—A ramjet test missile burning one of the new high-energy fuels, a boron compound, has flown more than three times the speed of sound, or faster than 2,000 m.p.h.

This was disclosed at the triennial inspection of the Lewis Flight Propulsion Laboratory of the National Advisory Committee for Aeronautics, the top Government agency in basic air research. The N.A.C.A. showed publicly for the first time a $2,500,000 rocket engine research facility which was completed last August.

BORON HIGH-ENERGY FUELS

REPORT
OF THE
COMMITTEE ON SCIENCE AND ASTRONAUTICS
U.S. HOUSE OF REPRESENTATIVES
EIGHTY-SIXTH CONGRESS
FIRST SESSION

[No. 39]

OCTOBER 13, 1959.—Committed to the Committee of the Whole House on the State of the Union and ordered to be printed

UNITED STATES
GOVERNMENT PRINTING OFFICE
WASHINGTON : 1959

In 1957 the new Boron High Energy Fuels were being reported as second only in importance to nuclear weapons in the US defence programme. However, by 1959 they were the subject of a Congressional Enquiry (right) asking why the expenditure of some $250 million had taken place before the programme was stopped

The start of the open-pit mine at Tincalayu, Argentina, at 13,000 ft

The borate of lime deposit at Salar de Ascotan, Chile (elevation 13,000 ft)

Pandermite ore leaving the Sultan Tchair mine in Turkey (c. 1890)

Across the fields of Anatolia to the mine (women were used for surface work on ore-grading)

Borates at work in the modern world

In glass fibre – used extensively for thermal insulation

In glass fibre reinforced with plastics – now widely used in boat-construction

In heat-resistant borosilicate glass

In all kinds of enamel-ware

In ceramic glazes on tiles, earthenware and china

In detergents containing sodium perborate bleach

railway is dependent mostly upon minerals. It passes through a nitrate district and taps a number of silver and copper districts as well as a number of borate fields.

JOURNEY TO ASCOTAN – we started from Antofagasta on 19th September in a private (rail) car, the party consisted of Mr. Lesser, Mr. Boyes, Mr. Marconi & myself. We were compelled to carry a week's supplies, which, with our party, comfortably filled the small narrow gauge special car of the Railway Company. The train consisted of one or two passenger coaches besides our car, and the balance was made up by goods carriages, hauled by single powerful locomotive. The ascent commenced immediately we left the Railway station of Antofagasta and, during the first 11 hours of the trip, namely from 8 a.m. until 7 p.m. our time was occupied between Antofagasta and Calama, a point a little over half way to Ascotan, being 238 Kilometers from Antofagasta and with an elevation of 2,266 metres, without passing over a bridge, through a tunnel, or through a cut of any importance. The railway is a gentle incline from start to finish and apparently in a level valley between mountains. The soil is a dark brown and very rocky: There is practically no vegetation, no cactus, no trees, and not much sand or dust. The green patches, which are to be seen on some of the hills, are evidence of the copper present.

After leaving Antofagasta, the weather gets very warm, the thermometer registering about 98 to 100 Fahr. . . . There are very few stations and they are simply to accommodate copper and other mines, which are in immediate vicinity of the road. There appears to be nearly the whole length of the line, the water pipe which not only serves as a supply for the Railway, but partially supplies water to Antofagasta. At Calama, the end of the first day's journey, the train stops and does not proceed until 6 o'clock the next morning, reaching Ascotan or more properly speaking, Cebollar, the present headquarters of the Company at about midday. . . .

Just before reaching Ascotan, we pass over an elevation of about 4000 metres, or about 13,200 feet. The nature of the country from Calama to Cebollar is considerably different from Antofagasta to Calama, and more interesting, while the whole distance may be considered to be of volcanic formation: the latter portion is much more plainly marked and the mountains surrounding the lake of Ascotan, say from 3 to 5,000 feet elevation about the lake, are almost all extinct volcanos. There are three or four of them, which are throwing out a little smoke but, at present, no lava. Some of the mountain tops, which are extinct volcanos, show the crater very distinctly and in some cases the colour of the crater and also down the mountain side, show blue, yellow, red and variegated colours. It is quite evident that the lake of Ascotan, and possibly other borate deposits, were formed by accident, being between volcanos surrounding it. Evidently the entire bottom of the lake, for possibly thousands of feet down, has been thrown out of the volcano and, as some of the mountains carry some considerable snow in the Winter, this melts and forms a lake as at Ascotan. If the ground under the lake were free from salt and other similar chemicals, the water would probably remain fresh and pure, but, at Ascotan, the ground under the lake must have not only contained sulphate of Soda, salt but boracic acid as well, and these chemicals, especially boracic acid, with their tendency to work up to the surface, have formed a deposit, which in the Summer time, when there is no water on the lake, has the appearance of dirty ice, with here and there a break, where the water is seen through. At a close examination the surface of this salar is rough and full of small holes as might be expected from the salt forming in a crust, after having worked its way through the soil. There is no evidence from Antofagasta to Ascotan of well defined ledges of rock of any sort. The hill tops are either good brown soil or lava, clear to their tops, and, if there is any rocky projection from the mountain at any point, this projection is more in the form of one of the many varieties of lava which has so far not disintegrated.

. . . On this trip, we reach an elevation of 14,000 feet by rail and the atmosphere is frequently very oppressive, many people having what is known as 'Soroche' or mountain sickness. People with any affection of the heart take the trip with great caution for the high altitude and rarity of the atmosphere produces more or less of a palpitation and the discomfort, caused through the struggle to breathe, is very great. So far it has had no effect on me, except to raise my pulse from about 60 to 85. Mr. Lesser's pulse was raised from about 75 to 105 and he has something of a headache, which he attributes to the elevation. Sleep is accompanied with great discomfort, for the necessity frequently to increase the amount of air in the lungs weakens one.

The Indians, who are working, appear to do a fair day's work, but some of our foremen are seriously affected by the altitude, the one at Ascotan particularly: he breathes like a man with a bad case of Asthma.

Then on to the company's non-operational property at

RIO GRANDE – Although Ascotan was practically our destination, we did not stop there but a few moments, but proceeded to Rio Grande in Bolivia, by the same narrow gauge railway as Ascotan, but about 100 miles further inland and practically at about the same level as the Ascotan lake. The first mile is across what would appear to be an old lava flow and we immediately came upon Carcote, one of our properties, which was purchased after Ascotan. The Railroad strikes almost through the middle of the lake of Carcote for about 15 miles: then continues either on the outskirts, or through the bed of smaller salt lakes, until Rio Grande is reached. In fact this whole upper level is a great salt lake plain, dried up in most of its parts, but containing salt water in the middle. The area is about 150 miles long with an average of something like 60 miles wide, but at some points it is as wide as 120 miles. Many parts of the lake that the Railway passes, have the appearance of being borate land. Mr. Lesser considers that it is only in appearance; that while there may be borate in small quantities almost universally over the whole, still the places where borate can be found in economically working condition are few.

At Rio Grande, we found in charge Mr. Jeronimo Aras, who appears to be a very good man, earnest, honest, careful and watchful and one of the few men, that I have so far found, who appear not to be affected in the least by the mountain sickness. From Rio Grande Station on the railway we took mules and went to the deposit over somewhat marshy ground for 6 miles.

At Rio Grande deposit ('Llipi Llipi') I was interrupted by 'Soroche' sickness, but the deposit seems to be in good condition . . . Rio Grande, you understand is the Railway Station on the narrow gauge road, 100 miles further inland than Cebollar. The town is situated on the outskirts of this great dried up salt lake and our deposit is 6 miles towards the middle of the lake.

Cuevitas, near Rio Grande, was later acquired by the Company but Locke wrote of it in 1901:

Now in regard to the appearance of Mr. Walker's deposit: The surface is covered in places with a thin layer of skimming or salt, crimped up by the sun, until it has the appearance of dirty white scale. Under this is 8 to 10 inches of what appears to be a most beautiful rich brown garden loam, fine in texture, and as nice as could be found in any one's garden. Under this there is 6 to 8 inches of borate, which looks of inferior character for the borate appears to be a conglomerate mass of little balls varying in size from a little pea to a marble, the surface of each one of the little balls covered with something resembling iron rust and at first sight you might think that it was dirt or an impurity but after being calcined most of the colour disappears, from which you might judge that it was only a stain. Under this comes a foot of greenish clay, under which is a layer of black clay from 3 to 4" thick, mixed with sulphate of lime in crystals.

From surface indications, one would expect to find borate just as quickly in Wimbledon Common as in Walker's property and there appears to be no part of the property that one can look at and say from appearance 'that borate should be found there' and the only thing that worries me is that if borate can be found under 12" surface of good garden earth, what security have we got against deposits turning up anywhere within the 9000 square miles of the salt lake, laying East of Ascotan? Mr. Lesser is of the opinion that some sort of arrangement can be 'got at', with Mr. Walker, not to take up further borate grounds in case we should buy Cuevitas; but I am afraid that Mr. Walker would operate through others, even if he apparently did nothing himself, provided he was fortunate to learn about the deposit through Indian women and half castes, as he did learn about the deposit at Cuevitas, and would turn up in some shape or other and give us trouble. The most plausible scheme seems to me to have the whole of the latter prospected, or such portion of it as Mr. Lesser and Mr. Macdonald would think proper. . . . If we could only satisfy ourselves as to the possibilities in this country, first by the inspection of places convenient to reach and gradually working into points of greater distance we would feel much stronger in our position I should assume. In the description of this country, which is most wonderful, I have no hope of being able to convey to you the results of my observation, for words can hardly express to anyone, not acquainted with such totally different conditions, the effect that this country has upon one. Generally speaking at this latitude, Chile is standing on edge, one side on the Pacific Ocean, and the other side (Ascotan) at almost the highest point in the Cordillera at this latitude.

The lake is almost 16 miles long and has the appearance of the Swiss lakes excepting the volcanic mountains surrounding it. The narrow-gauge railway from Antofagasta reaches a little Railway Station, called Ascotan, at the South end of the lake at nearly 14,000 feet above sea level. Looking down upon the lake at that point, one has the impression that the lake is entirely covered by ice, except at a few points where the ice appears to have melted and the water shews through. The surface of the lake is a little whiter than the melting snow and ice would be, but the appearance is not far different. Coming closer to the surface of the lake, it looks like a salt lake would, after the water has evaporated. The surface is not smooth, far from it, it is lumpy and full of small depressions as if cattle had been tramping about on it. From this high point, Ascotan, the little Railway follows around to the North the whole length of the lake, when we reach Cebollar 388 kilometers from Antofagasta. Cebollar is our own Town. Whilst it is not much larger than the Railway Station at Ascotan, still the buildings, which are made by covering wooden frames with corrugated iron on both sides and tops, are serviceable, being well painted and in remarkably good condition and the establishment, as it is called, is a credit to the Company.

The railway company have their own agent on the spot and all the labourers and managers, in charge of the lake and the workings, live at this point. Formerly extraction was carried on near Ascotan. Why the establishment is at Cebollar at present I do not know, excepting possibly a greater quantity of good borate is found nearer that point than at Ascotan. Then again Ascotan is 227 meters higher above the lake than Cebollar, which is down almost to the surface of the lake and situated directly on the outer edge of the lake. The ground from which the borate is being taken is possibly 2 miles from Cebollar, and is connected with the storing ground, or 'Cancha' at Cebollar by a light tram line, over which mules haul, in neat, well constructed iron tipping cars, the crude borate. This borate when it reaches the 'Cancha', which covers an area of say 12 to 15 acres, is stacked in lumps, in the shape of cones, possibly 4 feet in diameter at base and rising to a height of about 6 feet. These cones are allowed to stand for months, sometimes as much as 6 months: the air being very dry most of all the free moisture is taken.

As soon as the borate is thought to be sufficiently dry, it is shovelled directly into sacks, which are soon loaded and put on to a flat tram and hauled to the railway cars, which are standing on the siding, running down into the 'Cancha'.

The surface of this new borate section of the lake is possibly the same in appearance as much of the surface over the other parts of the lake, viz: on top, a corrugated crust of dirty white colour, composed principally of salt and other worthless chemical ingredients. This surface breaks up as a crust naturally would and the foot sinks through for a half to one and a half inches. With a spade a workman can easily dig down about 8 to 12" when he comes to a smooth white layer of borate. This borate on an average, as shown by examination, is usually about to 12 to 18" thick.

Under the borate is a white slaty looking crust of a more or less regular formation, which the workmen instantly detect immediately their spades reach it. . . . The test holes that we saw this morning, show a most excellent quality of borate and seldom was there any evidence of salt or sulphur of soda in the borate . . .

The labour at Cebollar is mostly drawn from the Indians in the neighbourhood, who, being born in the country, suffer probably less from 'soroche' or mountain sickness than foreigners or people from other sections of the country. Still they are unable to do violent work for any length of time without a great amount of puffing and blowing, on account of the rarity of the atmosphere.

There is very little to say about Cebollar and the Ascotan lake, except possibly to state that it is the largest and best supply of borate of lime in South America. . . . The property is managed directly by Mr. Rescalli at Antofagasta, whose many years connection with the Company has made him fully acquainted with every detail, but at present his health is so impaired as to make it impossible for him to give the attention to the Company, which the property should receive. His illness is somewhat mysterious. It seems to come on whenever he visits the property and prostrates him immediately. It is not an uncommon thing for people of his build to wear out their constitutions in this altitude. . . . Mr. Lesser thinks that some change in this department may be necessary; therefore it is necessary to get an able man to take the responsibility of the work conducted at Cebollar. He must know how to manage the workmen, have sufficient energy and ambition to watch our interests in choosing ground for extraction, manage the store and store keeper, and generally conduct the business at this point in a business like way. He must also attend to accounts, books and work generally done at Antofagasta, which is the seaport point, where the borate is calcined and shipped to Europe.

Locke went on to see the non-operational properties at Surire, Cariquima and Pintados.

Valparaiso, 2nd October 1901.

DEAR MR. BAKER,

IQUIQUE TO PINTADOS – Iquique, the great nitrate port of South America, is from 24 to 36 hours by steamer North of Antofagasta and reached in about 3 to 4 days by quick steamer from Valparaiso. Iquique is, in appearance from the sea, something similar to Antofagasta, namely it is a little above the water level, on a slope to a very high hill to the back of it. At nearly all times of the year, there are from 30 to 40 sailing vessels at anchor in the bay, awaiting their cargo of nitrate for European ports. . . . The borate properties, reached from Iquique, are 'Surire', some 3 days by mule back: 'Cariquima', a small property, where we have no man in charge at present, and 'Pintados', a point easy to reach and one where we have a good many matters demanding an early settlement.

The country between Iquique and Pintados is mostly nitrate ground . . . The country we passed through is barren of trees and very mountainous: although, like other railway roads in this country, there are no tunnels, as the makers of the Railway were quite satisfied to follow the formation of the hills until the nitrate fields are reached . . .

Continuing past the nitrate fields, we strike that great Salar in which Pintados is located. Pintados is so named for the reason that it is situated on the edge of the lake and near to the route originally followed by Indians in their passage from their own part of the Country to the other. In order that they might leave word behind that they had passed this point they made hieroglyphics in the side of the hill, such as crosses, circles, and so forth as signs to those that followed them that they had passed this point. The hills are now covered with these hieroglyphics at and near this point and the Spanish word for them is 'pintados' or 'painted'. This enormous salt plain, where the Pintados property is situated, is something like 30 miles wide by 180 miles long, and while I have had some experience in deserts, I never have seen anything that approaches this vast waste and language is almost unable to express its terrible condition. If you can imagine all the water being drawn off from an inland sea, leaving a perfectly level bed of the size above mentioned, with a crust from 8″ to 16″ deep, composed of salt, sulphate of soda, and sulphate of lime of a greyish brown colour, baked hard as a brick, you will get some idea of what it must have been originally. In the course of time, much of this crust, having become hard, has by some peculiar condition of the atmosphere, expanded on the surface, the result has been that ridges have risen up thickly all over parts of the salt lake to the height of 2 and a half and 4 feet, with cracks all over them.

At Pintados, as one stands near the establishment in the forenoon, one hears innumerable bursts of the surface, caused by the expansion of the uppermost portion by the heat of the sun. The sound is something similar to the noise made of ice being broken on large ponds. Of all God forgotten countries, this one seems to be the worst that I have ever seen.

The hills surrounding the lake are not very high and are usually composed of what appears to be a good rich brown soil. There is no vegetation apparent, and little if any rock in evidence. If there is anything in South America that would appear like Colemanite, or boracite, it would possibly be found in these hills, but nothing from their present appearance would lead one to suppose that any such formation exists. As before stated the elevation at Pintados is only 3,000 to 3,500 feet and people here generally consider that low elevation borate deposits are worthless. There appears to be in South America two ranges* of borate deposits, so far as elevation is concerned. One about 3,000, and the other range, including Ascotan, and adjoining properties, 'Surire', 'Salinas' and 'Azangaro', which latter deposits are in the neighbourhood of 14,000 feet. The high deposits are considered the better: the low deposits are almost always intermixed with salts and sulphates to a greater extent than the others, notwithstanding which there may be compensating advantages, as at Pintados, where Mr Lesser claims that, with cheap labour from the nitrate fields in close proximity, cheap coal and material from Iquique, which is within 4 hours by rail, good water in near vicinity, not only the salt, but about 70% of the sulphate of lime contained can be washed out. In fact a marketable article can be produced at as low free-on-board cost as from some of the better deposits from the Cordillera level further inland.

ESTABLISHMENT AT PINTADOS – This is a very low one storey dwelling house, and a wall surrounding a vast drying ground. Back of the house is a very extensive establishment, constructed originally for the purpose of making borax: but, like all similar establishments, was found to be unsatisfactory.

The manufactory of borax has been attempted at Antofagasta on our grounds back of Mr. Recalli's house, at Caldera, where the 'Maricunga' borate was formerly turned into borax, as well as at this point of Pintados, but failure has resulted in every instance. The Pintados plant consists of from 25 to 30 very large wooden tanks. 8 or 10 of them are lined with lead and they are variously arranged in rows to suit the purpose for which they were erected. As you know, they are not at present in use and their last service was in washing borate.

Mr. L. Ovalle Arrieta is in charge of the property, with a salary of 350 dollars a month. He and his wife live on the premises, and he occupies himself as best he can in various pen and ink work, carving and such other light amusement, until he can get something better to do. Formerly, he had charge of the Police force at Callao, but, through some political disturbance, lost his position and now he is in hopes of something better to do.

There is no rain at Pintados, excepting once in 3 or 4 years. The heat in summer is intense. At this time of the year it ranges about 90 degrees Fahrenheit.

The district appears to be prolific in earthquakes, the time between same being a week to 2 weeks, the shocks lasting about 2 days and sometimes even longer, namely good strong shocks occurring at frequent intervals during that period per day.

All houses are constructed so as not be interfered with and business progresses as usual. At Ascotan and some of the other points they have earthquakes, hardly as frequent, but sometimes much more severe. About three years ago they had an earthquake lasting for

---

*Locke later changed his views to a 'three-range' theory – at 3000, 7000, and 14,000 feet approximately.

about 3 days at Ascotan. This was practically one earthquake for there were from 300 to 400 shakes during that time and the quaking was so severe that one storey stone houses, where the stones were 18" to 2 feet square, were thrown off a good cement foundation and the whole house razed to the ground. The houses are in evidence today, which are nothing more or less than piles of rocks, though I learned of no opening in the earth, the worst feature being a general shake up. The whole country seems to be pretty well accustomed to earthquakes. At Santiago for instance, while I was there, they had 8 or 10 of them, most of them being slight shocks, like what might possibly come from heavy gusts of wind, or a jar from a railway train running under a house. None of them were mentioned in the papers as being anything worthy of notice.

Around Pintados there are no volcanos active or any craters within reach of the eye, but still the soil has somewhat of a volcanic appearance. From my knowledge of the Colorado Desert I should say it was a garden spot as compared with this saline dried up lake at Pintados. In the approach to this property, through the nitrate fields, it is sufficient to give one the horrors before one gets inside of it, for the nitrate people could give points to the devil in destroying the physiognomy of the earth in this section. There is not a redeeming feature in the whole formation of the country from Iquique to Pintados, except in its hidden mineral value, which in proportion to its horrible external appearance, must be great. There are silver, copper, iron and other mines en route and on one side of our Pintados property there is, as you know, a deposit of most beautiful sulphate of aluminium.

From there Locke went on to assess the practicability of operating at Arequipa/Salinas.

Arequipa, 10th October, 1901

DEAR SIR,

Mollendo, the so-called sea port, 900 miles North of Valparaiso, is probably as poor a port, in comparison with the business done, as there is in South America. The Ocean is usually very rough at this point: frequently the waves dash over rocks from 20 to 30 feet high. As it appears from the seas, the harbour is awful notwithstanding which, small boats by good management, can go to and from the Ocean steamers, which usually lie out about a quarter of a mile. The business done is chiefly in borate and minerals like copper, silver, nickel and tin, which are carried from the port to the steamers side in lighters. . . . Often the weather is so bad that steamers cannot take freight or discharge, and they have been known to stay there as long as 14 days, awaiting subsidence in the water.

Mollendo itself has one or two alleged hotels, several business streets, and a very steep hill, at the bottom of which is the railway station and the various warehouses, in one of which we have 1,000 tons of borate. As it seldom rains very hard in Mollendo during the rainy season of July, August and September the buildings are not very well constructed, and, while the leakage will not be great, a little of our borate appears to be wet. Everything throughout the town is in the most primitive style and it is not a place where anyone can live with much comfort.

FROM MOLLENDO TO AREQUIPA – At 8 o'clock every week day, there is a train over this broad gauge line from Mollendo to Arequipa. For the first hour or so, the train runs South along the Coast, through several small towns interested in fruit growing. The land is productive and, with irrigation from a small stream reaching the sea about this point, good crops are easily produced. The railway road here goes directly inland East and on a very steep incline. It follows hill top after hill top, remarkably crooked and the weather is usually exceedingly hot . . . . and there is very little soil. Apparently the mountains are piles of rocks broken up, while there is some soil between the rocks from decomposition of the rocks, which, during the rainy season, produces considerable grass.

In the dry season, it looks nothing but a pile of stone. The stone is either of the nature of sand stone, granite, or hard lava.

About midway between Mollendo and Arequipa, we strike a high level plain. It is about 87 kilos from Mollendo and about 4,000 feet elevation. This plain is something like other plains in South America on the first range of mountains at this elevation. It, however, has one peculiarity in its drifting sand, which I have not noticed elsewhere. This drifting sand is of a bluish grey colour and forms in mounds. . . .

The trip from Mollendo to Arequipa takes about 9 hours and the country is as described until within 10 or 12 miles of Arequipa. At this point the train runs along an old river bed and gardens, farms and homes have been planted on the banks of this 'arroyo' (ravine) and, where the ravine spreads out a little, pretty green patches of barley, corn wheat, and garden vegetables, as well as fruit trees and vines, are seen. From the Indian women, who are selling fruit there to the people in the train, I should judge the country to be very productive: figs, oranges, guavas, as well as strawberries, were very nice. Finally the train reaches Arequipa, a city of 30,000 population built very much after the style of all South American Spanish cities, only possibly 'more so'. There are no carriages at all in the city (Mr. Forga has recently got a carriage*). The sewage is carried away by means of an open ditch, through the street of the city. The water supply is defective although they have introduced a fair system of electric lighting, the motive power being derived from the little stream afore mentioned. The houses are mostly built of mud or a very light pumice stone lava and most of the houses, on account of earthquakes, which were formerly frequent, are of one storey high. Those being built today are rather more pretentious and are two or three storeys high and are mostly built of a light coloured volcanic soft rock. The people devote a great portion of their time to their Catholic religion and from 5 to 7 in the morning the churches let off fireworks in the shape of bombs, crackers of enormous size, until the morning hours are similar to the 4th of July in America and Guy Fawkes in England. . . . A few hours in the evening is also devoted to fireworks and the general hubbub is somewhat similar to that of the morning.

The town is full of pack mules and llamas, managed principally by Indian men and women, who bring in fire wood and minerals from the surrounding country. The city is possibly a few more hundreds of years behind the ordinary standard of civilisation than the cities nearer the Coast, where civilisation has penetrated.

AREQUIPA TO PROGRESO, SALINAS – From Arequipa the trip to Progreso is made in about 9 hours on horseback or muleback. The start is usually as early as possible in order to get out of the valleys and up into the higher mountains before the heat of the day. After leaving the town on horseback the traveller runs into volcanic sandy country through an old river bed. With the sand there are distributed a few rocks from the size of your fist to the size of your head. Mules and horses acquainted with the journey have no trouble with this part of the trip and can trot a little slowly. The incline is steady: it is very hot and dusty for the first two hours. Then the country becomes rougher and the rocks amongst the sand become thicker until the sand is hardly visible and for about 3 hours the country is about this description. The road gradually gets very much worse until it is very similar to the bed of an old stream and, in fact, this is exactly what it has been . . . As the road progresses it gets worse and the stones, instead of being the size of your fist or your head, become as big as your body. The horses have to climb over them and sometimes around them. . . . Vegetation, while not prolific, is general, and is in the shape of bushes every 5 or 10 feet apart. Little Indian huts are often seen along the route with a stone wall enclosure, where corn and other cereals are seen growing. During the passage along this little river, the traversed road crosses it 29 times. Most of the crossings are through the water. A few of the crossings at the upper part of the stream are over stone bridges, very substantial and constructed when the old Arequipa Company carts formerly came down the road part way with borate. The road might possibly keep getting worse but it cannot as nothing could be worse and how these little 'burros' or donkeys with 200 lbs on their backs, can pick their

---

*Forga was one of Borax Consolidated's lawyers.

way is marvellous and accompanied with great danger to them as is shown by the great number of dead animals on the way. . . .

These little 'burros' make the trip from 'El Progreso', Salinas to Arequipa, in about 2 days usually, though they can be pushed through in one day on a pinch; but the drivers prefer to stop and take off their load and let them run in a 'corral' and finish the trip the next day. The 'Llama' usually takes 5 days between 'El Progreso' and Arequipa, making about 3 round trips a month. The little donkey carries 200 lbs. and the 'Llama' 100 and mules 320 lbs. . . . After about 7 hours travel over the roads described, we reach the highest mountains, called Pichu-Pichu. Why it should be the water source is difficult to understand. There is, at this part of the year, no snow on any part of the Pichu Pichu, and the elevation is nearly 15,000 feet at this point, yet the whole country seems to bubble with springs (and I think that there are springs of various kinds, such as sulphur, iron and so forth). In passing up this river, one cannot help but notice what grand water power is here possible for electric purposes. The volume of water is considerable, the width of the stream being 10 to 12 feet wide and possibly 2 feet deep in the middle. The fall of water from the crater of Pichu Pichu to the wall below is possibly 5/6,000 feet. If it should be thought advisable to construct a monorail or any other mode of conveyance electricity, if wanted, could be found without the least trouble if this route was chosen. From the crater of Pichu Pichu to the Salinas deposit the route changes again (the air is bitingly cold and a good stiff wind usually blows.) The road bed is almost clean sand and from 2 to 3" deep with no rocks. The incline is gentle for 2 hours in which you drop down 500 to 1,000 feet to the lake. The appearance of the lake, when it first comes into sight, is grand. It has exactly the same appearance as a lake of ice on which an inch of two of snow has fallen and frozen. As one comes a little nearer, the surface does not change in appearance except possibly to become a little darker or dirty like. There is vegetation almost the entire way from Arequipa to Salinas lake. On account of the vegetation there are a few wild animals to be seen en route. We saw a couple of bands of Vicuna or small deer. . . .

In this trip I managed to hold together from the time I left Arequipa until Salinas, at which point I was severely attacked by the mountain sickness, and was entirely incapacitated for 12 hours with pains in the head, shortness of breath, nausea and vomiting, chills and fever, alternating every 5 minutes throughout the night.

WEATHER AT SALINAS – The rainy season here is from December 15th to the middle of March and during the present month the temperature during the day is 80 degrees Fahr. and during night 5 to 6 degrees below freezing. The humidity during the day, under cover in the house, is 27 to 28 while in the open air, in the sun, it is 17 to 18. Water boils, I believe, at this altitude at something like 90 Fahr, against 212 degrees at the sea level. This of course will bring prominently to your mind the question of calcining. The atmosphere at Antofagasta shows a humidity of 75 deg., the atmosphere here in the open from 17 to 18 deg. a difference of 58 deg. in the humidity alone. As water boils here at 122 deg. lower temperature than on the Coast, it would seem as though it would be an ideal place for calcining, i.e. a certain number of units of heat should have more working force by a great many times than the same units at sea level. We have no figures to prove this, but we hardly think that this is a question that needs proof. Water can more easily be driven off in a dry atmosphere than a damp one and water is more quickly evaporated on the mountain top than at the sea level and therefore the thing is self evident. At about 13,000 feet elevation 'Yareta' is found on the way from Arequipa to Salinas. This yareta, in its natural state, has very much the appearance of a rock, the size of your body, over which a green plush shawl has been thrown. A nearer inspection shows a growth of wood somewhat similar to honeycomb. As a matter of fact, by probing into this yareta you found that it has, as nucleus, a rock and yareta is somewhat of a fungous growth on this rock. Whence it gets its nourishment, except possibly from the air, must remain a mystery for each one of these little stems, laying together so close as almost to resemble honeycomb, is variously from 2" to 1 foot in length. This yareta is stripped off the rock, carted to our establishment, piled up or spread out, as may be thought fit or best to dry, after which it is used and apparently makes a very good fuel.

My comment to Mr. Lesser, in regard to the Salinas borate deposit, has been that the difficulties in transportation over 40 to 45 miles of rough country (i.e. over a road almost impassable) are so great as to make the business always unprofitable and at times almost impossible, but Mr. Lesser's answer, to the effect that the business is done and has always been done, largely neutralises my statement. Still it does appear to me that to carry on business in such a God-forsaken country, depending as we are upon a few erratic Indian muleteers and Llama drivers, is unbusiness-like and bound to be coupled with insurmountable transportation difficulties for all time.

There is a certain gambler's infatuation about the business, which is likely to lead a man with a great 'bump of hope' on and on at the present time; however, my bump of 'hope' is a hole unfortunately, and I cannot see how we can reasonably overcome the difficulties that are bound to crop up. True, if these difficulties were out of the way and if no further difficulties come up, if we could get good management, and if the borate was of the right quality and a few more 'ifs', we could do well, but how can we ever expect to compete with the deposit similar to Ascotan, which, I will here state, has practically spoilt me for any other deposit in South America? Theoretically Salinas works cheaply and successfully but, practically, the difficulties, which keep cropping up, overbalance the results of our most sanguine hope and make our work poorer than we had expected.

SALINAS BORATE DEPOSIT – This extensive and valuable deposit of borate of lime, situated about 40 miles East of Arequipa at an elevation of about 14,000 feet, is, like Ascotan, partly surrounded by extinct volcanos ranging from 3/5,000 feet above the surface of the lake. The formation is practically the same as Ascotan, except that the borate is found frequently a little deeper and possibly the strata may be a little thicker. Generally speaking, it is safe to say that there is usually found a whitish crimped salt crust, more or less dirty (from dust) of about 1" on top. Under this is about 1 foot of different stratas of sand, each strata being approximately 2" thick: under this comes about a foot of dirt, mixed with borate, rarely of any value and seldom utilized, and then comes from one foot to 2 feet, and on some occasions, as deep as 3 feet, a strata of borate.

Like Ascotan, the Salinas property (deposit) is more or less in and under strata of water viz: after digging through the top stratas and through the borate, water immediately rises to within something like a foot of the surface and there is something a little queer about this water. Naturally you would suppose that water in the whole valley, as you might call it, would be of the same level, but this is not the case. Small ponds are frequently seen on the outskirts of the deposit and the water in these ponds is frequently 4 to 5 feet above the level to which water will rise in the test holes of the lake and there seems to be no reason why this should be as it is.

There are certain parts of the lake where borate is found more or less of a yellow colour. Spots of similar borate have also been found at Ascotan and Surire. Mr. Lesser is of the opinion that this borate owes its colour to the presence of arsenic and is therefore not used. There are at Salinas, and also at Ascotan Lake, a few hot sulphur springs, usually about the border of the lake. I do not know that there is anything surprising about this excepting the fact that the water is hot that comes from the spring. There is one peculiarity at Salinas, which is different from Ascotan, in the fact that when making test holes in the lake, the ground emits a very disagreeable odour: in fact some of the test holes smell worse than a sewer. The smell is possibly more like the burning of horses' hoofs than of sulphuretted hydrogen gas.

ESTABLISHMENT AT SALINAS – This consists of a calcining plant, a house for the residence of the manager and officers of Salinas and a small store: also a corral for animals. The calcining house appears to be neat, orderly and well conducted as also the rest of the establishment.

Reviewing the trip, which I have just taken, I will state that Mollendo, as viewed from the sea, is a most repugnant looking port to visit and from the shore it is but little better.

... The trip from Mollendo to Arequipa is not pleasant and all people making it must feel a certain disinclination to be taken through such a hot, dusty, barren country, away from sea breezes and civilisation, and to a latitude very inconvenient to many. The valley of Arequipa is an agreeable contrast, with its fruit, green fields, and vegetation, in the proximity of the little stream, which trills through this enormous pile of lava. This feeling soon vanishes when a long stay in this primitive city becomes necessary, for the attempt at Civilisation has not progressed sufficiently to give one all that is required. Surely it has not yet reached the culinary department, that is to say, speaking generally, the only apparent use they make of food is to spoil it.

There are no satisfactory amusements in Arequipa and, in fact, the same remarks may apply to almost all the South American ports and particularly Iquique. Strong drink appears to be the only medium of real complete happiness and is much resorted to by all classes. The difficulty therefore in getting a good sober man or a good man, who will stay sober, must be apparent.

The trip from Arequipa to Salinas is coupled with so many hardships and inconveniences that it is safe to assert, without fear of contradiction, that no man ever went there twice, if he could conveniently avoid it.

Our Arequipa lawyer, Dr. de la Torre, although a young man, made all manner of excuses in order to avoid accompanying us and yet the elevation above this point is only half the elevation which Mr. Lesser and myself had to face. Now, if you can grasp the above statement, you can readily understand that it requires a man of most peculiar disposition to jump at his work at Salinas and do it with the same vim and without cessation that men work at in other places. . . . Still, I will assert that no civilised white man ever made the trip twice if he could avoid it and the men who live there in our interest are to be respected for their pluck, if they do their duty while they stay there. None but young men can stand it and they, only for a few years.

Mr. Goodair appears to be fairly well content with his lot but Mr. Lesser tells me that he (Goodair) has hopes of something better in the future. He is an ideal man for the place and just built like a Polo pony, nostrils distended like a native, lungs like bellows, skin either by nature or sun and alkaline air, a deep beef steak colour. 31 years old with a good business training as book keeper. Born at Preston, Lancashire and a gentleman. If he does his duty to Mr. Lesser's satisfaction it would be almost inhuman of me not to do what I could to say a word for him, unsolicited as it is, and to wish him well.

I remain,

Yours sincerely,

In spite of all the hardships described in these letters from Locke, this was still 'VIP' travel as far as Chile was able to provide it in the first decade of this century. A rather different viewpoint came in a letter to Lesser from Goodair, 'the ideal man for the place' mentioned in Locke's letter of 10 October 1901 (quoted above). This describes just as vividly something of the hardships endured by prospectors and other ordinary employees of the company, who had to pass months at a time in the High Cordillera, but it is not surprising that Goodair left for a less exacting post not long afterwards. The letter reads:

Cebollar, August 4th 1902

The General Manager,
Borax Consolidated Limited, Valparaiso.
DEAR SIR,
Since last addressing you under this heading I have been continually in camp, working at a distance from Rio Grande, so have had no opportunity of meeting any of the parties interested in the lawsuit or gaining any particulars as to what is going on. . . .

CARCOTE – Pozos de Ordenanza were all finished as mentioned in my last respects on the

20th ulto. I made them all 1.50m. square and sunk them 60 cm to 1 meter as the nature of the ground permitted.

EXPLORATION of likely Borate ground.

I have now examined all such grounds abutting on the Antofagasta and Bolivia Railway as far in as the oficina 'La Carrilana', which has been mentioned already in previous correspondence. Report and plan of work done will follow in due course.

As I understand it to be your wish that our parties again run over and report on all such known Borateras in passing, I was on the point of addressing myself to Mr. Kirkwood of Chattapata who is, I understand, one of the principal shareholders, if not the owner, of the place, asking his wishes in the matter.

Unfortunately a touch of rheumatism caught in my left hand assumed an acute form, and the pain and swelling were such that it became quite impossible for me to keep the field any longer. On top of this I caught a severe cold and lost my hearing completely.

Under the circumstances I broke up my camp and came back again to Cebollar. . . .

Meanwhile I wish to record my opinion that it is quite impossible for a party to keep the field for any length of time at this season of the year. I myself, who have up to the present been able to hold my own under such circumstances with all comers, am quite unable to continue.

Let me explain just what the life is in order that you may more fully appreciate the difficulties.

To do so the better let me take up the day at 5.00 to 5.30 p.m. when I get back to camp, pretty well done up with the day's work, having walked probably 25 to 30 kilometers over the worst kind of ground, sinking at each step ankle deep in sulphate, ashes, salt – whatever it may be. The wind as a rule has gone down, the sun is just disappearing, a thin chill atmosphere piercing to the bone is just settling down almost audibly over everything. I get into my tent, put on a couple of ponchos, wrap myself in two heavy rugs, and call for dinner. After dinner often the cold is so great that I pass the whole evening just tramping up and down to keep a semblance of warmth in my unfortunate limbs.

At 10.00 to 10.30 p.m. I turn in.

Turning in consists of crawling between the blankets after taking off boots and gaiters. Over my head, I don a woollen headcover, with just a gash in front for breathing purposes.

Over my head again I draw a doubled vicuna poncho and still suffer from the cold to the point of being kept awake by it. At 6.45, at sunrise, I get up; I swallow a cup of tea and a biscuit – the cold just takes away all appetite for anything else. Any white Chilian wine in my tent is frozen solid – a bottle of mineral water left exposed is invariably burst and frozen – and so on. At 7.30 I am in the pampa with my men and with them I tramp all day long, with just 45 minutes for lunch, until I again arrive in camp at 5.00 p.m. If you say 'Well, what of the things sent out from England?' I can only reply that on the frontier, where a man is liable to have his throat cut any night, he does not care about crawling into a bag and lacing himself up in it – when he is alone in a tent.

At noon the wind springs up, sometimes half a gale, coming down icy-cold from all the snow-clad peaks around. Then by 3 p.m. it is almost unbearable. On Friday last the men's tent, though well pitched, was carried away by a gust of wind and mine almost followed suit. Add to all this that a man is absolutely alone and never speaks for days at a time, except just to give the usual stereotyped orders to his men, and you have at once a condition of things which no ordinary man can endure, in health, for long at a time. In any case a man can run over as much ground in a week if pushing ahead conscientiously as he can work up in the office in 3 days, so that it would appear that 3 engineers or inspectors heading two parties could keep such parties in the field the whole time and still be able to have a little change of occupation, and keep their office work abreast of work in the field. More than this is impossible, and only men of exceptional physical strength could stand even this. These opinions I put forward with all due deference as the result of the experience I have had of the Cordillera. Now I hope to have the opportunity to talk the whole matter over at length with your goodself on my arrival in Antofagasta. For the present I can only say that, during the time I have been in the Company's service I have put their interests before anything else, but I also now see plainly that my physical endurance would not allow of my passing month after month without rest in the Cordillera on exploration work, and of my still doing that work to the satisfaction of the Company and of myself.

> I remain, Dear Sir,
>
> Yours very truly,
>
> F. Goodair.

The pity of it is that, after Locke and Goodair, no one else seems to have possessed quite the same flair for describing these far-off places in largely unexplored parts of South America, and there is no one to tell us what it was like on a summer's day at Hombre Muerto, Pajonales, Zenobia, Infieles or some of the other strange places that make a brief appearance in the records of the first decade of this century.

When Federico Lesser went to South America in 1898 he had the enthusiastic encouragement of both Smith[3] and Baker in planning to acquire any borate deposits that might be of commercial importance, and shortly after arrival in Chile he went to Arequipa in Peru to negotiate the acquisition of a deposit which might be of equal importance to the one at Ascotan (see page 81). News was anxiously awaited in London, but when the expected cable arrived it was received with mixed emotions, as it contained only the glad news of Lesser's marriage to a señorita of Arequipa.[4] However, news of successful negotiation soon followed, which all demonstrated a sound sense of priorities on Lesser's part.

After a spate of early acquisitions, Lesser's task became increasingly difficult, and years went by with no opportunity for him to discuss with Baker and Smith what should be done within this vast continent in relation to what was happening in the borax business elsewhere. None of them had realized how extensive these borate deposits would be, and how different the situation compared with that in California and in Turkey. Not many years elapsed before the glamour of the High Cordillera had faded for Smith. Summarizing, he wrote:[4] 'If we were to take all these properties at the original valuations of the owners, I am afraid we would have to stop paying dividends for the next twenty years and be working entirely in the interests of unscrupulous Chilians and Peruvians.' From about 1902 onward, Smith left South American affairs entirely to Baker in London.

In seeking to apply his strict London office standards to the staff in Chile, Baker never appears to have grasped the difficulties of running an organization in what was still a very remote part of the world, and Lesser received a continuous flow of often critical letters challenging costs and expenditures. Most of the staff had come from Europe to what must have appeared a God-forsaken country expecting at least reasonable generosity on pay, and possibly opportunities of making a little extra outside office hours. By 1905 Lesser was defending himself and explaining the situation to Baker in simple terms as follows: 'I know you consider the wages we pay here very high but this is not so. You can see how one man after another leaves our employ.' Unable to compete in pay with the prosperous nitrate industry in Chile, and with Baker insisting on being consulted on every detail, it is not surprising that Lesser was surrounded much of the time by staff who if they were any good soon sought prospects elsewhere, or if they stayed had to be sacked sooner or later for incompetence or some misdemeanour.

Lesser's letters give some idea of the local conditions that had to be endured. In June 1905 he wrote that he had been intending to visit Antofagasta from his headquarters in Valparaiso, 'but the smallpox epidemic here is so terrible that I dare not go away and leave my family alone – there is no smallpox hospital. . . .' In the following year he was again writing from Valparaiso:

> The tremors still continue, although three months have already passed. Now we have some days and nights without any at all, which is a blessing. My family and myself are still at an hotel. I am promised that the house I used to live in at the time of the earthquake and which is being rebuilt of adobe (mud) instead of stone will be handed over to me on February 1.

To be fair to Baker, it should be mentioned that he did arrange for compensation to all employees who had suffered loss as a result of the now historic earthquake of 1906.

As already outlined (pp.87–89), many borate deposits were acquired at mounting costs and at a financial outlay that caused increasing concern. From 1905 serious production was confined to two deposits: Ascotan in Chile, with its depot and new calcining plant at Cebollar; and Salinas in Peru, with its ovens on site. Salinas (at 13,000 ft) was worked only intermittently, as costs were always higher there, the quality of the borate produced was inferior to that from Ascotan, and there was also a difficult forty-mile haul on the backs of llamas to the railway at Arequipa.

In the early days it was realized that by sending South American borates to Europe, United States colemanite could be conserved for the growing domestic market, which Borax Consolidated at this time supplied from a single mine with limited reserves.[6] Later, however, it became clear that the delivered cost of pandermite ore to anywhere in Europe from the mines a few miles from the coast of Turkey was bound to be less than that of material produced in South America, given normal commercial dealings on all fronts. However, in these early years of the century dealings were often far from normal, and traders in South American ore constantly found opportunities among the numerous borax-refiners who sought to establish a position in the European markets and in the USA. The precarious and volatile business climate in which they worked has already been described in Chapter 8. For these reasons, even when rising costs made the viability of such operations increasingly doubtful, Borax Consolidated could not afford to let go of the grip it possessed in South America for fear of attracting competitive production which could be supplied from these virtually unlimited reserves of surface borates.

Borax Consolidated gave production in Chile every chance by working at high levels in terms of world production, as the following figures show.[7] In the five years before 1913, the United States and Chile had each an annual production averaging about 40,000 tonnes of borates, while the comparable figures for Turkey and Peru were 10,000 and 2,000 respectively. But by now Baker recognized that high costs were inherent in South America, and Borax Consolidated decided to concentrate in future on production in the USA and Turkey. As a result, it was agreed in 1913 that Lesser and his growing family should return to London, where he became manager of the newly created South American Department. He never wrote his memoirs, but it would be interesting to know which he considered the most trying of the many trying aspects of the sixteen years that he had devoted to South America – earthquakes, fever, uncomfortable conditions, lawyers, politicians, the abuse of alcohol by unruly staff and workmen, or the continuing criticism of detail by R. C. Baker, seated in London.

However, at this point Borax Consolidated policy was overtaken by history. The Balkan War of 1912, followed by the First World War of 1914–18, and then by the Mustapha Kemal rising removed the Turkish mines as a source of borates for more than a decade, and war also dislocated and soon terminated shipping of borates from South America. When hostilities in Europe ceased Turkey remained isolated, and the importance of the South American borate deposits was revived and prolonged until the discovery of the sodium borate deposit at Boron in California in 1926, after which even the most entrepreneurial efforts could not have overcome the disadvantages which faced South American production. However, the rising xenophobia of Japan and later Germany in the 1930s dictated a pattern of trade which became increasingly political as their preparations for war escalated, and was enough to provide some interest in borates from South America. Nevertheless, when war came in 1939 it was not long before South American borates ceased to move anywhere by sea.

After the Second World War, those same pre-war economic factors ensured that there was no place in world trade for borates exported from South America. However, the increasing demand for refined borates by South American industry and the need to avoid the use of foreign currency on imports, focused new attention on the use of an indigenous raw material.

By the end of the 1950s Borax Consolidated had translated its assessment of the future into investment, and a new company Boroquimica, S.A., was formed in Argentina. Production was based on a new tincal mine situated at an altitude of 13,000 ft in the province of Salta at a place to be named Tincalayu, on the northern side of Salar del Hombre Muerto. This was coupled with a new refinery built at Campo Quijano near the ancient city of Salta in the foothills of the Andes, some 250 miles by road from the mine and connected by railway to the industrial centres of Argentina. As events have proved, Boroquimica was also well situated to supply refined borax overland to Brazil, which had become the largest consumer among the South American countries and is without indigenous borate deposits. The internal industrial demand for borax in Chile – mainly for glass-manufacture – brought some revival of activity by Borax Consolidated between 1954 and 1966, but in the face of high costs and price controls it became unprofitable to continue.

With Boroquimica well established in Argentina, the continuing cost of maintenance of ulexite deposits in Chile and Peru could no longer be justified, and in 1967 after some seventy years, Borax Consolidated decided that a mining presence in these countries was no longer necessary.

REFERENCES

1. No. 234 of 23.9.1880.
2. M.D. to Lesser, 6.5.01.
3. F.M.S., 12.3.98.
4. Gerstley, J.: *History of Borax Consolidated* (unpublished). Also Lesser, A. B.: 'B.C.L. in South America' – Borax House Journal, Winter 1970, London.
5. F.M.S. to R.C.B., 25.6.02.
6. F.M.S., 28.1.98.
7. Yale, C. G. and Gale, H. S., *The Production of Borax in 1914* – Dept. of the Interior – U.S. Geological Survey, Washington, 1915.

PART THREE
The pursuit of borax in Turkey

The pursuit of borax in Turkey

It has already been related in Chapter 6 how the Turkish calcium borate mines began, which early in 1899 came under the control of Borax Consolidated, together with the two French companies to whom they had formerly belonged.

This mining complex, created by the merger of these adjacent mines, went out of production finally at the end of the 1950s, by which time Borax Consolidated was already taking active steps to continue mining borate material in areas farther south and east of this, its original location in historic Anatolia.

When the original mines came into existence about 1865 the reform movement which had been encouraged by Mahmud II after the destruction of the Janissaries in 1826 was rapidly petering out. But Turkey still possessed an empire around the shores of the Mediterranean comparable in importance with the scattered empires built up by the nineteenth century Great Powers in less civilized places, and far more mature in antiquity and culture. Although Western influences had long since touched the upper classes, the preponderating outlook in dress, custom and thought was still Islamic and oriental.

During the succeeding years the changes in Turkey were on the surface at least as great, and arguably greater, than those which overcame the three European-based empires of Austria, Russia and Prussia in the same period. To summarize, they were the Young Turk revolution of 1908; the deposition of Sultan Abdul Hamid the following year; the entente with Germany, resulting in defeat and loss of most of the empire in 1918; the Allied occupation of Constantinople (Istanbul); the invasion of Smyrna (Izmir) and part of Western Anatolia by a Greek army supported by the Western Powers in 1919 to 1922; the rise of a newer and wiser type of Young Turk in Mustapha Kemal and the amazing revolution which he brought about, driving out the Greeks, reconquering Izmir, forcing the Allies to cede back Istanbul, abolishing both the Sultanate and the Caliphate, and directing the Turkish homeland forcibly towards the pattern of a modern Westernized secular state. To put it undramatically, all these events, and others besides, provided Borax Consolidated with a succession of commercial problems over the years. Subsequently there were further problems of a less sensational type after Kemal Atatürk's death in 1938, and these really centred round the question of whether he had truly laid the foundations of a modern secular state or had merely succeeded in Westernizing surface appearances, leaving the old, aloof and unpredictable Turkey unchanged underneath. This question particularly concerned Western business-men trying to do business in the country or, like Borax Consolidated, trying to procure some return on capital sunk in developing concessions duly granted by some government of the hour. The events of this century up to Atatürk's death also bequeathed two realities to posterity which did not exist in nearly so strong a form in the Sultan's time – nationalism of a fairly pronounced anti-foreigner variety and the power of the Army, which is always close behind the political façade and sometimes, as in the recent past, out in front of it. This political uncertainty, together with surviving Ottoman habits of thought and procedure in the bureaucracy, did not always combine to provide that predictability about the basic facts of a situation which reasonably safe investment of capital normally requires.

This then was the historical background to the operations of Borax Consolidated in Turkey after the pandermite* (calcium borate) mines of the two French companies came under its direction in 1899. It was of course assumed from the beginning that the Borax Consolidated's Sultan Tchair mine would be run as one unit with the Société Lyonnaise's Azizieh mine, particularly as their workings were so close at one point that it became possible to drive galleries from one to the other.

However, there were personality problems to be overcome before this could be done. The obvious choice for the post of Borax Consolidated's agent in Constantinople (as Istanbul was then called) was Aristide Tubini, who seems to have 'known everyone'† at the Palace and among the business and other circles in the city, and had been doing the job most successfully on behalf of the Société Lyonnaise ever since the early 1890s.[2] He had also been responsible for supervising the work of the manager of the Société's Azizieh mine. When, however, he was asked, in May 1899 to take over also the Borax Company's agency at Constantinople the sitting tenant, a Frenchman called David, put up a prolonged passive resistance to handing it over. It was not until Daniell, another and more reliable Borax Company executive, unearthed David's thefts of company cash at Constantinople and Panderma[3] that the situation could be resolved.

Only three weeks later, and presumably on the recommendation of Tubini, a certain Pierre Coulbeaux was appointed 'Chief Engineer and Manager of the Mines in Asia Minor' on a contract for two and half years; but this immediately created another personality problem. The Borax Company mine had been managed since 1890 by Charles J. Bunning, 'Assoc. M. Inst., of C.E.'[4] – most obviously a forthright character, and one whose few surviving reports contain page after page of detailed exposition of his own technical opinions and merciless comments on the failings of others. Much more will be written about the adventures of Bunning later in this chapter, but for the moment it soon became obvious that poor M. Coulbeaux was in for a disagreeable time. Already, less than a year after his appointment, Tubini asked Baker if he could come to London to discuss

---

*See Glossary p.292.

†Obviously the Tubini family enjoyed considerable standing in Istanbul. As late as 10 November 1907 the English *Times* recorded the death, while hunting in England, of Mr Cecil Tubini, who 'will be laid to rest in Istanbul'.

BORATE AREAS OF NORTH WEST ANATOLIA-TURKEY

Map X

a scheme for reorganizing the mines 'chiefly on account of Coulbeaux and Bunning who, from his account, are not hitting it off'.[5] A letter from Bunning,[6] only three weeks later, does far more than surmise can do to reveal the man's character: 'I venture to write to you asking if my eleven years of good work at Sultan Tchair has no merit in your eyes. Am I destined to work much longer under a younger and less experienced man?' ... and so on, with much more in the same strain. No reply is traceable but perhaps it was less sarcastic than the usual Baker style as Bunning wrote again three months later: 'I regretfully beg that from 1 October this year I be given the same emoluments as your manager, whose work I now do.'

It is not surprising, therefore, that only four months later Coulbeaux was arguing out with Baker[7] the meaning of the terms in his contract relating to premature resignation. He was gone by 1902, thus dispensing with the need to solve the problem by transferring Bunning to the mines in America, as had been hinted in earlier correspondence. We shall never know in fact quite how large an explosion the mixture (of Bunning with F. M. Smith, Ryan and Zabriskie) would have produced.

Instead, from 1902, business in Turkey started to settle down for a few relatively peaceful years, with Tubini in charge of the diplomatic and commercial side of the work in Constantinople and Bunning in charge of the mines. With the dissolution of the temporary 'advisory committee' in France, the death of Lawrence, and the setting up of a new united company, Borax Français, in 1902, the old asperities between the personnel of the two French companies began to disappear rapidly. It became possible at last to regard Sultan Tchair and Azizieh as one mine instead of two, and to intensify a process of consolidating this complex with other promising concessions around it, which Tubini had been carrying on in a quiet sort of way ever since 1899.

There was indeed not much else to distract his attention during working hours. In July 1902 Daniell visited the Constantinople office and reported discreetly[8] to Baker on its activities. His general conclusion was that 'the organisation of the Constantinople office leaves the staff with very little to be done, but we know from experience that that 'little' was never done in the days of our predecessors'. He had a high opinion of Tubini, 'an agent whose position and prestige allow him to arrange matters which have long been in suspense', and of his influence with the Ministries, but noted that he did not 'occupy himself in anything concerning the mines' and showed 'some disinclination to work with Bunning'. This was understandable, 'but at the same time it cannot be said that the mines have suffered in any way through the present system of administration'.

> As regards the rest of the staff, Mr. Coyoumdjian may be considered as a present to Selim Pasha, pure and simple: he is useless to us as an employee [that is to say, he had been given the job to keep one of the Palace Pachas on the company's side].
>
> Monsieur Manolaki fulfils the duties of a permanent lawyer and intermediary with the Ministers. He may have been useful in the past, when the management was incompetent, but ... we gain nothing by M. Manolaki making a daily visit to the Ministry to ask if there is any complaint against us.
>
> Mr. Psalty, the clerk, looks after the work in the customs, purchase of powder, stores etc., in fact he does most of the work for which the office is required and does it very satisfactorily.

According to Daniell, Psalty had been connected with the boracite mines since 1884, which if this date is correct would have brought him into the business in the early days of Desmazures and Groppler, and before the first official concession was granted in 1887. Indeed, he might have told an interesting tale if this history had been commissioned half a century or so earlier.

The description of this office set-up would of course move any modern office consultant to mirth or indignation, depending on his particular temperament, but in either case, and even in terms of present-day Turkey, he would be wrong. When Bunning wrote to Baker on 3 August 1899 about the difficulty of securing a particular mining permit area he unintentionally summed up the atmosphere when he said: 'But could the Consolidated get this permis without raising the suspicions of the hungry Palace Pachas, who would be furious if they knew that a concession or permis could be got without it passing by their hands?' He quoted a particular example to illustrate this surviving idea of the 'spoils of war' which has lingered on in Turkey ever since the Ottoman army took Constantinople in 1453: 'We tried in 1890–91 to get Djivad Pacha's concession through Lafontaine (Manager of the Ottoman Bank) but the disguise was not veiled enough, the Minister of Mines refused it to Lafontaine but gave it to Djivad Pacha.'

Tubini, therefore, although he might not be a 'nine to five' office man, was doing a very good job precisely at the point where it mattered most – influencing Ottoman officials in the interests of Borax Consolidated and occasionally being forced into offering a sinecure in his office to a friend or relative of a Palace official in order to secure more permanent forms of influence. In the first year of the new company's existence there was particular need for his intervention. The Borax Company's mine and La Société Lyonnaise's mine were both surrounded by at least ten mining concessions granted to other people, and if one or more of these were to come into operation the value of the company's product and property might be correspondingly diminished. The situation was in fact similar to the one confronting the company in South America at this time. As Bunning's remarks also indicate, most of these concessions had been pegged out with the principal object of making a broker's turn by selling them to Borax Consolidated or to some other rich foreign concern. For Borax Consolidated, therefore, it was a question of deciding which, if any, contained payable mineral and then using Tubini to negotiate a reasonable price with the 'hungry Palace Pachas' and with quite a number of lawyers and other hangers-on, several of whom seem to have been interrelated.

The first and most important purchase was the 'Ahmed deposit' (otherwise known as Tchakmak Bair), which was in operation as a working mine at the time.[9] After inspection by M. Henri Lenicque, a mining engineer, and four months of bargaining and broken promises, this was eventually purchased from His Excellency Fuad Pacha, Aide-de-Camp General of the Sultan, His Excellency General of Division Ahmed Pacha, Aide-de-Camp General, etc., and two civilians, for a sum of Lires Turk 76,000. By the time the negotiations were concluded the Borax executives concerned were expressing, even on paper, opinions about the two Pachas very much at variance with their high-sounding titles.[10] Fuad was at loggerheads with Ahmed, and all four vendors, it was discovered, had charged the property in various ways with their debts. On top of this, Fuad was perpetually trying to negotiate behind the backs of the other three and to secure advances of purchase money before the agreement was struck by threatening to sell the property instead to the Germans. In this way he hoped to evade the prior claims of his creditors on the purchase money. Tubini added in the same report to Baker[11] that: 'If my informations are correct Fuad has been encouraged by Chivinian, Whittall's man, who is said to have proposed to him to sell you the mine for LT 90,000 in return for sharing the difference from the LT 76,000 which the Company has offered.'

In spite of the success with which Tubini had threaded his way through this web of Oriental intrigue, there is no evidence that the 'Ahmed Pacha deposit' was ever worked by Borax Consolidated after its acquisition.

In May 1899 the 'Whittall deposit', which had formerly been a concession with a different and now untraceable name, was also acquired, and became the 'Lawrence concession' in accordance with the Turkish mining law requiring concessions to be registered in the name of individuals. This fact has been mentioned because it added greatly to the difficulty of tracing Borax Consolidated's former properties in Turkey.

In addition, in this case, there was also an 'old Whittall deposit', which had been abandoned in 1893 and which was continually being dangled in front of likely buyers in these years. This former 'Demi-Capou' concession had acquired the reputation of containing no worth-while mineral, but all the same Borax Consolidated decided to buy it for £10,200 as late as 1909[12] 'because it is contiguous to our property'.

There were also three further Borax Consolidated concessions, registered in the name of de Friese, Daniell and Falcouz respectively, but which they were (out of the thirteen listed by Bunning, in a letter of 3 August 1899, under other and more Turkish names) is impossible now to determine.

In these transactions a firm called Edward G. Thompson frequently appear, trying to sell Borax Consolidated the 'abandoned Whittall concession' and trying, unsuccessfully, to push Tubini out of his job as agent of the company. Their supporting argument[13] was that the continuing French management of the mines should be replaced by Englishmen, as 'for the same reason that the Frenchmen are bad colonists they are bad managers of natives'. We are not told whether the repetition of these astonishing Victorian sentiments to the proud rulers of the Ottoman Empire contributed to the bankruptcy order under which the Thompson firm were then labouring. One of the Thompsons had married a sister of Edward Pears, a 'Civil Counsel' of Constantinople and legal adviser to Borax Consolidated in Turkey at this time, and was on the other side connected by marriage with Mahmoud Pacha, brother-in-law of the Sultan. He appears, therefore, to have been in an advantageous position to make mischief, but in spite of Thompson's intrigues and Fuad Pacha's hatred of Tubini, Borax Consolidated had the wisdom to retain Tubini as their agent right up to his death in 1908.

However, the immediate game of acquiring concessions was not played for long, as it was soon realized that there was a government decree forbidding the holding of concessions for more than a certain time unless they were worked. As Daniell reported in July 1902, the Council of State were getting suspicious that amalgamations would reduce output and hence revenue due to the State through royalties, the level of assessment of which was a frequent matter for argument with the Palace in these years.

This led Tubini in 1904 to arrange a compromise whereby Borax Consolidated was permitted to work only one of its five concessions, in return for payment of £500, and working was then confined to the former Sultan Tchair mine of the Borax Company. This permission was due to last for two years only, but it seems probable that in fact it continued indefinitely, as there is very little information about Borax Consolidated working any other concession than Sultan Tchair during these remaining years before the outbreak of the 1914–18 War closed the mines for the duration.

A scaled-down map of the workings at Sultan Tchair and Azizieh mines survives,[14] and there are of course references in Bunning's reports to the numerous galleries and areas that were being worked. These, however, were written for contemporaries with detailed knowledge of the background, and are unrewarding information eighty years later. It is therefore interesting to find two references, in letters written by Baker to Bunning in the early months of 1914,[15] that give some general idea of what plan was being pursued underground. Probably of the Sultan Tchair mine only, Baker wrote: 'The question of opening up the Northern workings can rest until we have exhausted the Southern workings'; and a few weeks earlier he had written:

> I note that, generally speaking, the ore in the Desmazures concession in the South of the Borax mine is being found in a restricted area which gets narrower as you go South and approach the Whittall, Reched and Tubini concessions and that, to the West of the fault running to the South-West side of the area you refer to, there is a field of ore but that exploration stopped owing to poor ore and water.

Only scattered and incomplete information survives also about actual production at the mines, but, such as it is, it provides a useful picture. In 1899 and 1904 Bunning wrote long and verbose reports.[16] These show that the mine had a shaft with a winding-house and two cages descending unstated depths to 'galleries', the number of which is also not stated. The principal method of working was payment by the cubic metre for mineral obtained in galleries six metres wide; but there were also narrow galleries two metres wide 'adapted to explore or to traverse sterile parts when met with in driving the large galleries'. 'The Borax (Company) has been working in poor rock in comparison with Azizieh (i.e., Société Lyonnaise), that is, in 19% richness, and the Azizieh in 30% richness'; but the Hon. Lawrence and Bunning had adopted a policy in 1897 of picking the richest parts out of their mine, which, Bunning admitted, might have endangered the mine with too many over-wide galleries if it had continued.

Bunning claimed, in his report for 1904, that he had introduced a new system in February 1900, which consisted of working the ore in galleries of great width (from eight to eleven metres wide), and in increasing the height from two metres to any height in which good mineral would render it necessary. This system was eventually to form one of the causes of the strike of 1908, when the miners refused in future to mine above two metres height, much to the annoyance even of Baker, away in the London office. At the time, however, Bunning justified the operation by pages of statistics in which, while there is mention of more effective use of powder in blasting, there is no mention of timbering or of the safety aspect. In fact, about the check-board system of working off pillars formed by narrow galleries eight metres apart which was formerly used in the Azizieh mine, he wrote in light-hearted manner: 'Where is the incentive for the miners? Powder was wasted, men were careless, explosions took place, accident to lives resulted, yield was lowered and cost price increased.'

He had not, of course, to deal with labour unions in these far-off days and places, but had other problems of a more unusual nature:

> The management of mines in this country is fraught with various difficulties arising from the many races of men employed. The Levantine has long been known for his

willingness. The Greek, of whom there are various, the Ottoman, Hellenic and those protected by several European nations, some catholics, some orthodox in religion, all are clever but designing and subdolous [cunning]. The Armenian, of sneaky, cringing character, who, behind your back is continually striving to circumvent your precautions in preventing yourself being swindled. The Circassian, obedient if rightly handled, clever without the cunningness of the Armenian, submits to discipline and strives to be honest. The Albanian, of whom we have none now, lawless even mutinous, but faithful to those he belongs to. The Nogai or Tartar, who is an unclean Moslem, lacking his sobriety. The Muhagir, the Moslem immigrant from Bulgaria or Macedonia, who works well, is submissive and disciplinable, to the Italian miner, who was well represented here in former days but of whom rest only two now. The indigenous Turk will not work at the mines, he prefers cultivating the soil.

Amongst all these you have the good, bad, and indifferent and especially the intriguant who waits his time to set all a blazing. Among the Circassians, there are always some idle hands, who have a chronic grievance against us, who are always grumbling and are quite ready, when bad advice is given, to get all the other workmen to strike, as happened in the Spring of 1903 when the Government Engineer came amongst them and made them believe that we were acting unfairly towards them. The Armenian renders all checking and controlling ineffective by his ingenious methods in hiding the truth and making money where he should not. We have this in the [mineral] dressing shed where, ever since I came here, we have been continually catching them if not in swindling then in hiding their profits.

Payment at the mines was obviously[17] based on piece-work, and hours worked averaged twelve a day, which compared favourably with those worked at the deposits in South America and was on a par with those worked at the factories in Paris and Vienna.

However, as Bunning wrote in 1904, 'I have passed through strikes and troublesome times here.' Perhaps the most notable of these came late in 1908, when the 'Young Turk' movement revolted against (and eventually deposed) Sultan Abdul Hamid and seized the government, which they held until just before Turkey's defeat in the Great War. Baker wrote to Zabriskie on 19 November:[18]

> We have just got through a strike of several months at Sultan Tchair and today have a cable that it is now ended. After a peaceful revolution the working people in Turkey considered that the Millenium had arrived and, lightly brushing aside all economics, demanded a 100% rise in wages and the dismissal of everybody in authority. Fortunately we had accumulated a large stock of boracite and it suited us very nicely to have the mines shut while the people were striking for the impossible.

Although no one, probably, realized the full implication at the time, 1908 was to mark the beginning of nearly twenty years of serious trouble, from outside causes, for Borax Consolidated in Turkey. At the time, however, the shape of things to come appears to have had remarkably little effect on the steady rise in production from the mines, which surviving statistics reveal. In 1895, according to Bunning, the two mines produced a total of 14,608 tonnes* of mineral, which had risen to 15,000 tonnes by 1898. For the twelve months ending 30 September 1903 production was 6,800 tonnes from the Sultan Tchair mine and only 772 tonnes from the Azizieh mine, making a total of only 7,572 tonnes.[19] This increased to 9,610 tonnes the following year, and varied rather erratically between 11,470 (1906–7) and 16,712 (1912–13) tonnes over the rest of the pre-war period.[20]

*1 tonne (or metric ton) = 1000 kilograms = 0·9842 of an Imperial ton.

No information at all survives of how the ore from the mine was treated in the mineral-dressing plant, but at the next stage – its transport from there to the depot and port at Panderma, about thirty miles away – some radical ideas were bandied about inside and outside the mine boundaries in these years.

Originally, as is obvious from a letter of Bunning written in 1899,[21] 'carriers' on contract to the companies 'carted' mineral down to the Panderma depot along a road in a shocking state of disrepair, but whether they used carts or donkeys or their own backs or a combination of all three is not clear. Then in 1902 J. W. Whittall & Co., grain merchants and purveyors of doubtful mining leases, as described above, tried to persuade the company to put capital into a scheme[22] for restarting transport on the river Simav, which 'for some inexplicable reason' the government had stopped in 1882. Needless to say, 'some high Turkish officials have obtained a concession for the navigation of the river that passes the borate of lime mines in Sultan Tchair' and 'H. & S. Pears, lawyers of Constantinople', also had interests in it, but needed to recruit capital from abroad. However, all these blandishments were resisted. Daniell (now company secretary of Borax Consolidated) pointed out shrewdly that the whole scheme depended on the proposed river service being allowed to carry all the trade, both export and import, of the Mihaldieh basin, in which the mines were situated. This was a very small area of Turkey, where in any case the near-by government in Constantinople and venal local officials had already killed any prospect of prosperity, the mouth of the river was likely to silt up, Panderma was a considerable distance from it, and no one could prevent the local inhabitants continuing to carry their own produce to Panderma themselves. He made an alternate suggestion that a light railway should be built along the road to Panderma, and this pointed the way to the next idea which, however, did not become practical politics for another nine years.

Late in 1911,[23] the Compagnie Chemin de Fer Smyrne–Cassaba had got some way with their proposals to build a line past Panderma to Smyrna (later Izmir), and it had become obvious[24] that the Turkish Government had been putting pressure on them to link the proposed line with Sultan Tchair mines. Needless to say, Borax Consolidated hastened to add their support to this proposal, and also to press for a 50 per cent reduction in the proposed freight rate for minerals as already fixed by the Government. The news of all this[25] had hardly reached London, however, when there was a change of plan by the military government. The 'Great Powers' had seized the excuse of the revolution of 1908 to continue the dismemberment of Turkey, the 'Sick Man of Europe'. Austria seized Bosnia-Herzegovina in 1908, Bulgaria declared its independence, and Crete declared itself a part of Greece. Then, in October 1911, Italy seized Tripoli and the Italian fleet bottled up the Turkish fleet in the Dardanelles and blockaded all Turkish ports from which reinforcements might be sent to Tripoli. This brought the Turkish Government quickly to the decision that the rail link between its capital at Constantinople and the key Aegean port of Smyrna must be completed with the maximum speed by the shortest and least costly route and, as a high official told Devisscher of Borax Consolidated early in 1912:[26] 'A route passing the Sultan Tchair mines would have been longer and considerably more costly than the route finally chosen.'

The company therefore fell back on thinking about various other ways of connecting the mines to the mainline,[27] and by the time the new railway line was opened, in December 1912,[28] expert opinion had come down in favour of constructing a ropeway to Omer Kerri station; but there is no evidence that the newly constructed ropeway was ever used before the outbreak of the First World War less than a year later.

The international situation had in fact given the company increasing cause for concern ever since 1908. It was known, of course,[29] that Germany had been building up her influence over the Ottoman Empire for at least thirty years – the *Drang nach Osten*. Indeed, early in 1903 Daniell had written to London, on the occasion of his negotiations to secure a reduction in mineral royalty, that: 'We are having great trouble with these Austrian and German hangers-on at the Ministry of Mines and the Germans, being all-powerful at the Palace, naturally hang as a drag on our negotiations.' The ultimate German objective of taking over all useful mines in Turkey became clear in 1915, as will be recounted later; but there was already a foretaste of it as early as 1899, when Bunning reported[30] that a commission, consisting of a German called Weiss and two Turks from the Ministry of Commerce, was 'visiting all mines and concessions in Asia Minor'. Its appointment had been instigated by the Deutsche Bank, to which the Sultan's government was heavily indebted, in order that, according to Bunning, they might 'have the preference to work any (concessions) should they be found payable'.

So, by October 1912, when Bulgaria, Serbia and Greece attacked the Turkish positions in Macedonia and the First Balkan War broke out, life in Turkey was becoming decidedly difficult for Borax Consolidated. As Baker wrote to the stockbroker, Emberson, on 21 October:[31] 'Yes, we are being bothered by this war – our miners are mostly Turkish reservists and have been called out and there is trouble anyhow in getting stuff away', particularly with the Dardanelles closed. Peace came, after several broken armistices and much double-dealing, in mid-1913, but the surprising thing is that there is not the slightest hint in Borax Consolidated's archives, or indeed in the general record of the times, that more than a few far-sighted people saw this peace as a pitiable prelude to a much greater holocaust starting in Eastern Europe just over a year later. Baker paid his first recorded visit to Turkey as late as July 1914, and, even after the First World War broke out on 4 August 1914, he could still write to Bunning: 'Although Turkey has mobilised her troops I see it stated here that she has given assurance that this is only for the protection of her frontiers and that she has no intention to embark on any enterprises. I trust this may be so.'

The bold and indomitable qualities of Bunning were shortly to show themselves as the very real asset that they had always been to Borax Consolidated in the outpost of Sultan Tchair. These qualities had hitherto irked rather than pleased Baker, the disciplinarian, ever since Bunning's brash but successful attempt to drive Coulbeaux out of the manager's chair at the mines in 1902. In 1908, when Aristide Tubini died, Bunning, along with Ernest Whittall and one or two other chance applicants, had been passed over in favour of Aristide's son, Hyacinth, for the job of Constantinople agent. This was partly because the Tubini family, in spite of all the rows with Fuad Pasha, were still more acceptable at the Palace than anyone else.

Then, in 1910, Bunning expressed himself very freely to his fellow-executive, Lesser (who was on a visit to Turkey) about the 'inexplicable apathy' of the London office in not securing the 'old Whittall' concession, and had been foolish enough to record the conversation in a letter to Baker, presumably while pressing his views on the concession's desirability. Inevitably he had received in reply[32] a mammoth and majestic rebuke, full of expressions about presumption and impertinence. When in 1913, having endured the increasing uncertainty and unrest in

Asia Minor resulting from the Young Turk revolution and the Balkan War, he asked for an increase in salary he received an outright refusal,[33] which, however, contained some interesting details. Baker pointed out that Bunning's salary had risen from 584 to 940 Turkish Pounds (LT) in fourteen years, and that he had also been granted an annual bonus and a travelling allowance, each of fifty pounds, a furnished house, free firing and lighting, servant's allowance, 'a riding and two carriage horses and carriage, and a special monthly allowance towards a further house and gardener'. 'These conditions', Baker added, 'do not point to any very serious underpayment for your services . . . and I must call your attention to the very heavy increase in cost which has taken place recently at Sultan Tchair mine' and must, of course, be reduced as soon as possible. To this cold comfort Baker added what he himself wished to believe about the political situation: 'There seems no reason to anticipate any disturbances in your locality likely to lead to serious trouble and I hope that they may not arise.'

Baker then proceeded to give Bunning what amounted in effect to an order[34] to break the three-year agreement reached with the miners about not removing mineral above a height of two metres: 'You go on to state that any change from the present system would probably lead to a strike of the Circassian miners. . . . To accept the position that the miners are to dictate how the mines are to be worked and whether we shall lose ore or not is one that I will not sanction.'

Probably his eyes were at last opened to the reality of Bunning's difficulties by his visit to Turkey in July 1914, and by events which occurred, and could not be shrugged off, as soon as Germany went to war on 4 August 1914 (leaving her future ally, Turkey, to sit on the fence for a few more months of armed 'neutrality'). He wrote to Bunning on 19 August:[35] 'By your letter of August 4 I see that our men between the ages of eighteen and forty-five have been called to the colours and that on the 4th all the workmen came to the office to be paid. We see the trouble that you have had with regard to finance,' but in spite of frantic exchanges[36] with Glyn Mills, the London bankers, no remittances of money had reached Tubini at Constantinople since the outbreak of war. He continued: 'The steps taken by the government will, as you say, presumably stop all work. Certainly, if the Christians are called to the army then you will only have a few old men left. Our business so far as continental deliveries are concerned is so much disorganised at present that it is difficult to foresee what our requirements of boracite may be', but it would still be as well for Bunning to go on producing all he was able to, and to take the opportunity to do some exploration work.

Hardly two months later Baker's hopes of 4 August – that Turkey's mobilization was purely precautionary – were disappointed. Russia's support of the aggressive anti-Turk nationalism shown by the Serbs in the Balkan War, and on the other hand a secret agreement between the Young Turks and the Great Power that they had called in to reorganize the Turkish Army after its defeats in the Balkans drew Turkey in on Germany's side at the end of October.

Baker summarized the probable effects to be expected by his company in a letter of 17 December: 'The Unspeakable Turk having joined in this War we are no longer able to produce from our Asia Minor mines, which again raises the question of obtaining further supplies from the States.' His hopes[37] that the Dardanelles campaign would soon put an end to this predicament were short-lived. On 18 May 1915,[38] when the Allied campaign at the Dardanelles was already bogged down, he reported to his Chairman, Lord Chichester:

In December last Mr. Bunning, our Engineer-in-Charge at the mines, was in Constantinople and had an interview with the American Ambassador there, who advised him to engage an American citizen to return with him to the mines, the idea presumably being to safeguard the mines to some extent by the presence of an American citizen in the service of the Company. A man named Militicks, an American citizen, was therefore, on the recommendation of the Ambassador, engaged, and returned to the mines with Mr. Bunning. In January (1915) Militicks informed Mr. Bunning he wished to leave as he had found more lucrative work: our Constantinople manager advised Mr. Bunning to offer some increase of pay – whether this was done or not I have no information: but in the early part of February Militicks denounced Bunning to the authorities on the ground that he had secreted rifles to distribute to the workmen, that he had used wireless, and that he had sent three cases addressed to the American Ambassador in the name of Militicks, and in consequence of this denunciation a perquisition was ordered at the mine. The result of the investigation was that it was proved there was no truth in the accusation of Militicks other than that the three cases had been sent to Constantinople and these contained only old books of account and vouchers. Nothing further was heard of the incident until the end of February when our Constantinople manager received a message from Nadji Bey, Chief of Political Section of Police, to attend and explain why the boxes were sent to Constantinople and what they contained. This was done and our manager was sent with a police agent to the Police Direction at Stamboul – the boxes were examined and, it being found that the contents were as stated, our manager and the boxes were released. The whole question now seemed finished: but on 4th April our Engineer-in-Charge at the mines, Mr. Bunning, was arrested and taken under escort to Baloukhissar. Our Constantinople manager sent to that place and found that Mr. Bunning had been arrested by order of Talaat Bey, Minister of the Interior, and was to be deported to Tchoroum. The local authorities, however, kept him at Baloukhissar as he was unwell. Steps were taken to approach Talaat Bey with a view to obtaining Bunning's release and on 19th April our manager in Constantinople was informed that Talaat Bey had wired that, unless the local authorities insisted on keeping Bunning, there was nothing to prevent his liberation. Up to the present time no news has been received that this has been effected.

About the middle of April [1915] [wrote Baker] two individuals, a German and a Greek, arrived at the mine with the local authorities and took all details regarding our stocks of mineral; and, about the same time, our manager in Constantinople was informed confidentially that the Minister of War had instructed the Minister of Commerce to requisition the stocks and to sequester the mines. On April 18 our manager saw Chevket Bey, Director of the Mines Section of the Ministry of Commerce, who admitted the order had been given and that there was little chance of withdrawing it but advised our manager to see the Minister of Commerce, Ahmed Nessimi Bey. This he did and informed the Minister that he had just heard that he had decided to requisition our stocks of mineral and to sequester the mines, that the Government hitherto had requisitioned everything that was necessary for their needs, but that there was never any question of requisitioning the mineral, which they could not want, and that in any case to sequester the mines was an utterly unjustifiable proceeding. The Minister asked whether we were working the mine, to which our manager replied that since we could not get the mineral to Panderma (the port) – nor could we ship it, the Dardanelles being closed, we were not carrying on the mining for the time being, but that, if the Government forced us to work, rather than allow the sequestration to be carried out and the mine thrown into the hands of people who would damage it, we would continue the mining. Our manager also called to the attention of the Minister the services rendered to the Government in the past and the profits (royalties) they had made out of the mine. The Minister stated that the remarks seemed perfectly just and that he would take them into consideration. On our manager informing Chevket Bey of the interview he was told not to count too much on the result.

Our manager is trying to appeal to the Minister of War, who, however, cannot be seen. He is continuing his efforts 'to have this appeal brought before the Minister'.

It appears to me that as a German went to the mines with the local authorities, there is probably some German intrigue at the back of this sequestration. All the German refiners took their supplies of mineral from us and know the value of the mines: and as they are presumably now out of raw material, it is possible that they look to this source to replenish their supplies or to holding the mines after the war. . . . I realise that while the war lasts but little can be done, but, when it is finished, it is of immense importance to the Company that they should not be deprived of this property.

Baker had previously remarked about one of the principal villains in this story: 'Militicks was evidently an utter blackguard and I have no information as to why he should have been recommended by the American Ambassador' but all the same a further approach was made to the Ambassador to assist in procuring the release of Bunning and the desequestration of the mines. It was not until December[39] that it produced any result:

We have heard from the Foreign Office during the past few days that they have been informed by the American Ambassador in Constantinople that our Mr. Bunning, our Engineer-in-Charge of our mines, who was under arrest and had been interned for sometime, has now been allowed to come back to the mines. I cannot conceive why the Turks should have done this, other than that they perhaps intended to work the mines during the war and have sent Mr. Bunning back for that purpose. It may be that this is the reason for this move as they may see their way to taking the material through Germany or Austria now that the route via Serbia is open; but, at any rate, it is satisfactory for us to know that, if the mines are to be tampered with, it will be under the supervision of our own engineer!

Nothing further was heard about the mines until the end of June 1916, when Baker wrote to Lord Chichester:[40] 'We are now in receipt of further information. The lawyer in Constantinople who looked after our affairs, Nihad Bey, has recently been in Lausanne and, through a mutual acquaintance there, communicated his desire to see somebody from this office, so that he might put forward important information.' Daniell had been sent out to meet Nihad Bey, but much of the information imparted by the latter appears merely to have filled out the story already related, including the fact that the remainder of the British staff had been deported as well as Bunning.

The more alarming part of the information, however, covered the twelve months or so since the correspondence of May 1915. It was a relief, of course, that

the steps taken to guide our protest into the proper quarters were successful. It was decided to drop the question of confiscation of the mine, that of the mineral was held in abeyance, and Enver Pasha, who had hitherto been intractable, stated that the mine and concession were of no interest from the military point of view and that he withdrew entirely from the matter. . . .

But,

fresh intrigues were then commenced by the Deutsche Orient and the Kriegsmetall-Gesellschaft, working under the auspices of the War Department, the former institution having taken charge of all commercial questions in Turkey so far as they concerned the purchase and export of minerals. Efforts were made by them to get hold of the mine either by confiscation or purchase; they and several other individuals were continually in our Agency discussing the purchase of the mine and the mineral: but their proposals lacked seriousness and our steps to counteract their influence were successful.

There was 'a certain Guebhart' from Berlin who tried to tempt the Turkish Government with a scheme for supplying them 'with certain raw materials required for their Munition factories' in exchange for boracite to be taken from Borax Consolidated's mines and exported to Germany. But he roused the opposition of his compatriots, Deutsche Orientalische, and 'met with a chilly reception at the hands of the authorities'.

On his heels appeared Dr. Mohr, who came to our office accompanied by Zellinger, a German merchant of Constantinople. This individual stated that he was an agent for Deutsche Orientalische and . . . was prepared to treat for the mines, or for 1,000 tons of mineral per month, and threatened us with the serious consequence which would ensue if we did not fall in with his proposals.

We then came into touch with one Groskop, a German hanger-on of the Government Offices, who (also) stated that he was acting for Rademacher, agent for Deutsche Orientalische, and intended to buy or acquire in some other manner our mines and stocks. He visited the mine and reported to the German authorities at Constantinople. We would have nothing to do with him, and shortly afterwards two German geologists visited the mine, made their report, and were followed by the Military Attache to the German Embassy.

While

at this period Dr. Mohr returned to Germany and the negotiations of the Deutsche Orientalische for purchasing a regular supply of mineral ceased – At the same time the Minister of Commerce – hitherto well disposed towards us – veered round completely.

As a probable sequence to the German Embassy report to Berlin, which was addressed to the War Department there, we were informed confidentially that the Minister of Commerce had submitted to the Sublime Porte a project of a 'provisional law' providing for the purchase by the State of the Company's mines for LT 400,000 on the grounds of their great value to the State and the undesirability of leaving them in the hands of an enemy subject.

As soon as we became acquainted with this very serious position steps were immediately taken to counteract it. A protest was prepared and submitted to the American Embassy, who, beyond approving the terms of it, did not appear able to give us any active assistance: this protest contained further references to the Mazbata assuring the rights under the existing Firmans, to the fact that the Firman contained no provision for the purchase by the State of the mine, that the concessionaire was in perfect order vis-a-vis the Government, and that a forced sale of this nature would be contrary to the fundamental laws of the Turkish Empire.

This petition was handed to the Sublime Porte and Ministry of Commerce . . .

The details had also been communicated to the Parliamentary Counsel, Whitehall and to the British Foreign Office.

Baker concluded this long report to Lord Chichester by summarizing the significance of these events:

You will see that my anticipations of intrigue by the Germans have unfortunately been only too well endorsed by subsequent events. The Germans are evidently now making a determined effort to obtain the mines: they did not succeed in inducing the Turks to cancel the concession and now, evidently, are trying to appeal to their cupidity by urging them to buy back the concession for LT 400,000, which the Germans will repay to the Turks and take over the mine. The sum mentioned we should probably never see and in any case we could never consent to a sale at any price. The Turkish mines are a keystone to the edifice of our Company, the whole reason for our existence as a Company was the consolidation of the various mines producing the borate mineral.

Baker then alleged that the Germans had made a great mistake twenty years previously when they had stood by and allowed Borax Consolidated to gather in the South American mines which made them dependent on the English company for their supplies of raw material for borax and boracic acid making. They are endeavouring to undo this mistake of theirs by obtaining our Turkish mines with which to compete with us in the world markets. It is absolutely essential to us that this should be prevented – we have nearly £5 millions of English capital invested in this business and the loss of the Turkish mines might seriously jeopardise this capital. Nothing can be done while the war lasts but when terms come to be discussed with the Turks, the handing back of our mines with compensation for damage and confiscation of plant should be made a condition.

Prophetically he continued:

If the matter is left for us to fight out after the war in the Turkish Courts, then we shall be helpless for many years. Experience of the conditions which such a course would entail convinces us that it might take 20 years to arrive at anything, and that we might find ourselves cheated out of a just decision. In the meantime the Germans would remain in possession of the mines to our very serious detriment.

Long after the Germans disappeared from the Turkish scene these words could also be taken as a general description of the Company's predicament during the last two decades of its operations in Turkey.

Meanwhile, for the rest of the War, odd snatches of news about the mines came in at irregular intervals, but the trend was on the whole reassuring. At least, 'the Turks do not appear to have confiscated the mines in the way the Germans were trying to engineer', although by May 1917[41] they had put a German in charge and sent[42] Bunning once more off to detention, this time to Konieh in the Southern Turkish plain but with his family as a consolation. It was obvious[43] that 'the mines are being worked for the benefit of the enemy', but it is also probable that by the time the War ended the Turks and the Germans had not really got as far as serious mining, but were merely picking over the dumps. A German chemical trade paper stated[44] that the price of borax in Germany at the end of April 1917 had reached £870 a ton, which seems to indicate that Turkish material was not having much effect on relieving the borax-shortage there. This was confirmed shortly after the end of the War, when Baker wrote:[45]

One of our people, who has been in Asia Minor since the beginning of the war, turned up yesterday. . . . It appears that the mine has been worked since last April with Germans there but the operation being under Turkish control. They have taken out about 6,000 tons of 1st class ore but my informant states that only about 80 tons actually left Turkey, the rest being either at the mine or at the nearest railway station or at Banderma, the port . . . circumstances were against them evidently so far as getting away the ore is concerned.

It therefore came as a shock when in March 1919 news arrived[46] from the USA that Turkish boracite was being offered to the Administration in payment for foodstuffs, and that this was from the remainder of the 6,000 tons mined at Sultan Tchair during the last year of the War. Baker's hackles once more rose – 'they (the Turks) have not the slightest right to deal with it'. Protest was made to the Foreign Office, and the claim was added to the demand for the return of the mines and for compensation for every scrap of damage done. Correspondence on these subjects between Baker, Lord Chichester and the Foreign Office reached a crescendo in volume and notes of indignation in the early months of 1919, until in April Daniell reported[47] that 'the mines and all the mineral were handed over to us on 1st April by the British Commission and from that moment the Turks have no power to deal with the ore in any way'.

All, however, did not live happily ever after. The minutiae of the negotiations, leading to the return of the Turkish mines to the Company after the First World War, seem now utterly irrelevant to the subsequent course of events, except for the bizarre coincidence that the mines were returned on All Fools Day (1 April 1919). The next decade was to prove the most violent and uncertain in the whole of Borax Consolidated's long history in Turkey, a state of affairs which must be attributed largely to the general history of the times. Tsar Nicholas I's opinion of Turkey as 'the Sick Man of Europe' has coloured unduly the outlook both of politicians and historians until comparatively recent years, but late in 1918 it seemed once more to be justified. The Young Turk Movement collapsed with the Germans, and Enver Bey and his henchmen fled across the Black Sea on a German gunboat. The new Sultan, Mohammed VI, who was a younger brother of Abdul Hamid, immediately showed a tendency to re-establish personal rule from the Palace and to go to all lengths demanded to appease the conquerors. An Allied military administration was set up in Constantinople, British forces occupied the Dardanelles, Samsun, and Ayntab, and the French advanced from Syria into Cilicia and the Adana district. Finally, and biggest insult of all, a Greek army landed at Smyrna (Izmir) on 15 May 1919 in pursuance of the 'Great Idea' of restoring the Greek Christian Empire of Byzantium.

Not unnaturally the general atmosphere was one of anarchy and smouldering resentment on the part of the Turks. Within the narrower world of borax the irrepressible Bunning re-emerged from internment and Turkish supervision and wrote to Baker on 30 December 1918:

We have been surrounded by brigands but never been much disturbed; on the 29 October they came to kill the Germans, one officer, two sergeants and 18 men, but they were happy to get 1000 liras in paper . . . they took also all the rifles and ammunition. This so frightened the Germans that they left on 31st October. . . .
The underground workings of the mines have not suffered from the Turkish occupation. . . . The machines, stores, buildings, have suffered very much, especially the buildings of the Azizieh mines where all the window panes and doors have been robbed. Also all the scrap iron has disappeared and been sold at Sussurluk. . . . Since last November there has been no production and, as the mine is still under the Turks, they drain the mine, it being the only maintenance they do.

In his report[48] to Lord Chichester Baker added: 'No damage appears to have been done at the mine other than the removal of the pipes and gear from the Azizieh mine, this mine being now full of water. The rope belonging to the ore ropeway, which carried the ore over the hills to the railway about five miles distant, was taken away to use for nets at the Dardanelles.'

Bunning, however, was far from being treated as a returning war hero. 'He appears to have been quite well treated by the Turks and I imagine has not suffered any considerable hardship,' wrote Baker[49] unsympathetically to Lord Chichester. While Bunning was in England during the summer of 1919 Baker seemed determined to put him in his place . . .[50] 'Your family arrangements cannot influence me in determining the salary the Company pays you for your services in Turkey . . . the outrageous extravagance shown by you in dealing with the funds of the company in Constantinople during the latter period of the war does not qualify you to ask for this.' Bunning, apparently, had been living it up with his family at the Pera Palace Hotel, the No. 1 luxury hotel in Constantinople in those days, for part of the War, and his periods of internment services at the mines in trying conditions did not excuse this offence. A successor, H. W. Faulkner, had been appointed to succeed him even before the War broke out in 1914, but had been sent out to the American mines to fill in time until it became possible for him to reach Turkey. Therefore Baker told Bunning, during his visit to England in mid-1919, that he might not be reappointed to his post at the mines; but all the same, by October it was clear that Bunning was going to return there, for Baker wrote:[51] 'I agree, that so far as your wife is concerned, it may be not advisable for you to take her back at the present time to Asia Minor but possibly when you arrive there you will find conditions better than anticipated.'

Apart from continuing post-war anarchy, the trouble now was that, on 19 May 1919, Mustafa Kemal Pasha, one of the few successful Turkish generals in the recent war, had landed at Samsun and had openly started to set up a rival government at the remote Anatolian village of Angora (Ankara) and to raise an army to oppose the Sultan's government at Istanbul, the Allied occupation and the Greek invasion alike. This was only four days after the Greek landing at Izmir, so that for the next three years Anatolia was convulsed by this tripartite struggle. Therefore, by the spring of 1920,[52] although Bunning had returned to the mines, Daniell found himself unable to get beyond Constantinople in an attempt to reach the mines and neither Mrs Bunning nor Faulkner were even allowed to attempt the journey to Turkey. At this time the port of Panderma and the mines were both in the hands of the Nationalists (Mustapha Kemal's followers), but they were to change hands twice later. Although Baker wrote,[53] in mid-June, to Faulkner: 'As to the mine itself we hear from Mr. Bunning that things are normal – the Nationalists are not interfering with our work, they are only charging us so much a ton on all the ore we send out,' he also had to write to the USA[54] 'Boracite supplies have been cut off owing to the present situation in Turkey. In consequence of this situation material, which we had reckoned upon from Turkey for the French works, will not be available; therefore we have to fall back on North American material to keep us going.'

The situation seems to have improved in the second half of 1920, as Faulkner had reached the mines early in September[55] 'after considerable delays both at Constantinople and Panderma owing to the infrequency of communications'. But after his arrival and during the next two years, almost to the day, this mining camp existed virtually in a vacuum from which no one could venture far because of the troubled state of the country outside, although it seems to have been practically free from Greco-Turkish war trouble within. However, it had its fill of a different type of trouble. Faulkner (BA Cantab., ARSM) arrived with a reputation for having been in Baker's words,[56] 'a competent and loyal' servant of the Company during five years in the USA although 'highly strung and apt to go off at a tangent'. He was now to understudy Bunning, some of whose outlook and reputation has already crept into these pages. A Turkish effendi, who was appointed to some minor post at the mines in 1925 and recorded a few pages of 'Memories',[57] has left a record of some of the reputation that Bunning left behind him:

Mr Bunning who led quite a care-free and pompous life to the extent of the possibilities and customs of the time, had been nicknamed among the workmen as 'The Great Director'. . . . his splendid way of living was mentioned and praised over and over again. His house, which he himself had built on the Eastern side of the block of buildings, consisted of eleven rooms, a large saloon serving as the library and the dining room at the same time, a wide glass verandah, two kitchens and a cellar where wine used to be stored. This building used to be divided into two sections allotted for ladies and gentlemen respectively. The section for gentlemen, which consisted of two bedrooms, a dining saloon, servants rooms, kitchen and verandah, was always prepared to welcome and entertain visitors in accordance with the Turkish tradition of the time. Among the frequent visitors were the governor, mayors, gendarmerie, Commanders and other Government officials as well as the Chiefs of the raiders who still continued their activities round about and who were said to be close friends with Mr. Bunning. The rumour went around that these Chiefs stopped at the Mines quite often and some nights they sat and drank 'raki' with Mr. Bunning.

Mr. Bunning, whom I met for the first time in 1926, was a cheerful and accomplished person who had retained his dynamism and vivacity although he was over seventy years of age at the time.

During the reign of the Ottomans, due to the slackness of the Government and the deficiency of discipline, the inner parts of Anatolia were forsaken and neglected and as a consequence public order and security were in a state of corruption in those parts. So much so that cases of robbery, raiding the villages and farms, and waylaying were considered matters of everyday. It is evident that Mr. Bunning, who had to live under these conditions, acted in accordance with the necessity of the time and very cleverly succeeded in managing his environment. As I have heard from the old people, no such cases were recorded at the mines in his time. No-one disturbed the mines and even the Courier who used to carry gold from Balikesir every month making his way on horseback in two days, faced no attacks or dangers of any kind. Mr. Bunning told me in person that he was indebted to the chief guardian called Kara Ibrahim who was in his service for a long period of time and who played an essential part in securing this security and confidence. Every summer Mr. Bunning used to go camping to the forest on the 1,500 ms. high mountain 'Cataldag' but, due to the remoteness of the distance, he could go there only over the weekends and the rest of the time Kara Ibrahim used to guard Mr. Bunning's family who lived at the camp the whole summer long.

It was no wonder that Faulkner, fresh from the close supervision of Ryan, Zabriskie, Baker and the other martinets who controlled the company's USA operations in an ever-vigilant hierarchy, found this Turkish scene incomprehensible. A month after his arrival in 1920 Faulkner wrote to Zabriskie:[58] 'Everything is new and strange, except that it recalls quite vividly the atmosphere of Mexico. I find the same undercurrent of officialdom and tradition which has to be considered before any new move is made.' He appreciated also that Bunning had one useful advantage over him – a working knowledge of Turkish, French and Greek, without which it was difficult at this period to get a proper grip on what was going on around the mines.

However, he soon started finding his feet and sending back to Baker reports[59] on what seemed really to be going on as seen through the fresh eyes of a newcomer – the first objective reports to come from these mines since well before 1914.

It is a very different ore-body from those at Ryan. It lies about 300 feet below the surface; varies from a foot or so to six feet or over, and has a gentle but not entirely regular dip. There are several shafts reaching to the ore body but two of them have fallen in during the war and we are now opening them up again. Practically no timber is used in the mine, pillars being left to support the roof, which is an excellent one – a good solid bed of gypsum, which reaches to a thickness of forty feet, above this is shale and clay: below the bed is more gypsum, but I do not know what the bed rock is. . . . We have no electric light or rock-drills but I hope that there will be an opportunity for introducing these; fortunately there is a tolerably good supply of water. . . . Owing mainly to the very cheap labour before the war, methods are in some cases primitive, and there is room for many improvements if conditions will stand it. . . . Much of the ore occurs in well-defined chunks so that it is capable of sorting to a good grade. The discarded ore runs about as high as the grade we struggle to maintain for the shipping ore at Ryan. This stuff was concentrated in former years, much on the same lines as at Death Valley Junction. It appears to have been successful technically but not commercially owing to the low rate at which ore could be sorted in those days. Labour is mostly Circassian – Turkish speaking. Common labour seems to be plentiful enough, but there seems to be a scarcity of the more skilled type such as miners. . . .

Baker, of course, responded in full[60] to this opportunity to bring the Turkish mines back into de facto Company control and the modern world. 'I am convinced by what I saw in July, 1914 that a big overhaul at the mines is necessary and that energy and more modern methods are now desirable, looking to the present conditions of the cost of everything.' Development work must now be expedited, even if this meant taking on more men, and unmined ore reserves must be assessed. 'It used to cost about £2 a ton to concentrate and with a cost of 18/- for 1st class ore of 36/48% it was manifestly not economic to do this. Conditions have now changed and it may be that it would pay us to concentrate 2nd grade ore of 20/22% B.A. content', of which some 300,000 tons lay in the dumps. It was therefore suggested to the American organization[61] that Stebbins, the well-known expert on processes, might be able to improve on the old concentration process for 'Seconds' (dumped ore), samples of which were sent to Bayonne for trial. Success might mean production of 120,000 tons of high-grade borate from a low-grade dump. Maybe, it was hopefully believed, this task would not be difficult, as the existing process had been introduced by one of the two predecessor companies who worked the mines in the 1890s.

As regards the current position of Turkish boracite in the commercial world, Baker wrote,[62] significantly:

Before the war we shipped about 18,000 tons of boracite which is three times the present production, partly to our factories in France and the United Kingdom and the balance to various refineries in different countries. It is not possible yet to see how much post-War Europe will require. Boracite is one of the most difficult crude materials to refine and, for borax making, requires special plant, having to be decomposed under pressure and perfect decomposition is very difficult to obtain: but we are experimenting to improve this.

Even allowing for the benefit of hindsight it is difficult to understand how Baker remained so much out of touch with the atmosphere in Turkey at that time as complacently to believe that Faulkner, with Bunning still at the mines and the Greek line advancing with overstretched communications against Mustapha Kemal's growing forces somewhere east of the mines between Eskishehir and Ankara, could settle down calmly and carry out this programme worthy of Death Valley operations in their best-regulated era. But in spite of this daunting outlook, Baker wrote in March 1921[63] to Boyd, Faulkner's successor in the USA: 'There has been a conference recently in London and I imagine that the Turkish difficulties will soon be settled and that part of the country entirely quieted down.'

Meanwhile, as early as October 1920[64] Faulkner had written:

With regard to the present position between Mr. Bunning and myself, I am convinced that the arrangement cannot continue satisfactorily indefinitely. I can see that his attitude towards the staff and heads of departments has a very adverse influence on them and consequently on their work, and that they strongly resent his manner towards them. I do not at all approve of his demeanour and methods in general, and I think his continued presence here would be prejudicial to the Company's interests, and would largely counteract any efforts I may be able to make to improve matters here.

There were many signs that Bunning resented Faulkner's presence and was withholding information from him but this uneasy takeover of the management was allowed to drag on over 1921 and well into 1922 before Baker and the Board decided to take action in response to the growing spate of letters from Faulkner about Bunning's behaviour. Faulkner accused Bunning[65] of drunkenness, mental decay, and quarrelling 'in a manner that can only be described as infantile' with the younger members of his staff, who were staying in his house. The section that seems to reveal the differences between the two most vividly reads:

I have myself heard him relate what good times they used to have, the holidays they had, and how the afternoons were given up to riding and so forth: all of which has made it rather hard to impress upon the younger members of the staff that they are here to work, and thus to counteract this adverse influence.

In fact, it was an irreconcilable clash of outlooks on life and work.

But Baker still moved cautiously. Daniell was sent out to the mines in June 1922 to 'make a report on existing conditions there', and reported to Baker on 11 August: 'Mr. Bunning was aware, shortly after my arrival, of the object of my visit, and was not likely to furnish me with any opportunity to complain of the particular shortcomings of which he is accused.' It was obvious, however, that the tension between Bunning and Faulkner could not be allowed to continue much longer and on 28 August the Board in London accepted Daniell's recommendation and decided that it was 'no longer in the best interests of the company that Mr. Bunning hold the position of Joint Engineer in charge of the company's mines in Turkey'. Bunning was due to go on leave in any case, and from that point he disappears from the record for the time being – but only for the time being.

He must have left Turkey only just in time. In this same month of August the third and final phase of the Greco-Turkish war began. The Turks won a crushing victory at Dumlupinar, and, driving the Greeks before them, reoccupied Smyrna (Izmir) on 9 September, thus completing the reconquest of Anatolia. The line of their advance included the recapture of Eskishehir and Bursa, and thus the mines and the port of Panderma – which was devastated by fire – could not escape from the war, as had also been the case during the preceding Greek advance. Faulkner, his wife and son, and the other British families at the mines eventually took the wisest and indeed inevitable step and returned to England; and it is due to his having to explain to Baker the reasons for doing so that we owe the discovery, in the drawer of a confidential filing cabinet in the company's Istanbul office, of Faulkner's vivid report[66] on the events of those tragic days in the autumn of 1922. It is long, but the key sections of it read:

On the morning of September 5 rumours were abroad that the Greek front had been broken and that the Greek troops were about to retire from Sousighirlik [Susurluk]. Although we now know that fighting had been going on for some days, this was the first intimation we had that anything unusual was occurring. I telephoned to the Station master at Omerkeui to ask whether he had any information concerning these rumours, and he replied that the situation was serious and that he had received orders to leave. The Greek soldiers stationed at the mine said that they had not received any orders to leave, and that they knew nothing about the rumoured withdrawal of the troops from Sousighirlik or the collapse of the Greek front. I, therefore, went down to Sousighirlik to interview the Captain of the Greek troops there. He professed to know nothing of the situation but said that they were expecting orders to withdraw at any moment; that we were under their protection, but that if we wanted them to continue their protection, we must all be in Sousighirlik in less than two hours. I told him that was impossible; my remark was met with a shrug of the shoulders. He also said that I should return at once and advise all the Armenians and Greeks to leave with them; he thought that if the English preferred to remain it would be safe for them to do so. I hurried back to the mine (meeting on the way the Greek guard who were being withdrawn from there) and got word to all Armenians and Greeks as soon as possible. Most of them did not want to go and I had to urge them strongly to persuade them to move. By three o'clock most of them had been paid off and left. Some, who were unable to get Arabas (carts) for their women and children and their goods, insisted on remaining. For ourselves, the only train had left early in the morning, we had no means of getting our belongings to Sousighirlik, and, in getting the Greeks and Armenians off, there was no time left to think of ourselves: in any case I considered it my duty to remain, at least until the situation was rather more clear.

The Armenians and Greeks who had left, found on arrival at Sousighirlik that the Greek garrison had left shortly after I had returned to the mine. It later appears that a special train, sent through from Panderma to Balikesir and back to pick up refugees, passed through Sousighirlik at ten o'clock at night. A few of those who had left boarded this train; they were Dr. Joannides, the storekeeper Vereopoulos and his family, and Manouk, the dressing shed foreman; I have heard that the Doctor arrived at Constantinople. I have not heard what happened to the others. The rest of those who left the mine, hearing that they would not be able to leave Panderma, stayed at Sousighirlik. On the 8th the Mudir of Sousighirlik wrote that he wanted these people to return to the mine and asked me to go down to tell them to come back. I went down and saw the Mudir and then told the Armenians what he had said. They refused, however, to come back . . . they wanted to stay in Sousighirlik until the return of the troops; this never happened. . . . Meanwhile, we had managed to get work started again, although of course, with a very reduced mechanical crew. . . . The Mudir again wrote that the Armenians must leave Sousighirlik. I therefore again went down and saw him and arranged with him to give them protection on the road . . . I also induced him to come up to the Mine and exhort the workmen and villagers to continue their work peacefully. . . . The Armenians then returned, with the exception of a few single men who escaped and joined the Greek soldiers at Al-Saqal, and work continued up till the 20th. During this time we were entirely cut off from all communication with the outside world. The last train that passed through Sousighirlik was on September 5th. The railroad had been torn up and bridges destroyed and the telegraph wires had been cut. All efforts to get in touch with Panderma or Smyrna were useless. On three occasions I sent messengers off but they failed to get there and returned. Rumours reached us that Panderma had been burned down and that all the inhabitants had left, and that the same fate had been suffered by numerous other towns and villages around us and more distant, at the hands of the retreating Greeks, and that unspeakable atrocities and wholesale massacres had been perpetrated upon the Turkish inhabitants. All these rumours, which have been later been found to be true, made the situation very uncomfortable and it was a period of considerable anxiety. At Sousighirlik and in our immediate neighbourhood everything was quiet. Bands of 'chettis' [brigands recognized by the military authorities, whose duty had been to harass the Greeks and Armenians and destroy as many of them as possible], who had been infesting our neighbourhood during the last two years, entered Sousighirlik and Balikesir and took over the policing of these places. . . .

. . . Matters continued fairly peacefully up to the 20th and work was continuing more or less normally. . . . On the afternoon of the 20th, however, the trouble started. At 2 pm a band of about 15 men rode into the mine, the leader called me out and questioned me as to the number of Armenians and Greeks at the mine, accused them of having rifles in their possession, and stated that he wanted the Greeks handed over to him; this I objected to, telling him that I could not hand over any of our men without some written authority that I could recognise; he replied that, in order to please me, he would not insist on this if I would give him permission to loot an Armenian shop, the owner of which had left. He said that he could not leave without his having something to show that he had been there. It seemed that resistance was useless and I told him that I could not give him permission to loot the place but must leave matters in his hands; and that if he insisted on taking the Greeks or looting the shop I would prefer the latter. He then proceeded to loot the shop. The leader then questioned our guardians as to rifles and revolvers in possession of the Armenians and Greeks and threatened to shoot them if they did not give them up. He then sent men round with the Guardians to collect those they knew of. After cleaning out the shop they went off in about an hour. . . . I learned afterwards that these men were one of the bands acting independently of, but recognised by, the military officials and known as Chettis; they proceed in front of the Army and are disbanded when the Army takes possession of their district; thus the Army recognises their work, at the same time being able to disclaim responsibility for their excesses.

At nine o'clock at night, another similar band of about 40 men rode into the mine; I was again called out and questioned as to the nationality of the ownership of the mine, the number of English there were at the mine, and the number of Armenians and Greeks. This man also stated that the Armenians and Greeks had rifles in their possession. I assured him that, as far as I knew, there were none left; he also stated that we ourselves had rifles but that, if we stayed indoors and did not attempt to hinder him, we could keep them and would not be molested. He then proceeded to 'look for the rifles', which meant looting every Armenian's and Greek's house. As soon as this started, the Armenians, most of whom had crowded into the houses surrounding our garden, jumped through the windows and rushed into our houses, and hid themselves where ever they could, under the beds and among the rafters: thus they spent the night in terror of their lives, while the band of marauders continued their pillaging. After about an hour the leader sent word to me by a guardian that the Armenians must give up all their money or he would enter our houses and take them out. During the afternoon most of these poor people had either hidden their money or turned it over to us. Afterwards we also found that some of them had hidden packages of money and papers in the drawers of our houses with a note as to what should be done with it, according to the fate that might befall them. As there seemed to be no help for it, I advised the Armenians to hand over their money. It was a difficult matter to collect it as they were so panic stricken that they hardly knew what they were doing and were afraid to move from their hiding places. All the time the leader was sending in messages that he must have the money at once or he would take them out and kill them. Finally we managed to collect Ltqs. 425, and the band then went away. Three bullock carts had been brought up and taken away filled with goods. I am sorry to say that this band was assisted by a number of villagers; after the departure of the band these men kept up the work: and all night long we heard the smashing of windows and doors and furniture. Next morning the place was a scene of devastation; the street and our garden were littered with debris thrown out from the houses and shops; every house had been broken into except our own and Schiave's: what they had not taken away they had destroyed. The laboratory had been raided and all its contents taken away; the telephone had also been taken away and the governor belt of the cable engine. The hospital had also been entered and a few instruments and small appliances taken; and the two Greeks who had remained at the mine had been killed. . . .

At about five o'clock in the morning word was brought to me by the chief guardian that another band was on its way to the mine threatening violence against the Armenians and demanding the girls and young women. I have never been able to discover the origin of this report but as there was no time to investigate it at the moment I went to the Armenians in our houses and urged them to go out and hide in the fields until I could get protection for them. With great difficulty I got them to do this; but instead of going into the fields some of them went down to the old Concentration Plant: some of them I never saw again. Omnik, a mechanic, who also drove the car, was later found with his young wife dead, with his revolver by his side: he had evidently shot her and then himself; this he had previously told us he would do rather than let himself and his wife be taken by the Turks. Carabet, a carpenter, and his young wife and mother I have never been able to hear of since.

Immediately after getting them away to hide themselves, I sent off a letter by horse rider to the General of the Division who was reported to be at Balikesir. At about five o'clock I received a message from the General of the 61st Division that protection would be given to us and the property, provided that I did not shelter the Armenians or interfere in anything that was done with regard to them. I did not reply to this message but at about 10 o'clock two officers and 20 cavalry came up: . . . these officers asked for information as to what had taken place and took notes. . . . I asked them to leave some of their men with us, which they did. I think about half a dozen. At about 1.30 pm another group of cavalry came in, in charge of a Major, of the 41st Division; he told me that he had been sent to stay at the mine until the main body of this Division arrived and he promised to protect the Armenians from further attack. As a matter of fact, it afterwards transpired that his object was to collect as many of the Armenians as possible before they had time to escape. He said that in order to protect them he must have them under guard. I therefore collected them from the old Concentration Plant or wherever I could find them: two of them, Agajahn, the old mason, and Kircor, the Dressing Shed foreman, my wife had found hiding in the manger of the stable at the back of the garden. I told them that this officer had given his word to protect them and in all good faith I urged them to go with the others; but, as it developed later, I was actually assigning them to their death. The Armenians implored me not to let them go outside the mine and I therefore arranged for them to be put in the warehouse. There were about 50 or 60 men, women and children. I also arranged for them to be given bread and water, which the officer gave me permission to do provided that it was at our expense, and posted two of our guardians there to do what they could for them; the officer also posted sentries. . . . Next morning he told me that he had been up all night as he dare not trust his own soldiers such was their hatred against the Armenians. He related to me the atrocities that he and his men had seen, which had been committed by Greeks and Armenians retiring before the advance of the Turkish troops – the burned ruins of mosques full of dead bodies of men, women and children, who had been put in the mosque and slaughtered by the mosque being set on fire and by bombs being thrown in among the victims. These and other atrocities that can hardly be described here are also common report and based on such evidence as leaves little room for doubt as to their truth.

. . . During the morning this officer and his party left and was replaced by another party in charge of a Captain. This man came into lunch with us and was quite friendly, as also the preceding man had been. . . . In the afternoon of September 22 two regiments came to the mine; these belonged to the 41st Division which was on its way to Soussourlou: they had been preceded by various aides-de-camp of the general commanding the Division who had come up to enquire what space was available for troops. The Colonel in charge of the regiments told me he would take over all the empty houses and buildings but that none of the Company's property would be damaged. He asked for quarters for himself and staff and said that they would be there for some time. I put him and four of his staff in Mr. Bunning's house; a major was quartered in Schiave's house and other officers arranged for themselves at the Azizie mine. . . . After our passports had been examined the Colonel called us together, told us that 'we were friends', that they were our guests and we were his guests. He told us that we were not to go outside the gates except with an officer. Sentries were posted around our houses, offices, and garden. . . . Next day an officer was appointed specially to look after our persons and premises and arrangements were made that only three people, consisting of two of the guardians and Mustapha the odd-job man, should be allowed to come inside the gates; these three were to make the necessary connection with the outside world for the purpose of obtaining provisions or summoning others to the gates for us to talk to them. We were thus virtually prisoners, although the colonel disclaimed such term and said that these arrangements were made for our protection. . . . We were told to put red bands on our sleeves to show that we were under protection. He told me that I must make arrangements for the water supply for the troops. . . . He also called for firewood for himself and staff for his cooking. Constanti also had to prepare baths for them all and also for numerous visitors every day. . . .

On the day after their arrival the Colonel questioned me with regard to the Armenians and said that they must all be handed over and that he held me responsible for this. By this time he had discovered that some of them were in the mine and others in the Azizie mine. He therefore ordered that they should be brought out and commissioned Marra to go down with guards to find them. He went down and saw a light that they had with them but, on calling to them, they ran away and he was unable to find them . . . and we had no means of getting food to them. How they existed I do not know. . . . Before those hiding in the mine could be found, the Armenians who had been placed in the warehouse were sent off: After five or six days the others came out of the mine and delivered themselves up. Before this the General of the Division at Soussourlou came up to the mine and asked to see me. He told me that I must see that all the Armenians were handed over. He said that they would be sent to Panderma and there put on boats and sent to their own country. As a matter of fact according to all reports they did not even reach as far as Soussourlou but were shot on the road. Up to this time we had been allowed to keep our servants, who, of course, were all Armenian, and the Colonel had told me that he would do what he could for us to keep them but it was impossible for him to let them stay without the permission of the general. I therefore begged the general to let us keep them. He said it was impossible and that he had his orders. . . . On September 25 the order came for them to be handed over; almost frantic with terror they clung to our knees and implored us not to let them go. We actually had to force them to leave the house, they were ill from lack of sleep and food, and terror, for days past they had lived in a state of panic, running away to hide themselves at every knock at the door; we tried to persuade them that the officers had given their word and that they would not come to any harm; they said they knew the promise would not be kept and that they were going out to their death; and no doubt they were right. It was a heart-rending task to send them away from us but there was no help for it. They were put in the warehouse overnight and next morning they were sent off together with Arias, Karnick and the others who had by now come up from the mines. What has happened to them I do not know; the women were reported as having been seen at Sousighirlik but the men do not seem to have reached there; the men are reported to have been shot on the road like the others: with regard to the women the people in the neighbourhood were very loath to report what they knew of the happenings around them but, knowing what I know now I do not imagine that the women ever reached Panderma. . . .

Housekeeping became arduous. One morning the Colonel saw me beating the carpets and told me it was not fitting that I should do this. I warmly agreed but pointed out that the work had to be done, and there was no-one else to do it; my wife could not do everything. He seemed genuinely grieved at the state of affairs but did nothing to remedy it, except that he sent Constanti [i.e., an ex-servant of Faulkner now 'requisitioned' by the Colonel] over to beat the carpets. It should be mentioned that all the officers during their 15 day stay were quite agreeable and friendly: they came to our houses frequently and played cards; we also tried to teach them tennis; much to my surprise they also drank an amazing quantity of alcoholic drinks; the precepts of the Koran forbidding strong drinks are evidently a dead letter with the military forces, notwithstanding the fact that it is forbidden by the Ankara Government to manufacture, sell, or have in the house alcoholic drinks. . . .

. . . Most of the thieving and pilfering that went on was obviously directed by, if not instigated by, local ruffians who knew the details of the mine and premises. The Colonel apologised on each occasion. . . .

When the Colonel came to the mine from Soussourlou he asked for the use of the Ford car to make a journey to Panderma and back; two days later he sent for the car and this is the last we have seen of it. . . .

I have omitted to mention that the day after their arrival our rifles and revolvers, that we had obtained from Constantinople, were taken from us and a receipt given for them. The guardians' rifles were also taken.

After the Colonel had got his arrangements made he told me that the General was anxious that work should continue and he had arrangements made with the Commandant at Demirkapor for passes to be given to Yahya, the haulage contractor, for the number of men required. I did not put any miners to work as the stopes had been left too full for further working them until they were cleared. Also I thought it better not to attempt to do much work while the future was so uncertain. We therefore continued the work of clearing up the stopes in as small a way as possible without running the risk of producing the impression that I did not want to work the mine, and on October 9th enough ore had been cleaned up for a half day's extraction. We also hoisted ore on several days after this. The ore was stored in the yard, except for a little that was sent to Omerkeui in order to bring back 30 tons of coal that had been lying there. We continued in this way up to October 13 when the men struck work, refusing to continue until they had been paid for September. It had been explained to them that we could get no money to pay them but the foremen assured us that the men wanted work and relied upon getting their pay as soon as communication could be established, which according to all reports at that time would be before very long.

... All this time I had been trying to get into touch with Constantinople through Sahadeddin Effendi, by sending couriers to him. All these letters and messages had to be censored by the Colonel. On October 4 I received a message from the Mudir saying he had received a telegram from Nihad Bey asking for news and that he would reply on hearing from me. I immediately replied to him. . . .

On October 7 we were aroused at 4 am and told that the troops would leave at 7 am. . . . After the departure of our two regiments other troops were continually passing along the road and we had numerous visits from their officers. Some of them camped at night in the plains below the mine; the people of Demircapou suffered heavily at the hands of these passing troops who demanded of them practically all their food supplies and helped themselves to practically everything they wanted; but fortunately we were left alone. . . .

On October 9 the railroad started to run as far north as Aq-Saqal; there was no direct communication with Panderma owing to the iron bridge North of Aq-Saqal still being down. It was reported, however, that a train also came south from Panderma as far as the broken bridge. I started to try to get a message through to Constantinople but was informed that no-one was allowed to leave the country without special permission and foreigners and Circassians were not allowed to leave under any condition . . . we were not aware, however, how strong anti-British sentiment was since, personally, we had been treated with all respect and courtesy by all the military officers and Government officials with whom we had to deal.

We were rudely awakened, however, to the realisation of the strength of this anti-British sentiment by the next incident that occurred.

On October 13 an orderly arrived at the mine from the General of a division of the 4th Army Corps staying at Soussourlou with a message that all foreigners at the mine and at Demirkapou were to go to Soussourlou immediately to be interviewed by him; passports were to be taken. Having travelled to Soussourlou in two carts the party of some dozen males of miscellaneous nationality waited about an hour and a half and were then interviewed by 'his august excellence the General of Division, one Kemaliddin Pacha', who proceeded to deliver a long political diatribe, directed mainly against the English. 'The English were not their friends, we must realise that we were in an enemy country; the English were the only ones standing in the way of the Nationalist aims . . . we should be allowed our freedom to move about the mine but we were under military orders and were not to leave the mine without permission . . . if Great Britain persisted in her policy of obstruction she could force Turkey into declaring war against her conjointly with the Bolshevists. . . . If, he said, war resulted, as seemed very probable at the moment, we should have to be exiled. I therefore asked him if we could leave the country. He replied that permission to leave would be granted by the General commanding the army, with headquarters at Balikesir, on my addressing a written request. . . .

There was, of course, substance in the General's remarks. British ships took a prominent part in the Allied force which sailed up the Dardanelles and occupied Constantinople at the end of 1918; British, French and American warships covered the landing of the Greek army at Smyrna in May 1919; and – final diplomatic blunder – at the end of the Greco-Turkish war, only two days before the interview between Faulkner and his colleagues and the General, the British had been the last of the Allied force in the Dardanelles to withdraw from active opposition to Mustapha Kemal's wish to cross the Straits to drive the Greeks out of European Turkey. It is no wonder that Faulkner now found it easy to make a rapid decision. 'I decided that, although this was the first instance I had encountered of such strong anti-British sentiment, yet it might be much more prevalent than I was aware of, and the only thing to do was to leave as soon as possible. . . . Next day, therefore, we started packing.'

Faulkner's long and exciting account of the endless difficulties, dangers and discomforts through which he, his family and his colleagues passed before finally getting away on 9 November from Smyrna (Izmir) in a Dutch boat bound for Samos is too long to include here, and not strictly relevant to the history of the borax mines in Turkey. He concludes: 'We arrived in London on November 24, just about a month since we left the mine. We had received no letters or other communications from England since September 4, these being letters dated August 18.'

This, of course, is one man's account of the events of those tragic months in 1922, but parts of it are corroborated by correspondence extant[67] between Tubini and Nihad Bey (the Company's representatives in Constantinople) and the London office, and by the recorded facts of history. There seems no reason to suppose that Faulkner was exaggerating, as in many sections of his long report he missed opportunities for pretending that the mines and the British were in greater danger than they actually were; but he pays a warm tribute to Nihad Bey – who must have been a greatly respected lawyer – for the continuing efforts that he made to extricate the small British party via Izmir, in face of the strong Nationalist mood of his own compatriots.

The war was followed by long negotiations, which began in Lausanne on 20 November 1922 and were concluded by a peace treaty signed there on 24 July 1923. Borax Consolidated's difficulties with the new Nationalist government at Ankara went on much longer – until March 1927. Possibly this may have been partly Baker's fault. At some time during the early summer of 1923 the new Turkish government started claiming[68] some £74,000 of royalties due from Borax Consolidated, and Tubini cabled asking permission to settle for £27,000. The background was that between the First World War and the outbreak of the Greco-Turkish war in 1920 the Company had put in a claim (eventually whittled down to 238,277 Turkish Lire) for compensation, mainly for mineral taken by the Germans between 1914 and 1918, and also for mineral which might have been mined and turned to profit by the Company but for Turco-German occupation of the mines during these years. This claim had been allowed in 1919, and to meet it the Company proceeded to withhold royalty payments otherwise due to the Turks. The Nationalist victory in 1922 wiped out the harsh settlement imposed on Turkey at the Treaty of Sèvres in 1920, and eventually among many other items, the withholding of royalties by the borax mines came also to be questioned by the Nationalists. Baker does not appear to have appreciated how much the atmosphere had changed in Turkey from the old days of weak and corrupt government under the Sultan, and tried to drive a hard bargain, using the reluctant aid of the British Foreign Office through Lord Chichester and his 'old boy' network in Whitehall.[69] Nihad Bey, in Ankara, had warned him[70] that the Minister of the Economy, 'upon whom depends the mines', was 'as a man very difficult, possessing a comprehension of affairs and the situation almost nil, very impulsive and intransigent'.

Nihad realized also only too well that the Turks were already using one of several ways open to them to compel the Company to settle on their terms – their power to withhold export permits for the boracite much needed by the Company's British, Continental, and US works for boric-acid production. But eventually it was only a mistake by Tubini and Nihad Bey over interpreting the intentions of the London office[71] that led to a compromise over this matter on terms which Baker would otherwise not have accepted, and even then it became a struggle to get the Minister to abide by the agreed compromise. Baker wrote to the Foreign Office:[72] 'Nihad Bey now reports that, on informing the Minister of National Economy, Mahmoud Essad Bey, that we accepted the conditions of the compromise as agreed by him and that we now considered the matter as completely settled, the Minister replied that he no longer accepted the arrangement come to, seeing that the Treaty of Peace had altered the conditions.' So the compromise was not signed finally until early in 1924.[73] Its details are no longer of any significance, but something of the course of the negotiations has been related as typical of the atmosphere in which the company had to conduct its business with the Turkish Government for the remainder of its years in Turkey.

During the course of these negotiations also Baker had reluctantly to admit[74] to the Foreign Office that there was substance in 'the fact that the Turkish government may carry out their threat to seize our property, which, though possibly illegal, there appears to be no means of resisting'. By the autumn of 1925[75] the Turks were trying to find ways of making this threat a reality. They started by complaining that several of the Company's concessions were being worked jointly as one mine, instead of each being served by a separate shaft in accordance with the Mining Regulations. When Tubini and Nihad Bey had persuaded the Council of Ministers to withdraw temporarily the decree cancelling two of the Company's concessions on these grounds the Turks then suggested[76] the formation of a company with 51 per cent Turkish participation to run the mines, an idea hotly contested by Baker and the board. Finally, realizing that the Ankara government intended, come what may, to wring some additional cash benefit out of the mines in addition to existing royalties, Tubini suggested[77] that the Company should offer a substantial 'loan' to the Government, and a settlement was reached on this basis on 30 March 1927.[78]

This agreement[79] provided that the Company should make a loan of £130,000 for eleven years at 8 per cent interest to the Bank of Turkey for Industry and Mines, the loan to be guaranteed by the Turkish government. No written details are traceable of what was conceded in return, but it seems from scattered references[80] that the Company's concessions were renewed for forty-five years from 1927 on the basis of a sliding scale of royalties which virtually compelled the Company to produce a specified minimum quantity of material each year to qualify for payment of royalty at the minimum rate. This obligation assumed an increasing nuisance value as the years went on, but at least there appears to have been no further major trouble from the Nationalist government right up to the Second World War.

Meanwhile at the mines, when Faulkner fled into the pouring wet night in a horse-drawn cart in October 1922, two Levantines, Marra and Schiave, had been left in charge. There is no evidence of what happened there until mid-1924, by which time Faulkner had returned and shipments of boracite had begun a month or two later. By July 1925 proper development and assessment work had been resumed, but the end of these valiant old mines was already becoming faintly visible in the dim and distant future. Faulkner wrote[81] to Baker:

The position, as I see it, is that for the present, and under present conditions, we have enough ore, enough working places, and enough labour to produce an average of 1,200 tons per month. As far as I can judge at present, we can reasonably hope that this state of affairs will continue for more or less three years; after that it does not seem to me that it would be safe to rely on more than an average of 1,000 tons per month. . . . Boring has now, to a large extent, defined the limits of the ore-body. . . . We therefore find the possibilities of discovering more ore eliminated around practically the whole circumference of the mine, with the exception of two small outlets where some prospects still exist.

Faulkner, however, did not last long. In October 1925 he dismissed Day, who had been one of the few who had been through the Greco-Turkish war troubles with him, for insubordination and 'holding too convivial gatherings with the workmen' in his house. Day wrote to Baker complaining about his dismissal,[82] which he alleged was due to his knowing all about Faulkner and Leaver (a new colleague) running on the side a private lead-mine somewhere in Asia Minor, and also all about Faulkner's affair with (as Baker later wrote) *une cuisinière ou ménagère d'origine française ou arménienne*. Baker, in a rather heavy-handed manner, sent Tubini to report. Tubini had to admit[83] that there was substance in both allegations, but thought that they should be treated lightly – admittedly Faulkner had been seen with the lady at the races, and had asked Tubini *de temps en temps d'envoyer à cette personne des articles de lingerie etc.* but *il doit être extrêment difficile pour notre personnel aux mines de se procurer de la domesticité convenable*! Faulkner admitted the allegation about the lead-mine, which, he said, was only a minute hole in the ground and merely a weekend hobby, but as Baker insisted that this was a breach of his contract, he put in his resignation on 1 January 1926.[84] His successor? – Charles Francis Bunning!

Bunning, however, was intended only to be a stopgap,[85] and there is unfortunately no information traceable about his final stint at the mines, or about what happened to him afterwards. By August 1931 a 'Mr. Macdonald, our engineer in Turkey'[86] was in charge, and remained there until the mid-1950s, when the mines closed down finally, although little is recorded about this long and commendable tour of duty in difficult surroundings. Information on what happened between 1927 and 1939 is extremely scanty, but it can be deduced from correspondence[87] that when the concessions were renewed in 1927 the minimum production required by the Turkish government was 5,000 tons a year. It follows that with the discovery of the great Boron mine in the USA in the mid twenties, this minimum obligation in Turkey became an increasing embarrassment. In July 1930 Baker wrote to Zabriskie:[88] 'At the present time we have only our own works at Coudekerque and Belvedere taking Boracite, besides some small consumers, and it would be a very great assistance if demand could be fostered for it in the U.S.A. – it is certainly a wonderful ore and there is nothing to compare with it for purity as a borate of lime.' Britain's going off the Gold Standard had reduced the exchange rate from 10 to 7 Turkish lire to the pound, and correspondingly increased production costs[89] but by 1936 it was hoped[90] that trade revival after the Depression would stimulate interest in Boracite, at least among special glass and ceramic customers in the USA.

Then, in February 1937,[91] the Turkish Government suddenly removed the mines' incentive to improve on the 5,000 tons annual minimum production by cancelling their sliding scale of royalties, which had encouraged higher levels of production. Demand for chrome and other metallic ores produced in Turkey had certainly increased enormously owing to European rearmament, but the Turks proved impervious to the Company's representations that this demand had not affected boracite also.

In the Second World War the Turks stayed neutral, in spite of Winston Churchill's blandishments, but this neutrality in itself, although an improvement on their attitude in the First World War, kept alive an uncomfortable feeling[92] that they might at any time requisition the Company's ore and sell it to Germany. Old leanings towards Germany die hard in Turkey. Fortunately, however, events turned out less dramatically than feared, although still inconveniently. Operations at the mines had to be suspended in January 1942 owing to the two-way effect of the War on transport through the Mediterranean. Necessary supplies and spares could not be imported, and there is mention particularly[93] of the impossibility of continuing operations owing to shortage of carbide which was used in the miners' lamps. Also, after April 1941, it had become impossible to export boracite from Panderma owing to hostilities in the Mediterranean, and to the fact that the principal markets for boracite were in the war zone in Europe. As a result only a few hundred tons a year could be sold, locally, from the stocks lying at Panderma until hostilities ended in mid-1945. However, a work force was kept in existence at the mines so that mining resumed fairly quickly and smoothly when the War ended. This is shown in particular by the fact that no less than 3,451 tons were produced in 1945 up to 30 September, during which period 1,500 tons also were shipped to London, and production rose rapidly to over 11,000 tons in 1951, which was the normal pre-war level.

Turkey 1945–1960

Mining at Susurluk in the post-war era continued at an increasing rate to meet the demand for boric-acid manufacture in the European plants. By 1955 the Pandermite ore in the Susurluk mine reached the end of what for a mine had been a full and long life. Lower-grade ore was extracted from the tailings stock accumulated over a period of ninety years and 'Pandermite' continued to be shipped until 1961, when final closure came.

Production of boric acid in Borax Consolidated's European plants was readily switched from Pandermite to colemanite, although the former had been easier to process and more consistent than the colemanite now available.

In the post-war era prospecting work primarily in the areas of Bigadic and Emet had located a number of borate deposits consisting of colemanite and ulexite. Several Turkish private individuals established themselves in the mining business. There was no market for ulexite, and as long as Susurluk was in operation the demand for colemanite was small. Borax Consolidated had some difficulty in obtaining satisfactory supplies, and in anticipation of the closure of Susurluk, took the opportunity in 1955 to obtain a small colemanite property at Beyendiklir in the Bigadic area. Experience showed that costs of operation at the relatively low level possible had been seriously underestimated, and after two years it was decided to cease production.

In the late 1950s Lardarello decided to discontinue their process for the extraction of boric acid from steam from the natural *soffioni* (geysers) and to replace it with conventional plant based on imported colemanite. This procedure, together with that of the Borax Group, provided a level of demand which stimulated colemanite-production in Turkey on a regular basis. The

Etibank acquired rights on a large colemanite deposit in Emet, which was later to add further to an overall capacity.

The first few years of purchasing colemanite exposed problems in obtaining regular supplies of high-grade material of the right quality. Colemanite deteriorated if exposed to weather, and the product received had often to be dried and blended, giving rise to serious cost increases and stock levels. Thus in 1959 Borax Consolidated decided there would be long-term advantages in establishing a source of high-grade colemanite, with mining and shipment under the Company's own control and supervision.

After the Second World War foreign companies investing in Turkey had some unfortunate experiences. Difficulties with essential imports, such as engineering supplies, vehicles and all manner of goods not available in Turkey but needed to support a manufacturing project, together with risks of currency devaluation and blocked funds, all created an uninviting picture for those contemplating new projects.

Concerned with the lack of foreign investors, the Turkish government revised its Foreign Investment Law, and in 1954 it was reissued as the Foreign Investment Encouragement Law No. 6224, with special provision for mining activities. In order to conform and obtain the benefit of this law, Borax Consolidated, Limited formed a Turkish mining company, Turk Boraks M.A.S., with Turkish members of the board and Cabir S. Selek (Chairman of the Guaranti Bank) as chairman of the company. Mr R. D. MacDonald, manager of the Susurluk mines from 1931 to 1956, was also a director. MacDonald served the company with loyalty and resource through the difficult times of the Second World War, when he and his wife were cut off and unable to return to this country. Mrs MacDonald was by training a nurse, but in the Susurluk countryside she was in practice a doctor, a missionary of all that was good, and regarded by the people as whatever the Muslim equivalent of a saint may be.

When in 1959 it was decided that Turk Boraks should commence a properly equipped geological minerals exploration programme, whose prime target was borates, the Turkish government consisted of the Democratic Party led by Prime Minister Menderes. At that time there was a new spirit of welcome and enthusiasm in the government for foreign investment, and particularly in regard to activity based on Turkish raw materials and exports. Western governments, particularly the USA and also Britain, were anxious to keep the good-will and confidence of Turkey within the Western Alliance, and were doing much to encourage private industry 'to go to Turkey'. In this climate of optimism Borax Consolidated, Limited made available substantial funds and experienced geologists, and were the first company to import modern diamond drilling equipment for use in the field for mineral exploration in Turkey.

Within a year a situation, now regarded historically inside and outside Turkey – except by extremists – as tragic developed with alarming speed. In 1960 the Government was swept from power by revolution, supported by military action. Following a bizarre trial, at which the trivial evidence given became a tragicomedy, Menderes was hanged along with others. Hundreds of persons, politicians, civil servants, business-men, including the General Manager of the State Mining Corporation (the Etibank), were imprisoned without trial, fearing for their lives. Many were sentenced to gaol for long periods.

A military government was set up, and the Republican People's Party came back to power, still led by Inönü, who had fought alongside Atatürk in the First

World War and in the revolutionary war that followed. A revival of nationalism resulted, and with it increasing emphasis on *Etatism* in industry and its development through the many state corporations. New appointments were made at all levels of civil service and government-controlled organizations. The Foreign Investment Encouragement Law was not repudiated, but the climate had changed.

These dramatic events coincided with Turk Boraks exploration programme getting under way; they also coincided with the geologists making a striking discovery of a new borate region in the Kirka area. A deposit containing sodium borate was found, and in order to determine the size and shape of the body it soon became necessary to drill in adjacent exploration permit areas not available to Turk Boraks. The owner of these adjacent permit areas was not available, and still languished in gaol on an island in the Sea of Marmora; he was a former Democratic member of parliament and successful business-man in borate-mining in the private sector, Mr H. Sirri Yircali. On his release an agreement was made to combine the interests of Turk Boraks and Mortas, which was Mr Yircali's borate-mining company. Exploration of the Kirka borate ore body was completed and the basis of a joint company was agreed, which incorporated such of Mortas's colemanite mines as had sufficient capacity to supply all Turk Boraks' requirements, and also a programme to study the possibility of developing the Kirka sodium borate deposit.

In compliance with the exploration and exploitation permits granted to Turk Boraks, the company had reported to the Ministry details of the exploration results for each drill-hole as the cores were logged. Thus the importance and significance of the discovery became understood within the Government, but the enthusiasm and encouragement of the use of foreign funds and technical skills had only lasted as long as nothing of importance had been discovered. Within the corridors of government resentment grew, and, as became clear some years later, the problem for them became one of how to stop foreign investment and translate ownership from Turk Boraks to the State.

An application under Law 6224 to bring in plant and equipment to establish a large-scale operation based on the Kirka ore deposit was submitted in 1964 and turned down in 1965 as 'politically undesirable', and Turk Boraks could do nothing at this time other than wait for a more favourable political climate. However, the State interest in borates generally was increasing. An application under Law 6224 by Turk Boraks to import a relatively small amount of equipment to mechanize and improve the mining and upgrading of colemanite and to install proper port facilities for its storage and export was refused in 1966, on the grounds that this could be done by other Turkish companies.

In London Borax (Holdings) Limited now saw the continued refusal of applications to invest in Turkey as a major obstacle to any future plans, but continued to hope that some of the spirit of co-operation which had existed earlier might return, as the stagnation of the Turkish economy increased and the balance-of-payments situation worsened.

The situation did not improve, the extremist Press took up the issue of foreign ownership of the Kirka borate deposit, and it became a publicized political issue. A Senate Enquiry was held in 1967, much of which indicated a rising tide of emotion used particularly to cloak the serious problems of the Turkish economy.

Thereafter the Government moved to dispossess Turk Boraks of the right to mine the Kirka property by questioning the validity of the early mineral permits taken out by predecessor permit-holders, from whom Turk Boraks had had to acquire them to start its own exploration programme. Lengthy legal procedures continued through 1969, and the highest civil court in Ankara, with five judges presiding, decided by a majority of one that Turk Boraks should lose its right to exploit the deposits.

To no one's surprise the Kirka property found its way to the ownership of the Etibank. In 1979 the colemanite mines in the hands of various Turkish-owned companies were nationalized.

It is perhaps not surprising since the events of the 1960s occurred that there has been no effective exploration for minerals by any foreign company, nor any foreign investment. The Democratic Party was banned in the post-Menderes era and the Justice Party emerged as a new opposition party and eventually found its way to power, but whatever its views about the means to bring foreign help to the Turkish economy, the 'in-fighting' had been too hot and heavy for there to be any change. Military government took over again in 1970 to pave the way for a new democracy. The cycle repeated itself in 1980, and for all concerned the situation remains delicate, uncertain and hard for the Turkish people.

Throughout the last decade Borax (Holdings) has had no investments in Turkey and no encouragement by the 'Encouragement Law' 6224. However, its European boric-acid plants have been able to continue production based on colemanite supplies from Turkey, where the potential to supply exceeds demand.

Since the Second World War substantial borates ore deposits over quite a wide area have been discovered in Anatolia, but it is noteworthy that no new pandermite deposit – not even a small one – has been identified, and the Sultan Tchair–Azizieh ore body seems to have been a unique geological feature of this part of the world.

REFERENCES

1. In a letter of 22 January 1903 Baker wrote that the name 'Pandermite' was given by the former owner of these mines the Borax Company to the calcined mineral to distinguish it from boracite or untreated mineral, 'they taking the name Pandermite from the Port of Panderma, to which the mineral was shipped'. Panderma, now called Bandirma, is situated on the Sea of Marmara, about forty miles north of this historic ore body close to Susurluk. Later pandermite became the accepted name of this Turkish ore, which later was shipped in bulk without calcination – i.e., roasting. See also Glossary, p.292
2. R.C.B.: Daniell to R. C. Baker, 30.5.99.
3. *Ibid.*
4. Bunning's report to London of 27.10.04.
5. R.C.B.: Daniell to Baker, 1.6.1900.
6. R.C.B., 20.6.01.
7. R.C.B., 27.10.01.
8. R.C.B., 18.7.02.
9. R.C.B. from Tubini, 17.3.99.
10. R.C.B. from Tubini, 15.9.99.
11. R.C.B. from Tubini, 9.3.99.
12. Minute Book, 15.6.09.
13. R.C.B. from Thompson, 17.5.99.
14. In Mining section at Borax House, London SW1.
15. M.D., 26.2.14 and 19.3.14.
16. R.C.B.: Bunning's reports of 3.8.1899 and 10.8.1899 and 27.10.04.

17. M.D. to Locke, 19.11.08.
18. M.D. to Zabriskie, 19.11.08.
19. M.D. to F. M. Smith, 15.1.04.
20. M.D. to Engineer in Charge, Sultan Tchair, 26.2.14.
21. R.C.B.: Bunning to Baker, 3.8.99.
22. R.C.B. from J. W. Whittall & Co., 27.1.02.
23. M.D. to Devischer 29.12.11.
24. *Ibid.*, 16.1.12.
25. R.C.B.: Devischer to R. C. Baker, 12.1.12.
26. *Ibid.*
27. M.D., 29.8.12.
28. M.D., 18.12.12.
29. Eversley, Lord and Chirol, Sir Valentine: *The Turkish Empire (1288 – 1924)* 3rd Edn. (1924), p.369 (Unwin, Ltd., London).
30. R.C.B. from Bunning, 10.8.99.
31. M.D. to Emberson, 21.10.12.
32. M.D. to Bunning, 13.1.10.
33. *Ibid.*, 10.2.13.
34. *Ibid.*, 19.3.14.
35. *Ibid.*, 19.8.14.
36. M.D. to Glyn Mills, 7.8.14.
37. M.D. to Faulkner, 12.3.15.
38. M.D. to Lord Chichester, 18.5.15.
39. M.D. to Faulkner, 15.12.15.
40. M.D. to Lord Chichester, 30.6.16.
41. M.D. to Pacific Coast Borax, 29.5.17.
42. M.D. to Zabriskie, 24.4.17.
43. M.D. to Pacific Coast Borax, 29.5.17.
44. *Ibid.*
45. M.D. to Lord Chichester, 7.1.19 (Ms.).
46. M.D. to Lord Chichester, 8.3.19.
47. R.C.B. from Daniell, 2.4.19.
48. M.D. to Lord Chichester, 7.1.19.
49. *Ibid.*
50. M.D. to Bunning, 17.10.19.
51. M.D. to Bunning, 10.10.19.
52. M.D. to Zabriskie, 21.5.20.
53. M.D. to Faulkner, 24.6.20.
54. M.D. to Pacific Coast Borax, 3.6.20.
55. R.C.B.: Faulkner to Borax Consolidated, 4.9.20.
56. M.D. to Zabriskie, 21.6.18.
57. 'Some Memories of What I Have Seen and Heard 30 Years Ago at Sultan Tchair Mines' by N. Tumaz – copy in Borax House, SW1.
58. P.C.B.: Zabriskie from Faulkner, 4.10.20.
59. R.C.B. from Faulkner, 7.10.20, and Boyd from Faulkner, 17.10.20.
60. M.D. to Faulkner, 22.9.20.
61. M.D. to Zabriskie, 2.2.22.
62. M.D. to Faulkner, 22.9.20.
63. M.D. to Boyd, 16.3.21.
64. R.C.B. from Faulkner, 31.10.20.
65. R.C.B. from Faulkner, 9.4.22 and 22.6.22.
66. An undated report in Istanbul office filing cabinet.
67. Tubini's reports to R. C. Baker of 26.10.22 and 1, 2 and 6.11.22, also Faulkner's letters from Smyrna of 26 and 31.10.22 and M.D. to Pacific Coast Borax of 23.10.22. All these are in the records at Borax House, SW1.
68. M.D. to Daniell and Tubini, 6.6.23. Also M.D. to Lord Chichester, 27.6.23 etc.
69. M.D. to Lord Chichester 25 and 27.6.23, and to Under-Secretary at the Foreign Office, 18, 19, 21.7.23 and 14.8.23.
70. M.D. to Under-Secretary at the Foreign Office, 18.7.23.
71. M.D. to Tubini, 23.7.23.
72. M.D. to Under-Secretary FO, 14.8.23.

73. M.D. to Zabriskie, 18.1.24.
74. M.D. to FO, 27.7.23.
75. M.D. to Zabriskie, 14.9.25.
76. M.D. to Zabriskie, 16.10.25 and 10.12.25.
77. M.D. to Under-Secretary FO, 6.1.27.
78. R.C.B. from Lesser – cable of 30.3.27 *et al.*
79. Minute Book.
80. R.C.B. from Lesser – cable of 28.3.27.
81. R.C.B. from Faulkner, 8.7.25.
82. M.D. to Tubini, 30.9.25.
83. R.C.B. from Tubini, 6.10.25 and 7.11.25.
84. M.D. to Faulkner, 1.1.26.
85. M.D. to Pacific Coast Borax, 1.2.26.
86. M.D. to Zabriskie, 5.8.31.
87. M.D. to Pacific Coast Borax, 12.8.42.
88. M.D. to Zabriskie, 11.7.30.
89. M.D. to Mitchell of Sterling Co USA, 23.2.32.
90. M.D. to Jenifer, 29.3.36.
91. Anon. to Pacific Coast Borax, 23.2.37.
92. R.C.B. from Elias Raben of Hamburg, 28.2.27, Gerstley to Pacific Coast Borax, 22.11.39 and M.D. to Pacific Coast Borax, 8.7.42.
93. M.D. to Pacific Coast Borax, 12.8.42.

Glossary of chemical and mineral names

Alum
By ancient usage the name is attached primarily to a double sulphate of aluminium and potassium ($Al_2(SO_4)_3$, K_2SO_4 $24H_2O$) – potassium alum – or a similar combination of aluminium and ammonium ($Al_2(SO_4)_3$, $(NH_4)_2SO_4$ $24H_2O$) – ammonium alum. Alum owes its importance historically to its role as a fixer of dyes (a mordant) in the textile industry.

Anhydrous Borax
Borax (sodium borate) with water removed.

Ascherite
A mineral containing boron in the form of magnesium borate ($2MgO.B_2O_3.H_2O$).

Biborate of Soda
A nineteenth-century name given to Borax – now obsolete.

Borate of Lime
See under Ulexite.

Borax
This usually refers to sodium tetraborate decahydrate ($Na_2.B_4O_7.10H_2O$), but is also used to describe other forms of hydrated and anhydrous borax.

Boric Acid
Formerly called sedative salt, and boracic acid, H_3BO_3.

Boron
The chemical element (symbol, B) contained in all borate minerals and compounds.

Borosilicate Glass
A heat-resistant glass containing the element boron.

Calcium Borate
This substance occurs in different minerals – *see* colemanite, pandermite, priceite and ulexite.

Cinnabar
A mineral, mercury sulphide (MgS).

Colemanite
A calcium borate mineral – $2CaO.3B_2O_3.5H_2O$.

Datolite
A natural borosilicate of calcium, occurring in rocks in crystalline form – $2CaO.B_2O_3.SiO_2.H_2O$.

Green Vitriol
Ferrous sulphate, $FeSO_4.7H_2O$.

Gypsum
Calcium sulphate ($CaSO_4 \cdot 2H_2O$).

Kernite
A sodium borate mineral, $Na_2O \cdot 2B_2O_3 \cdot 4H_2O$.

Malachite
A basic copper carbonate, $CuCO_3 \cdot Cu(OH)_2$, occurring in crystalline form in the oxidation zone of copper deposits.

Marine Acid
Hydrochloric acid, HCl.

Marine Alkali
See under Sodium Carbonate.

Natron
Impure natural soda (sodium carbonate). Also referred to as nitrum.

Nitre
Saltpetre.

Nitrum
See under Natron. The term is sometimes confused with nitre.

Pandermite
A calcium borate mineral, $4CaO \cdot 5B_2O_3 \cdot 7H_2O$ (also called Priceite).

Pearl Ash
Potash – potassium carbonate (K_2CO_3).

Petuntze
This secret constituent accounted for the superiority of Chinese hard-paste porcelain, and for the failure of many early European efforts to imitate it. The word means 'little bricks', the material being sent to the potter in this form. It is a partly decomposed granite (consisting of felspathic minerals and quartz) mixed with kaolin. It is sometimes referred to as China-Stone.

Potash
A word applied originally to the alkaline ash of land vegetation (usually wood ash), which contained potassium mainly as carbonate (K_2Co_3). Today the term potash is applied to the potassium salts used in fertilizers, chiefly potassium chloride (KCl) and sulphate (K_2SO_4).

Priceite
See under Pandermite.

Sal-Ammoniac
Ammonium chloride (NH_4Cl).

Salt Cake
Anhydrous sodium sulphate (Na_2SO_4). The crystalline form is known as Glauber salt ($Na_2SO_4 10H_2O$).

Saltpetre
Potassium nitrate (KNO_3), an essential ingredient of gunpowder.

Sandiver
Fused salts removed as a scum in glass-manufacture and used as a flux in pottery glazes.

Sedative Salt
An early name given to boric acid (H_3BO_3), but now obsolete.

Soda
See under Sodium Carbonate.

Sodium Borate
As sodium tetraborate decahydrate ($Na_2B_4O_7.10H_2O$), it is known as borax, but see under Tincal and Kernite.

Sodium Carbonate
$Na_2CO_3.10H_2O$; also called marine alkali, soda or soda ash. These names were applied originally to the ash of sea vegetation.

Sodium Pentaborate
$Na_2B_{10}.O_{16}.10H_2O$.

Sodium Perborate
$NaBO_3\ 4H_2O$, an important bleaching agent in detergents.

Tincal
A sodium borate mineral containing crystals of borax ($Na_2.B_2O_4.10H_2O$).

Tourmaline
A complex silicate mineral containing boron, dispersed in crystalline form in granite-type rocks.

Trona
The most frequently occurring sodium alkali mineral – dihydrate of sodium carbonate and bicarbonate ($Na_2CO_3.NaHCO_3.2H_2O$).

Ulexite
A sodium calcium borate mineral, $Na_2O.2CaO.5B_2O_3.16H_2O$; also called borate of lime.

Verdigris
A green copper acetate, made by treating copper with vinegar.

Index

Abbasid Caliphs, 8
Abbe, Ernst, Jena glass works, 33
Abdul Hamid II, Sultan, 89, 90, 257, 264, 272
Acide borique, 11
Advertising, 45, 66, 67, 68, 115, 116, 140, 202
Advisory Committee in France, 85, 91, 260
Aerospace, boron in, 2
Agricola, Georgius, *De natura fossilium, De re metallica*, 6, 29
Agriculture, borax in, 159, 160, 172
Agricultural lobby on potash in USA, 159, 160
Ahmed deposit (Tchakmak Bair), 261, 262
Ahmed Nessimi Bey, Minister of Commerce, 268
Ahmed Pacha, H.E., 261
Ahmedabad, 16
Aircraft fuels (see also 'Boron High Energy Fuels'), 229
Air raid damage, 211, 212
Alameda refinery, 60, 62, 64, 65, 76, 79, 111, 112, 114, 117, 142, 147, 169
ALBRIGHT, HORACE:
   Asst. to, and then Director of National Park Service, 164–167
   General Manager and then President of U.S. Potash Co., Medal of Freedom, 200, 201
   Misc. references, 194, 203, 208, 209, 211, 218
Alchemy, 4, 8, 9
Alien Property Custodian, the, 214, 215, 222
Allen, Receiver for Western Borax mine, 222
Allen and Howard, chemical works, 14, 15
Al Razi, physician, 4, 8
Alum, 6, 7, 9, 32, 291
Amargosa, USA, 49, 118
   Canyon, 130, 132
   Desert, 127, 130
   Hotel, 192
   River, 53, 130
   Works, 53, 54
American Ambassador in Constantinople, 268, 269
American Borax Company and mine, The, 40, 118–21, 175
American Chemical Society, The, 160
American Civil War, 40, 125
American Cyanamid, 198
American Missionary Society, The, 112
American Potash and Chemical Corporation (see also 'California Trona Company' and 'American Trona Corporation'), 181, 182, 186, 193–198, 200–202, 204, 208, 214, 215, 217, 219, 222, 224, 225
American railways, investment in, 132
American Red Cross, 211
American Trona Corporation, 156–160, 166, 171–173, 176, 180, 181, 186, 193
Amsterdam, 14
Anaheim Research Laboratories, 231
Anatolia, 257, 273, 276, 288
Anderson, Chairman of Pfizer & Co., 143
Anderson, Frank, Bank of California, 143
Andes, The (see also 'Cordillera, The'), 27, 87, 88, 254
Anhydrous borax, 202, 224, 225, 229, 230
Ankara Court Decision on Kirka, 287–8
Anniversary Mine (Calville Wash), 175
Annual General Meetings, 139
Anti-British feeling in Turkey, 281, 282
Antiquities Act, The (USA), 164
Antiseptics, borax in, 34
Anti-Trust investigation, 217, 222
Anti-Trust Laws, US (see also 'Sherman and Clayton laws'), 192
Antofagasta, 27, 87, 237, 240–241, 243, 244, 245, 252
Antofagasta and Bolivia Railway, 251
Apothecaries, 70
Aquas Calientes, Chile, 98
Arab glasses, 30, 31
Arab trade routes, 7, 30
Arabic, writers in, 4
Aras, Jeronimo, Rio Grande, 242
Arbor Villa, Oakland, 65
Arbuthnot, William, refiner, 26
Arequipa, Peru, 81, 87, 119, 237, 246–248, 249, 250, 252, 253
Argentina (The Argentine), 27, 86, 88, 91, 204, 209, 224, 254
Armenians, 264, 277–280
Armenian borax, 3
Armour, Bernard, Chairman of American Potash and Chemical Corpn., 225
Armstrong, Professor H. E., 107
Army, Turkish, 258, 267, 278
Arrieta, L. Ovalle, Pintados, 245
Artists Drive, Death Valley, 49
Asbestos, replaced by glass fibre, 224

Ascherite (a magnesium borate), 113, 291
Ascotan Agreement, 80, 83, 87, 91, 96–98, 102, 103
Ascotan deposit, Chile, 79–82, 87, 89, 103, 237, 240–246, 249, 252, 253
Assam, 23
Assaying of precious metals, 29
'Associated' refiners, 97, 98, 102
Atchison, Topeka, and Santa Fe railroad, 125
Atlantic & Pacific (later Santa Fe) railroad, 62
Atwood, Albert, writer, 164
Auditors, 141
Aussig factory, 96, 170, 209, 210, 212–214, 226
Aussig Verein, 213
Australia, exports to, 148, 205
Austria and Austrians, 91, 95, 96, 151, 208, 209, 212, 214, 226, 257, 265, 266, 269
Austrian refiners' syndicate, 96
Autoclaves (digesters), 186–188, 224
Avicenna, physician, 6
Ayers, Dr William, 39, 54
Azangaro, S. America, 245
Azizieh mine, Turkey, 90, 258, 260, 263, 272, 279, 280, 288

Babcock, W. L. & Co., 44
Babuk, Nepal, 19
Babylon, 3
Badelona refinery, Spain, 172, 210
Badwater, Death Valley, 49
Baker, F. C., President of A.P. & C.C., 194, 219
BAKER, RICHARD CHARLES:
 Joins Redwood & Sons (food-preservatives), 70
 Upbringing and personality. Founds Burton, Baker and Co., 70, 71
 Expands Redwoods' sales, 72
 Plans for borax refining, 73
 Encounters F. M. Smith in England. Arranges merger between Redwood & Sons and Smith's borax interests. Subsequent enlargement into Borax Consolidated, 75–84
 Comments on merger. Able administration of new group, 86–87
 South American policy, 88–89
 Difficulties with French Advisory Committee, 93
 Approaches to Lardarello, 94–95
 Austrian difficulties, 96–97
 Belief in loyalty of German refiners, 98
 Restrains Smith from raising English capital through Borax Consolidated to finance non-Borax projects in USA, 100
 Relations with refiners, 101–104
 Passing revival of interest in Tibet and China, 110–112
 Financial worries caused by Smith's railroad ambitions, 127, 131–136
 On Smith's financial collapse, Baker succeeds in preventing Smith's controlling Borax shares from passing to outside group, 139–143
 Endeavours to restrain Smith's hostility to Borax Consolidated, 146–147
 On wartime uses of borax, 148
 Shipping problems, 149
 Attitude to labour problems and exports, 150, 151
 On the defensive about new potash and borax industry on Searles Lake, 156, 158, 159
 Discusses including Death Valley in National Parks system, 165, 166
 Involved in details of his Company's colemanite mines and mining camps in USA, 168–170
 Raw material contract with Stauffer, 172–173
 Declines to join consortium to purchase American Trona and another to acquire Smith's West End Chemical Co., 174–176
 Makes difficult decision to advise Board to acquire promising Sections of Kramer area, 179–183
 Problems over identification, extent, and treatment of new Kramer mineral and over keeping off interlopers, 186–191
 Decision to go in for hotels and tourism in Death Valley, 192–193
 Fails to come to terms with American Potash on stabilizing borax prices. Decision to compete with them in potash as well as borax. Comments on mysterious sale of American Potash interest by Consolidated Goldfields, 193–195
 Difficulties in way of resurrecting Searles Lake potash venture, 197
 Accepts Rasor's recommendation to seek participation in United States Potash venture at Carlsbad. Meets great difficulty in obtaining finance required. Main burden left with Baker and his company but venture succeeds, 198–202
 Works until two months before death in January 1937. Many tributes, 203–204
 On Senator Key Pittman's attack on borax and potash industries, 218
 South American affairs, 237, 252, 253
 Turkish problems, 258, 260, 263, 266, 267
 First visit to Turkey, 267, 268, 270
 Opposition to German takeover of mines, 271–275
 Miscellaneous references, 208, 222
Baker's letters, 203
'Baker' mine, 187
'Baker' shaft – see 'Kramer'
Balkan War, First, 254, 266, 267
Ballistic missiles, 232
Baloukhissar, Turkey, 268
Bank of New York, 65, 66
Bank of Turkey for Industry and Mines, 283

Bankers' Committee of San Francisco, 143, 146
Banque Franco-Américaine of New York, 140
Banque Privée, 92
Barba, A. A., *El arte de los metales*, 7
Barndt, Victor, 157
Baron, T., chemist, 9, 10
Barter trade, 205, 209
Bartlett mine, Calico, 120
Bastard emeralds, 5
Bates, Sir Percy, shipping, 148
Bates, company promoter, 193
Bath tubs, 120
Batten, Arthur van, Coudekerque, 211
Bayonne factory, New Jersey, 79, 114, 116, 149–151, 171, 186, 187, 275
Beatty, 127, 133, 135
Beik, Fred, engineer, 169, 202
Belgium, 91, 93–95, 148, 153, 225
Bell, Jacob, apothecary, 69, 70
Belvedere refinery, Kent, 72, 74–76, 81, 87, 101, 105, 111, 112, 150, 153, 187, 211, 213, 226, 284
Bengal, 18, 23
Bennet, Bellerin' Teck, 51, 52
Benson (Frederick J. Benson & Co. financiers), 135
Berlin, 80
Beyendikler colemanite deposit, 285
Bhots, 22 (footnote)
Bhutan, 22, 23
Biborate of soda, 39, 292
Biddy McCarthy lode, USA, 62, 135, 136, 168
Bigadic, Turkey, 285
Billwaeder, Hamburg refiner, 171
Biringuccio, Vanoccio, *Pyrotechnica*, 6, 29
Bishop (W. A. Bishop & Co., Warrington), 104, 105
Blanco Vale Borax Company, 42
Blane, William, traveller, 17
Blumenberg, Henry, 119–121, 144, 176, 190
Bob Tubbs' saloon, Death Valley Junction, 170
Bodington, Borax Français, 93
Bohm, Herr, of A.G. für Chem. Rheinau, 97
Bolivia, 27, 88, 236, 240, 242
Boraces, 5
Boracic acid (boric acid), 10, 11, 34, 241, 271, 291
Boracic crystals, 1
Boracic lint, 34, 148
Boracium, 11
Borate and Daggett Railroad, 63, 125, 137
Borate Mine (see also 'Calico'), 62–64, 120
Borate of lime (sodium calcium borate), 27, 149, 189, 205, 237, 240, 284, 291
Borate of soda, 4
Borate mineral resources, study of, 229
Borated vaseline, 153
Borates in Death Valley, discovery of, 51
Boraxaidal soap powder, 116
Borax bead test, 32
Borax, chemical definition, 1

Borax Company, The, 28, 73, 81–83, 85, 89, 91, 92, 258–264
Borax Company of California, The, 40
Borax Consolidated Limited:
  Preliminaries, 53, 69, 74–78, 81, 82; Incorporation, 82–83; Assets acquired, 85, 87, 88, 90–106; Sterling Borax Co., 122; Tonopah and Tidewater Railroad and ensuing financial problems, 126, 130–137; problems created by Smith's financial collapse, 142, 143, 146–148; First World War, 148–152; first response to Searles Lake rivalry, 157–160; final colemanite struggles in USA, 169, 171–174; new mining centre and borax mineral in USA, 176, 179–182; accompanying problems, 186–192; hotel business in Death Valley, 192–193; failure to come to terms with Searles Lake, 194–195; entry into potash mining, 197–202; changing borax scene and war preliminaries, 202–205; Second World War, 208–214; Anti-Trust case, 217–220; post-War policy problems, 222–225; open pit and new refinery in USA. Restructuring of Group, 228–230; South American problems, 237, 238, 242, 252, 254; details of involvement in Turkey, 257–288
Borax Consolidated Act (1907), 133
  Debenture issue, 135–137, 140
  Share capital, 133, 140, 143, 148
Borax crystals, 15
'Borax' – derivation of word, 4
Borax, discovery of, in the USA, 39
'Borax engineers', 188
Borax '5 mol' and '10 mol', 229
Borax Français, 93, 94, 211, 260
Borax: From the Desert . . . Into the Home, S. Mather, 67
Borax (Holdings) Limited, 230, 231, 287, 288
'Borax King', the, 140
Borax of Zarawand, 5
Borax refiners, number in UK, 72
Borax 'rushes', 39, 44, 52, 175, 181
'Borax Smith', 74
'Borax Trust', the, 105
Boraxo, 116, 171
Bore, 11
Borge, Wendell, US Asst. Attorney-General, 219
Boric acid, 2, 9, 11, 22, 24–26, 34, 45, 68, 69, 72, 78, 88, 91, 94, 95, 101–104, 106, 107, 115, 116, 119, 120, 148, 149, 152, 170, 210–212, 217–219, 225, 283, 285, 288
Boric acid powder, 153
Boric oxide, 33, 119, 176, 179, 229, 232
Boron, 1, 11, 291
Boron chemicals, 229
Boron chemistry, 11
Boron compounds, 2, 229
Boron hydrides (boranes), 229, 231, 233

Index

Boron Mine (see also 'Kramer'), 187–191, 203, 205, 214, 223–225, 228–231, 254, 284
  Open pit and new refinery project, 228–231, 233
  Policy background, 228–229
  Financing, 229–231
  Boron products, 229
Boron Products Company, Stroud, 26, 83, 101, 102
'Boron', the, railcar, 131
Boron tribromide, 2
Boroquimica Samical company, Argentina, 254
Borosilicate, 2, 33
Borosilicate glass (see also 'Heat-resistant glass'), 30, 31, 33, 223, 291
Borosolvay, 157–159, 197
Boverton (village), Glamorgan, 69
Boxer Rebellion, 112
Boyd, Major Julian, mining engineer, 275
Brabourne, Lord, Chairman, Consolidated Goldfields, 193
Brazil, 254
Bread borax, 5
Brettschneider, Dr, Manager, Aussig, 209, 213
Breyfogle, Jacob, 50
Brighton Chemical Company (Thorkildsen & Mather), 122
Brine, 154–158, 160, 161, 218, 219
British Commission in Turkey, 272
British-owned companies in USA, 217
British Foreign Office, 149, 269, 270
British Treasury, 216, 217, 230
Brock, John, entrepreneur, 132, 133
Brownrigg family, 101
Brunner, Mr, Banque Franco-Américaine, 140
Brunner, Mond & Co., 112
Buley, A. M., oil prospector, 189, 190
Bullfrog, 128, 131–134
Bullfrog Goldfield Railroad, 132–135
Bul-Tso Lake, 19
Bulwer-Lytton, Edward (Lord Lytton), 3
Bureau of Mines, Geological Survey Dept., 198
Burton, Baker & Co., 71, 72, 76
Burton, Sir Richard, 71

Cadet, L. C., chemist, 9
Cahill, 'Wash', 63, 64, 121, 127, 128, 130, 135, 204
Calama, Chile, 237, 241
Calcined Rasorite (C.R.), 189, 224
Calcining ovens, Antofagasta, 240
Calcium borate, 291
Calculators, boron in, 2
Calcutta, 16, 23, 111
Calico Mine (see also 'Borate'), 60, 62, 64, 117, 119, 125, 126
Calico Mountains, 60, 62, 64
California Eastern Railroad, 126
California Railroad Commission, 136
California Trona Company (see also 'American Trona' and 'American Potash'), 155, 156, 158
Callery Chemical Co., 232
Calm, Max and Charles, 47, 119, 120, 122
Calm Mine, 120, 122, 123
Calville Wash (Anniversary Mine), 174
Campo Quizano refinery, Argentina, 254
'Cancha', the, Cebollar, 243
Candeleria, 44, 114, 117, 164
Canterbury Tales, 7
Canton, 18
Capuchin Missions, 17, 18
Carbide for miners' lamps, Turkey, 285
Carcote, S. America, 81, 237, 242, 250–251
Cariquima, S. America, 244
Carlsbad, New Mexico, 198, 201–204, 226
Carlsbad Caverns National Park, 200, 202
Carlston, J. F., financier of F. M. Smith, 157, 176
Carnallite (potash), 198
Carrere, Dr, 27
Cebollar, S. America, 87, 237, 241, 243
Central Pacific Railway, 54
Ceramics, borax in, 2, 29, 31, 284
Ceylon, 24
Chandler site, San Pedro – see 'Wilmington'
Chapin, friend of F. M. Smith, 157
Chaptal, J. A., chemist, 4, 5, 14, 23, 24, 111
Chardin, Jean, 24
Chase Manhattan Bank, 229, 230
Chase National Bank, 199
Chau Ju-kua, writer, 30
Chaucer, Geoffrey, 7
  Prologue, 7
  Canon's Yeoman's Tale, 8
C.C.F. (Compagnie Chemin de Fer) Smyrne-Cassaba, 265
Chemical analysis, methods, 32
Chemical composition of borax, 9
Chemical Trade Journal, 201
Chevket Bey, Mines Section, Turkey, 268
Chicago sales office, 67, 68, 114
Chichester, Lord, Borax Consolidated Chairman, 149, 193, 204, 267, 269, 272, 282
Chickering, lawyer, 74
Chilcaya, S. America, 87, 103, 104, 135
Chile, 27, 35, 86, 88, 104, 149, 205, 225, 240, 243, 250, 252–254
Chilean export tax, 89
Chillicolpa, Peru, 81, 87, 237
China, 8, 16–17, 18, 23, 30, 31, 109–113, 148
China glaze, 34
China Lake, 154
'Chinese borax', 111, 112
Chinese Labour Corps, France, 150
Chinese labour in USA, 53, 55, 63, 117
Chinese Mandarins, Viceroys, officials etc., 112
Chivinian, Whittall agent, Turkey, 262
Christians in Turkey, 267
Chrysocolla, 5, 7, 9, 24
Chu Tsale (water borax), 19

Cibot, P. M., missionary in China, 8, 18
Cinnabar, 291
Circassians, 264, 275, 281
Citrus-fruit preservation, 172
Civil suit against Borax companies, 218
Clare, George (Clare & Co.), 143
Clark's railroad (the San Pedro, Los Angeles and Salt Lake), 126, 127
Clark, W. A., Senator, 126–128, 133
Clark, Tom C., US Attorney-General, 218
Classified information, 231–233
Clay, for oil refiners, 191
Clayton Anti-Trust Act, 217
Clear Lake (see also 'Borax Lake'), 28, 39
'Closed shop', the, 151
Clough, Alton, 46
Coal allocation, UK, 226
Coffin Redington Co., 66
Coghill, Frank, 88, 101–103
Coghill & Sons, 26, 73, 83, 101, 102, 104, 105
Colefax, Peter, President, A.P. & C.C., 225
Coleman, Morton, financier, 229, 230
Coleman, William T. (& Co.), 42, 44, 47, 48, 52–56, 60, 62, 65, 114, 118, 154, 168
Colemanite (a calcium borate), 1, 60, 62–64, 111, 114, 115, 117, 119–121, 125, 149, 168, 169, 171, 174, 175, 179, 181, 185, 188, 189, 192, 219, 223, 245, 253, 285, 291
Colemanite production in Turkey, 285–288
Colombia Chemical Divn., Pittsburgh Plate Glass – see Pacific Alkali
Columbus Borax Company, The (Calm Bros. – later U.S. Borax Co.), 47, 120, 122, 123
Columbus Marsh, Nevada, 40–42, 46, 51, 52, 54, 64, 76, 131, 155
Committee of San Francisco bankers, 143, 146
Compagnie France-Autrichienne de Produits Chimiques, La, 82
Compagnie Internationale de Borax, 88, 91, 93, 94, 97, 101–103
Computers, boron in, 2
Connah's Quay borax works, 78, 101
Connell, George A., 185
Consent Decree, 218, 220, 222, 223
Consolidated Goldfields, 155, 156, 159, 160, 186, 193, 194
Constantinople (later Istanbul), 90, 257, 258, 260, 262, 265, 266, 267, 268, 269, 270, 272, 273, 277, 281
Continental Bank, 200
Cook, prospector, 121
Copper propositions, 142
Coptic Papyrus, 3
Cordillera, The (see also 'Andes'), 243, 250–254
Corkill, Fred (I), mining engineer, 47, 63, 120, 122, 125, 148, 151, 168, 173
Corkill, Fred (II), 168, 188, 189
Corkill Hall, Death Valley Junction, 170
Corning, 33
Cornish reducing flux, 29
Cornwall works, Kennington, 71, 76, 87

Cornwall, Mrs, writer, 202
Corum, Ralph. M., driller, 179–181, 185
Corum syndicate, 181
Cosapilla, Peru, 81, 87, 237
'Cottonball' (ulexite-sodium calcium borate), 40, 41, 50, 53, 60, 62, 64, 114, 117
Coudekerque factory, France, 93, 94, 101, 148, 149, 187, 209, 211, 226, 284
Coulbeaux, Pierre, Turkey, 258, 260, 266
Courtaulds, 217
Coward, Noel, 1
Coye, H. L., 47
Cramer, Johann, assayer, 29, 30
Cramer, Tom, chemist, 147, 169, 180, 185, 188, 189
Crédit Liègeois, 93, 94
Criminal indictment of borax companies, 218
Crops – see 'Agricultural crops'
Crusades, the, 6
Crystal glass, 30
Cuevitas deposit, S. America, 242, 243
Cuevitas Trading Company, 105
Cunningham, Alex., Major, Bengal Engineers, 22
Custodian of Enemy Property (see also 'Alien Property Custodian), 225
Cuthbertson, Sir John N., refiner, 83, 103
Czechoslovakia, 171, 208, 209, 212, 214, 226

Daggett, 54, 60, 62–64, 76, 119–121, 125, 126, 128, 176, 179
Daniell, H. T., executive, 93, 94, 97, 258, 260, 262, 265, 266, 269, 270–271, 273, 276
Dante's 'Inferno', 24
Dardanelles, The, 90, 265, 268, 272, 281, 282
Datolite, 2, 291
Daunet, Isadore, 51, 52
David, Borax Co. agent, Turkey, 258
Davis, Bess, schoolteacher, 170
Davy, Sir Humphry, 10, 11
Dawes and Myler, 120, 121
Day, employee in Turkey, 284
De Aluminibus et Salibus, 6
Death Valley, USA, 28, 44, 49–55, 57, 60, 62, 64, 76, 117, 118, 122, 125–128, 136, 154, 155, 165, 166, 168, 170–172, 179, 188, 189, 192, 193, 216, 219, 223
Death Valley Days, radio programme, 202
Death Valley, flying school, 216
Death Valley, hotel business, 192, 216
Death Valley, literature, 202
Death Valley, National Monument, 166, 201
Death Valley Junction, 126, 130, 132, 135–137, 151, 168–171, 187–189, 192, 275
  Baseball team, 170
  Civic Centre, 170
  Roasting plant, 137, 151, 168, 169, 171, 274
  Wet concentration plant, 169
Death Valley Railroad Company, The, 136, 137, 168

De Guigne, Christian (Stauffer Chemical Co.), 119, 143
Defense Department, US, 229, 231, 232
Dehydrated borax (see also 'Anhydrous borax'), 219
Delameter, J. R., blacksmith, 56, 57
Demachy, G. F., chemist, 15, 17
Democratic Party, Turkey, 286, 288
Depletion allowances, US, 230
Depreciation, 141
Depression, the Great, 160, 188, 191, 192, 199, 202, 203, 284
De Rovato, Joseph, Capuchin Father Prefect, 17
Desert Culture peoples, 49
Desmazures, Camille, Turkey, 28, 81, 91, 261
Detergents, borax in, 2
Deutsche Bank, Berlin, 79, 265
Deutsche Borax Vereinigung, 190, 208, 210
Deutsche Orientalische, 269, 270
Devisscher, M. Jean, 93, 94
Devisscher, Jr., engineer, 93, 94, 265
Diamond drilling, Turkey, 286
Diehn, Herr, Kali Syndicate, 195, 197
Dioscorides, 5
Dividends, borax companies, 47, 48, 62, 66, 141, 233, 252
Djivad Pacha's concession, 261
Dolbear, C. E., chemist, 155
'Dollar gap', the, 225
Donkeys and 'burros', 248, 265
Dossie, Robert, *Handmaid of the Arts*, 30–32
Dowsing, W. M., driller, 179, 180, 185
Dresden, 30
Dress of fibreglass, 224
Drilling teams, 154
Dru, Count, American Potash, 186, 193, 194
Drug Markets, periodical, 203
Drummond, US Commissioner of Mines, 41
Duffield v. San Francisco Chemical Co., 44 (footnote)
Dumlupinar, Battle of, 276
Dunkirk, 211
Du Pont Company, 193
Dupont, F. M., 97
Dupont, M., 96, 97
Durval, 95
Dutch borax, 111
Dutch borax monopoly, 14
Dutch refiners, 10, 15
Dumont, Mr, 147
Dumont, Harry, 172
Dwyer, Mr, 197

Eagle Borax Company and works, 52–56
Earth's crust, boron in, 2
Earthmoving equipment, 228
Earthquakes, South American, 245–246
Eastern Trading Company, Shanghai, 112
East Indies, 14
Eberstadt (F) & Co., financiers, 230, 233

Eberstadt, Ferd, 233
Eddy County, New Mexico, 198
Edenburg, driller, 179
Edwards Air Force base, 231
Edwards, Ben, 117
'Effort', F. M. Smith's sloop, 144
Egypt, 3, 24, 223
Eighteenth-century chemists and borax, 9
Electricity, 105
Electronic capacitators, borax in, 2
Electronics, boron in, 2
Ellis, Evelyn (2nd Mrs Smith), 128, 144, 176, 192
Ellis, George, 176
Ellis, prospector, 53
Embalming, 3
Emberson, Alfred, stockbroker, 131, 132, 137, 266
Emberson and Hughes, stockbrokers, 133, 135
Emet Mines, Turkey, 285, 286
Empresa d'Ascotan Co., Chile, 79
Enamel, 2, 31–33, 116, 148, 151, 171, 172, 187, 202, 223
Enamellers, German Union, 98
Engestrom, G. von, chemist, 17
England, borax prices, 35
England, borax refining in, 14, 15
English Company Law, restraints of, 100
'English ore', 64, 65
'English Union', the, 73
Enver Bey, Turkish Minister of War, 90, 269, 272
Eppinger, prospector, 121
Epsom salts, 103
Ercker, B. L., metallurgist, 29
Ernst, Jack, Searles Lake, 158
'Essential Industries' classification, 212
Etibank (State Mining Corporation), Turkey, 286, 288
European Economic Community, 226
European Kali Syndicate – see Kali (German Potash) Syndicate
European market, 172, 202, 205, 214, 225, 253
European travellers in the East, 16
Examiner, The (Oakland newspaper), 127
Exchange rate, 226
Explosives, use of borax in, 148
Export duty, 149
Exports to Iron Curtain, US restrictions, 226
Ezcurra, Señor, 86, 88

Fairlie, H. C. & Co., Falkirk, 103, 105
Falcouz, M. Auguste, 82, 92, 262
Faraday, Michael, 32, 33, 69
Faraday's heavy glass, 32, 33
Faulkner, Herbert (mines manager) – USA: 169, 192. Turkey: 273–284 *passim*
Faulkner's report on Greco-Turkish War, 273, 274, 275, 276–281, 282, 283–284
Federal Homestead Act, 174

Federal Oil and Gas Permit, 198
Ferro-boron, 172
Fertilizer industry of USA, 201, 209
Field test for borax and borate minerals, 9
Financial News, the, 201
'Financial Panic', the (United States, October 1907), 116
Fine Glass (see also 'Glass'), use of borax in, 148
Fire retardants, borax in, 2
First World War (1914–1918), 143, 146, 150, 151, 156, 157, 159, 160, 171, 172, 188, 210, 212, 215, 217, 224, 252, 263, 264, 266, 271, 272, 281–282, 285, 286–287
Fish Lake refinery, 44, 64
Fleming, Ernest L., 102, 104, 105
Flux, use of borax as, 1, 28–31, 112
Food and Drug Act, USA (1907), 107
Food-preservation, boric acid in, 34, 69, 75, 106–108, 115, 148, 150, 212
Food rationing in Second World War, 107
Foot powders, 34, 153, 212
Forbes (asst. to 'Howe'), 193, 197
Foreign agents, 106
Foreign currency difficulties, 205, 254
Foreign Investment Encouragement Law, Turkey, 284, 285
Foreign Mines Development Co., 154, 155
Foreign Office, British, 149, 269, 270, 271, 282, 283
Forga, S. American lawyer, 247 *and n.*
Fossi, partner of Lardarello, 95
France, 14, 91–95
'Francis', locomotive, 137
Frazier Borate Mining Co., 120
Frazier Mountain mine, Ventura County, 114, 115, 120–122
Freight rates, 149, 188
French, Dr Darwin, 50
French office, Borax Con., Paris, 209, 211
French refinery, an early, 15
Fries, de, German financier, 122
Friese, Lafayette Hoyt de, 81–83, 141, 144, 262
Frit, 31, 33, 34, 223
Frodish, W. J., A.P. & C.C., 215
Fuad Pacha, H.E., 261, 262, 266
Fumaroles of Tuscany, 39
Funeral Mountains, Death Valley, 49, 50, 118, 135, 137
Furnace Creek, USA, 47, 49, 51, 52, 118, 122, 192
Furnace Creek Inn, 166, 192, 216
Furnace Creek Ranch hotel, 192
Furnace Wash, 62

Gale, Hoyt S., US Geological Survey, 155, 156
Galileo, G., 8
Gallipoli, 90
Gallois, John T., West End shareholder, 176
Gasseri, Professor, 25
Geisenheimer, M., 85, 91–93, 95

Gem borate property, Calm brothers, 119
General Land Office, US, 156–157
Geoffrey, C. J., chemist, 9
George, Dr S. G., 50
German Agrarian Party, 106
German Air Force, 213
German Army, 212
German blitzkrieg on West, 210
German borax refiners, 26, 72, 87, 97, 148, 172, 190, 205, 209, 269
German Control authorities, the, 209
German designs on Turkish mines, 270
German Embassy, Constantinople, 270
German I.G. Farben, 193
German law on boric acid in food, 106
German Reich Commissioner, 213
German Solvay Company, 197
German Union of Refiners, 72, 73, 80, 91
Gernsheim refinery of Stauffer, 97, 122
Gerstle, Lewis, 74
GERSTLEY, JAMES:
 Redwood and Sons (from c.1890) and Burton, Baker & Co., 69–73
 Association with F. M. Smith, 74, 75
 Unpublished 'History' of Borax Consolidated, 74, 101
 South American problems, 103
 'Chinese borax', 112
 Volunteers in First World War, 150
 Buys shares to ease U.S. Potash's financial burden, 199
 On Baker's death takes over as Joint Managing Director with Lesser, 204
 Views on worsening international situation, 204, 205, 208, 209
 Borax group problems in Second World War, 209–216
 Anxieties of the Anti-Trust Case, 218–220
 Retirement, 222
 On the A.P. & C.C. affair, 225
 On rationing, 226
 Miscellaneous references, 79, 80, 84
Gerstley, J. M., President, Pacific Coast Borax, 204, 211, 219, 222, 228
Gerstley mine (Shoshone), 177, 181, 187, 188
Gibbs Engineering Co., 126
Giles, W. B., FIC, chemist, 78
Glasgow refinery (Townsend), 105
Glass (see also 'Fine Glass'), 2, 29–33, 95, 148, 171, 172, 202, 203, 223, 226, 254, 284
Glass fibre, 2, 223, 224
Glass-makers, 224
Glauber salts (salt cake), 103, 104, 292
Glaze, 2, 29, 31–32, 223
Glazebrook (W. T.) & Co., Liverpool, 102
Glyn Mills, Borax Consolidated's bankers, 226, 267
Goethe, Wolfgang von, 14
Goffs station, 126
Gold Centre station, 133
Goldfields American Development Co. Ltd., 215

Index

Goldfield Mining Camp, 127, 132, 135
Gold rush of 1849, 42, 49, 50
Gold Standard, the, 199, 284
Goldsmith's borax, 5
Goodair, F., prospector, 250–252
Government contract work, 231
Gower, Harry, 170, 171, 192
Gower, Pauline, 171
Grand Canyon National Park, 165
Grand View Mine, Death Valley, 168
Great Basin, 49, 50
'Great Idea' of the Greeks, Turkey, 272
Greco-Turkish War, 276, 281, 282, 284
Green-flame test for borax, 9, 51
Green River, Wyoming, 155
Green, Thomas, 78
Green Vitriol, 291
Greenwater mining camp, 133
Griffin, Leslie, driller, 173
Griffin & Wyman, Fish Lake, 42
'Grime Off', 116
Groppler, Henry, Turkey, 28, 81, 261
Groskop, Constantinople German, 270
'Ground Afire' (i.e. Death Valley), 50, 51
Grover, Dr F. L., chemist, 158
Guebhart, Berlin, 269–270
Gunsight silver lode, legend of, 50
Guton de Morveau, L. B. et al, *Méthode de nomenclature chimique*, 10
Gypsum, 27, 28, 30, 180, 274, 292

Hachinhama (Little Borax) Lake, 40, 41
Hainan, 18
Hale telescope glass, 223
Halsey, Edward, 110
Hamburg, 80, 208, 209
Hannan, John L., prospector, 179, 185
Hanson (Charles) & Co., Constantinople, 81
Harding, Works Manager, Belvedere, 211
Harmony Borax Company, The (1884), 47, 53
Harmony Borax Works, 47, 52–54, 57, 60
Hartman, cowboy and prospector, 175
'Hauoli' railroad car, 128
Headlamp glass, borax in, 223
Heat-resistant glass, 2, 33, 172, 223, 290
Heat-resistant ovenware, 33
Hecker, Dr G., 97, 156, 158, 160, 197, 209, 212
Hell & Sthamer, 96
Hellmers, Henry, chemist, 161
Heyden Chemical Corporation, 225
High Maremma, Tuscany, 24, 26
Hill, John, translator, 17
Himalayas, 22
Hitler, Adolf, 204, 205, 209
Hoeter, Francesco, 10, 24
Hohne, Dr, 215
Holland, 93, 94, 153
Homberg, W., 9, 10
Hombre Muerto deposit, S. America, 252, 254

Hoover, Herbert, President USA, 166, 195, 201
Hope Syndicate of Amsterdam, 215
Hopkins, prospector, 121
Horton Junction, 136, 137, 168
Horsfall of Mear & Green, 78
Hospitals, borax etc. in, 148
Hotel business, Death Valley, 192, 216
House of Representatives Ctte. on Science and Astronautics, 232
Howard, John, 119
Howard, Luke, 14
Howards of Ilford, 15, 26, 83, 104
Howe, ex-Kali sales agent, 193, 197
Hughes Bros., Birkenhead, 103
Hughes, Connah's Quay, 101
Hughes, stockbroker, 133, 137
Hummer, Nazi Verwalter of Stadlau, 213
Humphries, Dr, 119
Hungary, 210, 214
Hunt, Captain Jefferson, 50

Ickes, Harold L., Secretary of Interior, USA, 201
'Illustrated Sketches of Death Valley', J. R. Spears, 57, 66
Import duty, USA, 42, 45, 65, 68, 75, 86, 105, 120, 154, 210
Imports of borax into USA, 39, 68, 79
Imports of potash into USA, 156–159
Inchbrook borax works, 101, 102
Independence (Inyo County) legal judgement, 157
India, 8, 14, 16, 23, 24, 86, 110–112, 148, 205
India Office, 23
Indian and General Investment Trust Ltd., The, 77
Indian refined borax, 23
Indian Wells Valley, 154
Indians, American, 49–51, 237, 242, 244, 247, 249
Industrial Surveys Corporation, Detroit, 210
Infieles deposit, S. America, 252
Insulating tiles, boron in, 2
International Borax Union, 73, 75, 80, 82
International Minerals and Chemicals Corporation, 226
International Union Contract, 73
Inter-State Commerce Commission, 216
Inyo Independent, 51
Inyo Register, 170
Italian boric acid, 24–26, 35, 72, 94, 95, 102, 103, 117, 149, 211
Italian drillers in USA, 186
Italy and Italian producers (see also 'Lardarello'), 16, 23, 24, 32, 35, 72–74, 86, 91, 102, 210, 211, 265
Iquique, S. America, 27, 237, 244, 245, 250
Ivanpah, 126, 127
Izmir – see 'Smyrna'

Jabir Corpus, 4
Jabir ibn Hayyan, 4
Jacquier, M., 85, 91, 92
Jagadhi, 23
Japan, 103, 111, 148, 187, 202, 204, 205, 214, 254
Japanese labour in USA, 130
Jasper, Henri, Loth works, 210
Jena glass works, 33
Jenifer, Frank, Vice-President, Pacific Coast Borax, 191, 202–204, 209, 210, 214, 216–220, 222, 228
Jenifer Mine, Boron, 223, 228
Jersey City, USA, powder plant at, 45
Johnson, Dr Samuel, 9
Johnson, Borax sales manager, 211
Jones, John, refiner, 83
Jones, Dr L. C., chemist, 158
Jones-Grover patents, the (carbonization process), 158, 197
Jumlate, Kingdom of, 17
Justice Party, Turkey, 288

Kali (German Potash) Syndicate, 155, 157, 159, 160, 194, 195, 197, 198
Kara Ibrahim, Chief Guardian, Turkish mines, 274
Karassi mines, Turkey, 82
Keene, Mr, chemical engineer, 159
Kemal Atatürk – see 'Mustapha Kemal'
Kemaliddin Pacha, General, 281
Kenyan I, Japanese artist, 31
Kenyon, Mrs, 40
Kepler, J., 8
Kern County, California, 185, 218, 219
Kern County Land Co., 223
Kernite ('Rasorite'), 185, 187–190, 218, 219, 224, 292
Kew, Royal Botanical Gardens, 112
Key Route System, The, 139
Kidsgrove borax works, 78, 101, 105, 150, 187
Kipling, Rudyard, 58
Kirka sodium borate deposit, Turkey, 287, 288
Kirwan, Professor Richard, 17
Kleiner, agent of Suckow, 176
Kleinwort Benson, merchant bankers, 230
Knean, President Santa Monica Paving Company, 179, 180
Knight, H. P., chemist, 160, 185, 197
Koko Nor, Lake, China, 113
Konieh, Turkey, 271
Korean War, The, 223, 228
Kramer district and deposit, Mojave desert (see also 'Boron Mine'), 174, 176, 179–186, 189, 204
    Detail: Kramer, S.22 – 174, 176
        S.18 – 179, 181
        S.19 – 179, 187
        S.24 – 179–182, 185, 187, 189, 192
        S.17 – 179
        S.29 – 182
        S.13 – 187
        S.14 – 190, 215
    Baker shaft (formerly 'discovery shaft') – 187
    Osborne (main) shaft – 187
    West Baker shaft and mine (formerly Suckow mine) – 191, 223
    Baker mine – 187, 190, 223
    financing – 187
Kriegmetallgesellschaft, 269
Kunckel, Johann, 29, 31
Kwantung Province, China, 112, 113

Laboratory glass, 223
Laboratory or School of arts, The, 30
Labour problems, 150–151, 171, 204, 212, 263–264
Ladakh, 19, 22, 23
Lafontaine, Ottoman Bank, 261
Laird & Adamson, Liverpool brokers, 103
Lane, F. K., US Secretary of Interior, 163, 164
Lang Mine, 121, 122
Lardarel, Francesco de, 24–26, 94
Lardarello, 25, 26, 75, 91, 94, 95, 97, 146, 285
Las Vegas, 127, 175, 216
Las Vegas and Tonopah Railroad, 127, 133, 135
Last Days of Pompeii, The, 3
Latin translators of Arab scientific writings, 6
Laundry, borax in, 2
Lausanne Peace Treaty, 282
Lavoisier, A. L., *Traité de chimie*, 9, 10
Lawrence, Hon. H. A., 81, 85, 260, 263
Lazard Frères, financiers, 230, 233
Lead borate glass, 32
Leather, 148
Leaver, employee in Turkey, 284
Leblanc, T., 94
Lebreton, Maurice, senior clerk, Paris, 211
Lee, 'Cub' and 'Phi', 51
Lee Higginson & Co., financial advisers, 230
Leghorn, 26
Le Havre, 92
Leibrich, Professor (see also 'Food Preservation'), 106, 107
Lemery, N., chemist, 9
Lend-Lease Act, 1941, 216, 217
Lenicque, Henri, mining engineer, 261
LESSER, FEDERICO:
    Seconded to Pacific Coast and Redwoods from Rosenstern for S. American duties, 80
    Joint Managing Director, Borax Consolidated, 204
    Second World War worries, 214–220
    Death, 222
    Work in South America, 237–254 *passim*
    Miscellaneous references, 140, 201, 209

Lesser, Frederick, Managing Director, Borax group, 204, 222, 223, 225, 228
Levantines, 90, 263–264
Leven, Lord, Chairman, Borax Consolidated, 166, 193, 199, 200, 203
Leytonstone laboratory of W. B. Giles, 78
Lhasa, Tibet, 17–19, 110
Libavius, Andreas, *Alchemia*, 6
Libby, E. D., glassmaker, 224
Lick Springs, California, 39
Lila C. lode and mine, 62, 64, 117, 118, 120–122, 125–128, 130, 132, 135, 136, 168, 182
Lila C. branch line, 132, 137
Liquor traffic during railroad construction, 131
Lister, Joseph, 34
Little, Ed, Shanghai, 112, 113
Little, Owen, Shanghai, 112
Little Placer Claim, 219, 223, 225
Liverpool Borax Company, 102, 103, 105
Liverpool, Old Swan Refinery, 26, 73, 102
Lizzie V. Oakley lode, 62
Llamas, 247, 253
Loan to Turkish Nationalist Government, 283
Locke, W. R., 105, 106, 237–250
Locomotives, 63, 128, 131, 137
Lode deposits, 44 (footnote), 121
London Chamber of Commerce, 226
London Stock Exchange, 140
Lorca, 8
Loth factory, Belgium, 93, 94, 148, 210, 226
Lovell, cowboy and prospector, 175
Lower Biddy McCarthy mine, 168
Ludlow, railroad town, 127, 128, 130, 134, 135
Ludlow, dance at, 135
Lui-li (glass), 30
Luke Howard & Sons, 15
Lyons Association Placer Mine, 155
Lyons Refinery, 28, 73, 82, 91, 92, 94

McCann-Ericson, advertising agents, 202
MacDonald, R. D., Manager, Turkish mines (1931–1958), 284, 286
MacDonald, Mrs, 286
Mackay, Sir Joseph, Transport, 148
McNutt, V. H., U.S. Potash, 198
Macquer, Pierre, *Dictionaire de chimie*, 9, 10, 15, 32
McSweeney, H., U.S. Potash, 198–201
Magadi, Lake, Kenya, 155
Magnetic separation process, 224, 229
Mahmoud Essad Bey, Minister of the Economy, 282, 283
Mahmoud Pacha, 262
Mahmud II, Sultan, 257
Maisons Lafitte refinery, 28, 73, 81, 82, 91, 92, 94
Malachite, 292
Manchuria (now Lianoning Province), 113, 204, 205
Mandelso, Sir John, 16, 24

Manolaki, lawyer, 260
Manteri, engineer, 25
Manvel, mining town, 126
'Manure salts', 198, 200
Marco Polo, 7, 17
Maremma, Tuscany, 10
'Maricunga' borate, 245
Marine Acid, 9, 292
Marine Alkali, 9, 292
Marion calcining plant, 64, 65
'Marion' locomotive, 137
Marseilles refinery, 26
'Marsh borax' companies, 42, 62, 117
Mastodons, 49
MATHER, J. W.:
  Vice-President of Pacific Borax, Salt, and Soda Co., 46
  Helps Smith find funds for expansion, moves to new Smith Office in N.Y., and boosts borax sales. Disagreements with Smith, 65, 66
  Resigns on Smith linking Pacific Coast Borax with British rather than American firm, 77
  Miscellaneous, 114
MATHER, STEVE:
  Journalist. Becomes part-time sales adviser to Smith. Initiates Twenty Mule Team publicity, 66
  Full-time advertising manager, 67
  New Chicago sales office, 68
  Temporary transfer to N. York and return to Chicago, 114
  Nervous collapse after overwork. Resigns Smith's employ. Joins Thorkildsen in joint borax company, 115
  Success in placer claims at Lang. Thorkildsen and Mather receive 20% in new Sterling Borax Co. to exploit Lang Mine. Mather expands sales and with Thorkildsen gains 60% interest in Sterling Borax Co., 121
  Interest in Sterling sold to Borax Consolidated for handsome sum, 122
  Continues refining and marketing of borax. In 1914 invited to run National Parks by Sec. of Interior. Goes to Washington and establishes National Park Service on permanent basis. Takes on Albright as Asst. Buys Thorkildsen out of Thorkildsen & Mather Co. Dies in 1930. Succeeded by Albright. Places named after Mather. Memorial plaques in 23 National Parks and 33 National Monuments, 163–167
  Miscellaneous references, 116, 200, 202
Mayer, André, Lazard Frères, 233
Means, driller, 198
Mear and Green Ltd., Staffs., 73, 78, 83, 101, 105
Mear, Stephen, 78
Meat, borax as preservative, 106

Mechanical miner, 228
Medical recipes and uses, 7, 116, 148
Medieval Europe, 5
Memo. of Association, 142
Menderes, Adnan, Turkish prime minister, 286
Mercantile Trust Company of San Francisco, 139
Meridian Borax Co. (1884), 47, 53, 60
Merlie Chemical Co., 156
Merrett, Christopher, 29
Merrill, Colonel, 156
Mesquite, 53
Metalwork, use of borax in, 1, 148
Method of production, 2
Mexican labour in USA, 130
Mihaldieh basin, Turkey, 265
Militicks, American citizen, 268, 269
Mills Tariff Bill, USA, 48, 65
'Mining and Scientific Press', The, 147
Minister of Commerce, Turkey, 270
Minister of Mines, Turkey, 261, 266
Ministry of Health Committee on food-preservatives, 1926, 107
Model, chemist, 9, 17
Model, Roland, and Stone, finance house, 229
Mohammed VI, Sultan, 272
Mohr, Dr, 270
Mojave, California, 44, 54, 55, 57, 62, 125, 126, 154, 156, 166, 174, 182
'Mol', 229
Mollendo, S. America, 27, 87, 246, 247, 249–250
Mona Lake, 40
Monchicourt, M., 85, 91, 93, 95
Monte Blanco, 62, 118
Monterotundo, 10
Montgomerie, Capt. T. G., 19
Moore Company, the, contractors, 199
Moradabad, 22
Morgan, J. P., Wall Street financier, 199
Mormons, 50
Mortas Co., Turkey, 287
Morton, Coleman, financier, 229, 230
Mosheimer, Joseph, 44
Mudd and Partner, 222, 223
Muddy Mountains, 174
Muir, John, conservationist, 163
Mule Team Canyon, Calico Mts., 63
Mule team treks and mules, 53, 54, 57, 63, 66, 125, 126, 154, 202, 244, 247
Multauf, R. P., 160
Multiple stores, packaged borax in, 212
Mulvey, driller, 198
Mumford, R. W., American Trona, 160
Muriate of potash (potassium chloride), 198
Muscovy, 17
Muspratt, Professor Sheridan, 15
Mustapha Kemal (later Kemal Atatürk), 171, 257, 273, 282, 286–287

Nantes, 211

National Advisory Committee for Aeronautics, 231
National Borax Co., The, 122, 123
Nationalism in Turkey, 257, 258, 273, 282, 287
National Park Service Act, 165, 200
National Parks, 163–166, 200, 216
Natron (or Nitrum), 3, 4, 5, 292
Nazis, 208, 209, 213, 225, 226
Nazi Party of California, 215
Nebraska Potash, 156
Necbal Lake, 17
Neel Consolidated Borate Mine (see also 'Borate' and 'Calico'), 76
Neill, Marion, prospector, 62
Nepal, 17–19, 22, 23
Neri, Antonio, *L'Arte Verraria*, 29, 31
Nether, 4
Neuschwander, R., prospector, 60
Nevada caves, 49
Nevada Chemical Company, 156, 157
Nevada Salt and Borax Company (1882), 47
New Brighton refinery, Pennsylvania (Dawes & Myler), 118, 120, 121
New Consolidated Goldfields, 215
New Deal, the, 204
New Jersey refinery (Calm brothers), 122, 123
New Mexico, 197, 198, 200
New Ryan mining camp, 136, 137, 168–171, 187, 192
New Ryan tourist hotel, 192
New York Office of Pacific Coast Borax, 65–67, 104, 114
New York Times, the, 231
Newman, Superintendent, Bayonne Works, 186
Nian Singh, Pundit, 19
Nicholas I, Tsar, 272
Nihad Bey, Turkish lawyer, 269, 281, 283
Ninzei, 31
Nitre, 292
Nobel Prize, 11
'Nolo contendere' pleas in Anti-Trust cases, 219, 220
North European Oil Company, 194
NRA Codes for borax and potash industries, 204
Nuclear reactors, borax in, 2
Nylon manufacture, borax in, 2

Oakland Traction Company, The, 139, 142
Oakley Mine, 168
Office of Chief Ordinance at Washington, 228
Old Dinah, traction engine, 63, 126
Old Swan refinery, Liverpool, 26
'Old Whittall deposit' (formerly 'Demir Capou') 262, 266
'Olinda', s.s., 210
Olin Mathieson Chemical Corporation, 232, 233
Omer Kerri station, Turkey, 265
Open-pit mining, 228, 229, 231
Optical glasses, 32, 33, 148, 172
Order in Council, 211

Index

Oregon Property, the, 62, 76
Osborne, L. D. (Roy), Lang Mine, 122, 182, 186, 188, 190, 191
Osborne shaft – see 'Kramer'
Otelia Suckow Placer Claim, 174
Ottoman Empire, the, 90, 262, 266, 274
Ovenware glass, 223
Owens Corning Fibre Glass Corpn., 224
Owens Lake, 53, 159, 225
Owens Valley, 153
Oxshott, wartime H.Q. of Borax Consolidated, Ltd, 211

Pacific Alkali, Owens Lake, 219, 225
Pacific Borax Company, The (1872–1876), 40, 42, 44
Pacific Borax Property, the, 46
Pacific Borax, Salt, and Soda Company, The (1886–), 46, 47–49, 62, 65
Pacific Borax and Redwood's Chemical Works Limited, 76, 78–80, 82, 83, 85
Pacific Coast Borax Company, The (1890–), 46, 49, 53, 62, 66, 67, 74–79, 100, 107, 115–117, 119–122, 139, 140, 144, 146–148, 153, 155–158, 160, 163, 165, 166, 170, 172–176, 180, 181, 185, 188–190, 192, 197, 214, 217, 222, 225, 229, 230
Package borax, 44, 67, 115, 116, 140
Pahrump Valley, 126
Paiute Indians, 50
Pajonales, S. America, 251
'Palace, The', in Turkey, 258, 266, 272
'Palace Pachas', 90, 260, 261
Palaeoindian settlements, 49
Palm Borate Co., 120
Panama Canal, 54, 117, 171
Panamint Range, USA, 49–51, 55, 193
Panderma (Bandirma), Turkey, 82, 258, 265, 268, 271, 273, 276, 277
Pandermite, 27, 28, 81, 258, 285, 288, 292
Parker, adviser on chemicals, 117
Parkinson, 'Borax Bill', 115
Parks, prospector, 53
Peacock, Sir Edward, British Treasury, 217
Pearl ash (potash), 292
Pearl Harbour, 210
Pears, Edward, lawyer, 94, 262
Pears, H. & S., Turkish lawyers, 265
Pearson, Norman, 208, 213, 214
Pennock, J. D., Solvay, 159
Pera Palace Hotel, Constantinople, 273
Perkins, prospector, 175
Perry, J. W. S., 56, 57
Persia, 14, 16, 17, 24, 215
Persian Gulf, 216
Peru, 27, 35, 86, 88, 89, 225, 252, 253, 254
Petuntze, 31, 292
Pfizer & Co., 45, 66, 75, 77, 117–119
Pharmaceutical Journal and Transactions, the, 70

Pharmacy, 34, 70
Philippines (Mindanao) project, 142
Philosophical Transactions of the Royal Society of London, 17
Phosphate, 150
Picardy, 101
Pichu-Pichu mountains, Andes, 248
Pintados deposit, Chile, 80, 87, 237, 244, 245
Pittman, Senator Key, Nevada, 218
Placer claims, 44 and footnote, 121
Played Out lode, 62, 167
Pleistocene, American, 49
Pliny the Elder, 5
Poland, 96, 172, 208, 210, 214
Pollak, Dr R., lawyer, Prague, 213
Porcelain, glazing of, 31, 32, 202
Pong-cha (pheng-sha), 8
Post-war Europe, 275
Post-war reconstruction, 220
Potash, 156, 158–161, 173, 193, 194, 197–202, 218, 224, 225, 230, 292; potassium carbonate, 4, 30; potassium chloride, 156
Potash Company of America, Carlsbad, 200, 206
Potash imports into USA, 201
Potash Leasing Act, 1917, 157, 158
Potash market in USA, 198
Potash process, 157, 158, 160, 197
Potash prospecting permit, 198
Potassium, 155
Potero, the (San Francisco), 119
Powdered borax, 210
Power generation, Lardarello, 26
Prague, 213, 214
Presdeleau, N.Y., 65, 144, 147
Preservide, 103
'Preservitas', 72
'Preservitas' Co., The, 72
Preston, 'Ma', 64, 128
Priceite (a calcium borate mineral), 27, 60, 291
Price of borax, 32, 33, 35, 39–41, 44, 45, 48, 68, 75, 80, 86, 93–97, 118, 120, 160, 166, 172, 173, 175, 176, 188, 191, 193, 197, 201, 202, 204, 214, 218, 219, 225, 251, 271
Price of potash, 197, 201, 204, 205
Pridham, C. F., 103, 112
Producing areas of the world, borax, 26, 35
Productora de Borates, La, 172
Profits, 76, 83, 95, 115, 139, 141, 172, 181, 191, 201, 202
Project ZIP, 229, 231
Proksh, Aussig, 207
Prologue of the Canterbury Tales, 7
Prospecting, Turkey, 285, 286, 288
Psalty, Mr, Clerk at Constantinople, 260–261
Puga, hot springs of, 22, 23
Purbet, Province of, 16
Pyne copper-smelter, 117, 142
Pyrites, 117

'Queen of Borax' Soap, 116

Ragtown, Nevada, 40
Railroad land grants, 179
Railway building and construction, 125
Raku (low-fired faience pottery), 31
Ramjet fuels (see also 'High Energy Fuels'), 229
Ransome and Smith, 79
Rasor Brothers, 173
RASOR, CLARENCE M.:
   Early life. Rasor Bros., consulting engineers. Deputy Mineral Surveyor. Helps build Tonopah and Tidewater Railroad. Becomes Chief Engineer. Then succeeds Corkill (I) as head of Pacific Coast Borax's Land and Exploration Dept., 173
   Otelia Suckow Placer Claim, 174
   White Basin, Calville Wash (Smith's Anniversary Mine), Shoshone (Gerstley Mine), 175
   Meetings with Blumenberg and Kleiner, Suckow's agent. Suckow mine acquired. Drill holes in Kramer district, 176
   Sodium borate discovery at Kramer. Secures agreement with drillers and Board. Discovery shaft sunk, 179–181
   Analysis of nature and extent of new material (trademark 'Rasorite'), 185–187
   Trouble with interlopers at Kramer, 189
   Seeks potash possibilities to compete with American Potash. Opens way for Borax Consolidated to join United States Potash Co. venture at Carlsbad, 197, 198
   Retires in 1936 and dies in 1946, 203
   Miscellaneous references, 127, 128, 136
Rasor, Edwin, 173
Rasor, Lew, 174
'Rasorite' (registered trade mark), various products, 185, 202, 212, 229
Rationing in UK, 226
Realty Syndicate, The, 139, 141, 142
Red Cross, 148, 149, 153
Redwood, Sir Boverton, 70
Redwood, George, 70, 71, 76
Redwood, Iltyd, 76, 97, 101, 105, 110
Redwood, Professor Theophilus, 34, 69–71, 107
Redwood, Theophilus Horne, 70, 71, 76
Referee in Bankruptcy, 190
Refining process redesign, Boron, 228, 229, 231
Refining metals, use of borax in, 1
Refining process of borax, 14, 15
Reichert, Dr, 86
Reichstelle für Chemie, 213
Reid, Arthur, Managing Director, 222
Reid, Colonel, Deputy Chairman, Borax Consolidated, 152, 168
Republican People's Party, Turkey, 286
Rescalli, manager at Antofagasta, 240, 245
Research laboratories, 231
Residual calcined shale, 189
Rhazes – see 'Al Razi'
Rhodes & Wesson, Rhodes Marsh, 44
Rhyolite mining camp, 133

Richards, Professor, 169
Richardson, Sir Benjamin Ward, 107
Rio Grande ('Llipi Llipi'), Bolivia, 242, 250
Ripley, President of S. Fe Railroad, 131
Robertson, relative of Coleman, 60
Robottom, Arthur, *Travels in Search of New Trade Products*, 16, 19, 81, 237
Rockefellers, the, 230
Rocket engines, 231
Rockies, The, 133
Roman times, use of pandermite, 27
Room and pillar stopes, 228
Roosevelt, Franklin D., President, 200, 204
Roosevelt, Teddy, President, 165
Rope, borax crystallized on, 112
Ropeway to Omer Kerri station, Turkey, 265–266, 272
Ropp, Baron Alfred de, 156
Rosenstern & Co. (F.), 80, 97, 208
Rothschilds, the, 157
Rotterdam, 209
Royal Commission on food-preservation, 106
Rubner, Dr (see also 'food-preservatives'), 106
Russell (W. H.) Mine, 122, 123
Russia, 73, 91, 95, 96, 148, 149, 156, 200, 201, 213, 214, 216, 257, 267
Russian Revolution, The, 150
Ryan, John, 46, 49, 76, 112, 115, 121, 122, 125–128, 130, 131, 136, 147, 168, 173, 260, 274
Ryan mines, the, 168, 169, 175, 179, 181, 188, 192, 274

Saldero potash project, Utah, 157
Sales of Borax Consolidated, 105–107, 115–117, 121, 172, 202, 212, 214, 223
Sal-ammoniac, 5, 6, 7, 9, 29, 292
Salicylic acid, 106
Salinas, Peru, 81, 87, 89, 236, 244, 247, 248–250
Salinas calcining plant, 249
Saline laws of Nevada, 41
Salt cake, 292
Salt Lake City, 50, 126, 127
Salt Wells, Nevada, 40
Salta mine and refinery, Argentina, 254
Salzdetfurth A.G., 215
San Bernardino Borax Mining Company, The, 47, 79, 154–158
San Francisco bankers, Committee of, 143
San Francisco Borax Board (1887–1889), 47–49
San Francisco Chronicle, 130
San Francisco Committee of Vigilance, 42
San Francisco earthquake and fire (1906), 132
San Francisco refinery of Stauffer, 120
San Francisco sales office, 46, 114
Sandiver, 293
Santa Fe Railroad (formerly Atlantic & Pacific R.R.), 54, 126–128, 131, 132, 134, 135, 187, 191
Santiago, Chile, 238, 240, 246

Index

Saunders, Mr, surgeon, Tibet, 18
Savage Landor, Walter, poet, 110
Savage Landor, H., explorer, 110
Saxony, 32
Schaffer and Budenberg, 213
Schering, Berlin refiners, 80
Schiave, Levantine, Turkish mines, 278, 279, 283
Schlagentweit, Dr H. von, 22, 23
Schott and Sons, 33
Schott, Otto, 33
Schott & Genossen, 210
Schroders, bankers, 194
Searles & Co., 44
Searles, Dennis, 44, 50, 154
Searles, John, 44, 47, 50, 154
Searles Lake, 28, 44, 79, 148, 154–160, 166, 172, 173, 175, 180–182, 188, 190, 191, 193, 195, 197, 201, 214, 219
Searles Lake, chemical analysis, 155, 160, 161
Searles Lake, Federal leases, 148, 157
Searles Lake potash, effect on borax industry, 158, 159
Second World War, 107, 113, 146, 151, 192, 205, 208–220, 222–226, 228, 229, 254, 283, 285, 286, 288
Sedative salt (boric acid), 9, 10, 293
Seitz, Dr, Stadlau, 209, 213
Selek, Cabir S., Turkey, 286
Semiconductors, 2
Senate Committee to investigate potash industry, 218
Sequoia National Park, 165
Sèvres, 32
Shadley, William, 57
Shankland, Robert, biographer, 164
Share capital of Borax Consolidated, 133
Shareholders' protective committee, West End, 176
Shaw, Peter, *Chemical lectures*, 14, 30
Sheep, transport of tincal by, 19, 22
Shelter Island, 65
Shepherd, prospector, 121
Sherman Anti-Trust Bill (later Law), 48, 49, 86, 194, 217, 218
Shipping, Ministry of, 148, 149
Shipping of borax, 2, 149, 150, 208, 210, 214, 215, 224, 275
Shoshone (Gerstley) Mine, 175, 181
Shrinkage stopes, 228
'Sick Man of Europe', 265, 272
Sierra Nevada, 154
Sikkim, 23
Silicon chips, 2
Simav River, Turkey, 265
Smith, Evelyn – see Ellis, Evelyn
SMITH, FRANCIS MARION:
    Early life. Teels Marsh venture and refinery, 41
    Link with Storey Bros. and William T. Coleman & Co., 42

Acquires assets of Pacific Borax Company. Builds Fish Lake refinery, 44
Package borax venture, 45
End of partnership with Julius Smith. Sole owner of Teels Marsh and Pacific Borax Property. Incorporation of Pacific Borax Salt and Soda Co., 46
Organizes San Francisco Borax Board, 47
Sole sales agent for the Board, 48
Coleman bankrupt, Smith acquires his borax properties. Floats Pacific Coast Borax Co., 49
Turns from working marsh and dry lake deposits to mining colemanite in Calico Mts., 62
Starts to live stylishly, 65
Financial and advertising disagreements with the Mathers, 66–67
Visits England, 74
Arranges with R. C. Baker the merger of Pacific Coast Borax with Redwood and Sons, 75–78
Acquires control of San Bernardino Borax Mining Co. on Searles Lake. Initiates new refinery at Bayonne, 79
Supports steps to control competition from South American and Turkish borates leading to formation of Borax Consolidated, 79–83
Relations with R. C. Baker, 83–84
Comment on Lawrence, 85
Attitude to merger, 87
South American policy, 88
Comments on Belgian affairs, 93
Approaches to Lardarello, 94–95
Ambitions to use English capital to finance non-borax enterprises in USA restrained by Borax Consolidated board, 100
Comments on Baker's idea of sole new borax refinery in Britain, 101
On borax in food-preservation, 106–107
Passing interest in Tibet and China, 110–112
Expands borax sales in USA. Gives up idea of diversifying into other chemicals but pushes borax soap and package borax, 114–117
Struggle to dominate colemanite supply in USA. Succeeds with takeover of Lang Mine. Mines of main rivals close down, 117–123
Ambitious struggle to build Tonopah and Tidewater railroad, 125–131
Resulting financial problems, 131–136
Smith's non-borax interests bring financial downfall. Struggles to prevent sale of his borax shares by bankers' committee. Resigns all appointments in Borax group, 139–144
Turns against Borax Consolidated and intends founding rival business, 146–147
Obtains lease at Searles Lake and starts West

End Chemical Co., 148
West End obtains successful process for producing borax from lake, 161
Outbids Rasor to acquire Calville Wash and re-enters borax mining with Anniversary Mine, 'Twenty Aeroplane' brand borax, and Stauffer as sales agent. With price fall in mid-twenties closes mine and falls back on Searles Lake production, 175
Shareholders' protective committee force him to relinquish chairmanship of West End Mining. Falls ill and dies in 1931, 176
Miscellaneous references, 154, 157, 168, 169, 173, 192, 202, 203, 251, 260
Smith Brothers, 42, 45, 46, 64, 67
Smith, Julius, 41, 44, 46, 64, 67
Smitheram, Billy, Supdt. Borate Mine, 117, 169
Smyrna (Izmir), 257, 265, 272, 273, 277, 281, 282
Snowden, Mr, U.S. Potash, 198, 199
Soap, borax, 115, 116, 140, 172
Soap plant, 116
Société Desmazures et Compagnie, 81
Société Lyonnaise de Borate de Chaux, 28
Société Lyonnaise des Mines et Usines de Borax, 73, 82, 83, 85, 89, 91, 92, 96, 258, 261, 263
Société Poudre, Russia, 96
Society of Arts, 10
Soda, 293
Soda ash (sodium carbonate), 4, 155, 158, 160, 161, 224, 293
Soda production, 158, 160, 196
Sodium, 11, 124
Sodium borate, 8, 180, 182, 185–187, 189, 225, 253, 293
Sodium carbonate – see 'soda ash'
Sodium Leasing Act (1920), 181
Sodium pentaborate, 161, 172, 293
Sodium perborate, 2, 172, 215, 224, 293
Soffioni, 24–26, 285
Solar evaporation, 119, 120, 157, 161
Soldering, use of borax in, 1, 6, 9, 28
Somptner, The, *Canterbury Tales*, 7
Source of borax, 2
South Africa, exports to, 148
South America, 26, 54, 72–75, 79, 80, 86–88, 90, 94, 97, 102–105, 119, 134, 135, 146, 148, 149, 189, 204, 205, 225, 237–254, 261, 264, 270
South America, borate of lime districts, 27
South American borate, imports into UK, 102, 103
South American Spanish cities, 247
Southern Anti-Aircraft Command, 153
'Southern borax', 30
Southern Pacific Railroad, 121, 125, 135, 156, 179, 187
Space missiles, 231
Space shuttle 'Columbia', 2
Spain, 6, 172, 200–202, 210, 225

Spangles, 1
Spanish potash producers, 201
Spears, J. R., *Illustrated Sketches of Death Valley*, 57, 66
Speculation in financial circles, 231
Spiller, Harry, 51
Spitfire 'Boron', 212
Stadlau factory – see 'Vienna'
Standard Oil Company and works, 79, 149–151, 157
Standard Sanitary Co., 120
State Line Divide, 126
State Line Pass, 126, 127
State Range, 44
Stauffer and Stauffer Chemical Co., 75, 115, 117–123, 143, 161, 171–176, 179, 189, 190, 217, 219, 225
Stebbins, Dr, 169, 188, 189, 275
Stedman, Mr, 140
Sterling Borax Co., 121–123, 135, 163
Stevenot, Emil, 44
Stevens, Hugh, prospector, 60
Stevens Consolidated Borate mine, 76
Stiles, Ed, teamster, 54, 56
Storey Brothers, Chicago, 42
St Petersburg, 23
Strikes, 151
Sublime Porte, German influence on, 89, 90, 270
Suckow, Dr John K., 174, 176, 179, 181, 190, 215, 217
Suckow borax refinery, Los Angeles, 176
Suckow Chemical Company, 174
Suckow Company, The, 190
Suckow mine (S.14 Kramer), 190, 191, 215, 223, 224
Suckow mine (S.22 Kramer), 175, 176
Sudetenland crisis, 209
Sultan Tchair mines and works, Turkey, 27, 82, 90, 149, 258, 260–269
Supply, Ministry of, British, 149
Supreme Court, USA, 204
Surat, 16
Suriri, S. America, 87, 244, 245, 249
Susrata Samhurta, 8
Susurluk (Sousighirlik), Turkey, 27, 90, 272, 277, 280, 285, 286
Sweden, 17, 32
Sylvite (potassium chloride ore), 156, 198
Syndicate attempt at takeover of Borax Consolidated, 229, 230
Syria, 30

Talaat Bey, Turkish Interior Minister, 268
T'ang dynasty, 8
Tankana, 8
Tariffs, Protective (see also 'Import duty, USA'), 48, 119
Tartaric acid, 172
Tartary, 17
Tasian culture, 3

Index

Tavernier, Jean B., 16
Teels Marsh, 41, 42, 44–47, 49, 51, 52, 54, 62, 64, 69, 76
Teeple, Dr John, 160, 176
Tehama County, California, 28, 39
Telescope Peak, Death Valley, 193
Television, 202
Tengri-Nur, Lake (Numar-Tso), 18, 19
Terrace Borax Mine, 155
Teshoo Lama, 18
Textiles, glass fibre, 224
Theophrastus, *On Stones*, 5, 17
'The Technical World', 136
Thermal and acoustical insulation, glass fibre, 224
Thibet (Tibet), 16
Thomas, M., Paris, 211
Thompson, Edward G., Turkey, 262
Thompson Mining Properties, Death Valley, 219, 223
Thorkildsen & Mather Company, 115, 122, 164
Thorkildsen, Thomas, 114, 115, 121, 122, 163, 164, 166
'Three Elephant Borax Corpn. of N.Y.', 219
'Three Elephant Brand', 160
Tiberio Vasques, bandit, 121
Tibet, 17–19, 22–24, 35, 39, 86, 109, 110, 113
Tincal, 1, 5, 7, 14–18, 22–24, 27, 54, 110–113, 179, 180, 185–187
Tincalayu mine, Argentine, 254
Tincal, sources of, 14, 24
Tincar, 5
Tissoolumbo monastery, 17, 18
Titus Canyon, 49
Tobeler, Dr Otto, 215
'Tomesha' (Death Valley), 50, 51
Tonne, definition of, 264 (footnote)
Tonopah and Tidewater Railroad, 127, 130–134, 136, 137, 146, 170, 171, 173, 191, 202, 204
Tonopah and Tidewater Railroad, financing of, 131, 132, 134, 135, 191, 192, 216, 226
Tonopah and Tidewater Railway Company, 134
Tonopah mining centre, 127, 132, 148
Torre, Dr de la, Arequipa lawyer, 250
Tourmaline, 2, 293
Townsend (Joseph Townsend Ltd.), Glasgow, 73, 83, 102–105, 149
Townsend, C. W., 103, 104
Traces of borates, 2
Traction engines, 63, 120, 126, 127
Trade and commerce, 14
Trade routes for borax, 2, 16
Tranquebar, 17
Transport problems, 54, 118, 171
Travels in Search of New Trade Products, Robottom, A., 16, 237
Treaty of Sèvres, 282
Tres Morros mine, Argentine, 93
'Trona', mineral, 5, 53, 154, 293
Trona Railway, the, 156
'Trona reef', 157

Troup, William, 40, 41
Trusts, 48, 218
Tubbs, Bob, 170
Tubini, Aristide, Constantinople agent, 90, 260, 262, 266
Tubini, Hyacinth, 266, 267, 282, 283
Tullner, Dr, Aussig, 209
Tunstall borax works, 78, 101
Turk Boraks M.A.S., 286, 287
Turkey, 27, 35, 54, 72–74, 81, 87, 89, 90, 149, 169, 171, 205, 211, 212, 225, 252, 254, 257, 288
Turkish courts, 271
Turkish economy, the, 287
Turkish mining concessions, permis and firmans, 28, 81, 90, 261, 266, 283, 284
Turkish mining law, 262
Turkish pandermite (calcium borate) mines of Borax Consolidated (see also 'Azizieh' and 'Sultan Tchair'), 254, 257, 258, 267–271, 276–282, 285
    Mining, production and labour relations, 263, 264, 271, 284, 285
    Mineral royalties, 266, 268, 282, 284
    Important events:
        1912 and 1914 call-ups, 266, 267
        German designs on mines, 269, 270
        Threatened Turkish sequestration, 268–271
        Mines handed back, 272
        Isolated by Nationalist rising, 273, 277
        Faulkner's new ideas, 274–276
        Bunning leaves, 276
        Greco-Turkish war, British escape, 276–282
        Compromise with Nationalist Govt. on royalties etc., 282–283
        Renewal of concessions and operations, 283
        MacDonald takes over operations until closure after Second World War, 284
Turner, Samuel, Lieutenant, R.E., 18
Tuscany, 24, 26, 54, 94, 95, 149, 163, 211
'Twenty Aeroplane Brand' ('F. M. Smith Borax'), 175
'Twenty Mule Team' episode, 54, 57, 66, 115, 126, 132, 202
'Twenty Mule Team' package borax, 115, 164, 172, 202
'Twenty Mule Team' soap chips, 116, 202
'Twenty Mule Team' trademark, 53, 66, 116, 160, 166, 175

Ulex, George, 27
Ulexite (sodium calcium borate), 1, 24, 27, 53, 169, 176, 179, 181, 285, 293
Underground mining methods, 228
Union Pacific Railroad, the, 127, 192
Unions, Trade, 150, 151
United Provinces, India, 111

United States, first borate discovery, 28
United States Geological Survey, 155, 156, 189, 229
United States production of borates, 35
United States Potash Company, 198–203, 216, 218, 226, 227, 230
Upper Biddy McCarthy mine, 168
US Air Force, 231, 232
U.S. Borax and Chemical Corporation, 200, 229–233
U.S. Borax Company (Calm Bros.), 122, 123
U.S. Borax Research Corporation, 231
US Defense Department, 228, 230
Uses of borax, boron, borates, 2, 26–35, 72, 106–108, 148, 223–224
US Navy, 232
US Surveyor General of Mineral Claims, 173
US Treasury, 217
Utah, 198

Valparaiso, 238, 240, 246, 250, 252, 253
'Val Jean' railroad station, 130
Van der Smissen, Loth, 210
Vatican, The, 211
Veatch, Dr John A., 39–41, 54
Vegetation, borax and, 2
Venetians, 6, 7, 14
Venice, 14, 29, 30, 224
Ventura County, California, 117, 120, 173
Verdigris, 5, 293
Vermier, partner of Fossi, 95
Vesting Order No. 249, 214
Vial and Pradel, MM., France, 85, 91–93
Vicuna, 140, 248, 251
Vienna refinery:
    256 Brigittaplaz, 28, 73, 82, 91, 96
    Stadlau, 96, 97, 148, 171, 208–210, 212, 213, 226
Vienna, Russian zone, 213
Viscose Corpn. (Courtaulds), 217
Vogelsang, Alexander, US Asst. Sec. Interior, 157
Volcanoes, 237, 241, 243, 249

Wadsworth, Nevada, 40, 54
Wages, 171, 204, 212, 252
Walker, Norman, 79, 88, 242, 243
Wall Street Journal, 231
Warrington Borax Company, 105
Washing crystals, 104
'Washing industry', perborate in the, 224
Washing machine soap powder, 116
Webster, J., *History of metals*, 7
Wehe, Frank, lawyer, 44 (footnote), 181
Weiss (German in Turkey), 266
Wells, President of Solvay, 194, 197
Weringer property, Ventura, 120
West Baker shaft – see 'Kramer'
West End Chemical Company, 148, 157, 161, 176, 197, 217, 219
West End Consolidated Mining Company, 148, 157, 176
West Kerman (Iran) discovery of ulexite, 24
Western Borax Company, The, 189, 190, 219, 222, 223, 225
Western Borax refinery, 190
Western Mineral Co., 120
Westminster Gazette, The, 104
Wheeler, John H., of Stauffer's, 119
White Basin, 175
Whittall, Ernest, 90, 262, 266
Whittall deposit (later Lawrence concession) (see also 'Old Whittall deposit'), 262, 263
Whittall, J. W. & Co., 265
Widdess, George, driller, 179–181
Widow lode and mine, the, 62, 168, 169
Wiley, Dr. H. W., 107
Wilkinson, Guy, 155, 156
Willcox, Sir Henry, 107
Wilmington works, USA, 169–171, 179, 180, 185–187, 189, 198, 224, 225, 228
Wilson, Belvedere manager, 153
Wilson, Sir Alexander, 77, 111
Wilson (Tariff) Act, 1894, 68
Winckler, Herr, 90
Windy Gap (Wingate Pass), 56
Winters, Aaron and Rosie, 51
Wintershall A.G., Germany, 214
Wire-drawing, borax in, 2
Wisconsin glacial epoch, 49
Woodman, Ruth, writer, 202

Yamdok Cho, Lake, Tibet, 19
'Yankee Girl' mine, 133
Yard, Robert, New York Herald, 164
Yareta, 248
Yellowstone Park, 164, 165, 200
Yircali, H. Sirri, Turkey, 287
Yosemite Valley and National Park, 163, 165, 166, 200
Young Ottoman movement, 89
Young Turk revolution, 90, 257, 264, 267, 272
Younghusband, Colonel Sir Francis, 23, 110

ZABRISKIE, CHRISTIAN BREVOOT:
    Joins F. M. Smith's Pacific Borax, Salt, and Soda Co., 46
    Moved to N.Y. Sales Office, 114
    On fall of Smith, Zabriskie becomes head of operations in the USA with Baker as President, 202
    Advises joining in Suckow Chemical Company, 174
    Reports activity at Kramer and together with Rasor recommends acquisition of mineral rights and land. Recommends sinking 'discovery shaft', 179–183
    Involved with Baker in solving problems

    created by discovery of the new mineral, in establishing Death Valley tourism, and in trying to identify the new (undisclosed) owners of American Potash, 185–195
    Health deteriorates. Succeeded by Jenifer in 1933. Dies in 1936, 203
    Miscellaneous references, 115, 122, 131, 144, 148, 156, 164–166, 169, 170, 175, 176, 200, 260, 264, 274, 284
Zabriskie Point, Death Valley, 203
Zabriskie railway station, 132
Zaffer (cobalt oxide), borax as flux for, 31, 32
Zeiss, Karl, 33
Zellinger, Constantinople German, 270
Zenobia deposit, South America, 252
ZIP project, 229, 231